AQUACULTURE DESK REFERENCE

R. LeRoy Creswell

Harbor Branch Oceanographic Institution, Inc.

An **avi** Book
Published by Van Nostrand Reinhold
New York

An AVI Book
(AVI is an imprint of Van Nostrand Reinhold)

Copyright © 1993 by Van Nostrand Reinhold
Softcover reprint of the hardcover 1st edition 1993

Library of Congress Catalog Card Number 92-24983

ISBN 978-1-4684-7117-5   ISBN 978-1-4684-7115-1 (eBook)
DOI 10.1007/978-1-4684-7115-1

Van Nostrand Reinhold
115 Fifth Avenue
New York, New York 10003

Chapman and Hall
2-6 Boundary Row
London, SE1 8HN, England

Thomas Nelson Australia
102 Dodds Street
South Melbourne 3205
Victoria, Australia

Nelson Canada
1120 Birchmount Road
Scarborough, Ontario MIK 5G4, Canada

16 15 14 13 12 11 10 9 8 7 6 5 4 3 2 1

Library of Congress Cataloging-in-Publication Data
Aquaculture desk reference/ by R. LeRoy Creswell.
     p. cm.
     Includes bibliographical references.

     1. Aquaculture--Handbooks, manuals, etc. I. Title
SH135.C74 1992
639'.8'0212--dc20   92-24983
                                                              CIP

CONTENTS

iii

III. ENRICHMENT FORMULATIONS

IV. HATCHERY SYSTEMS AND METHODS

V. PLUMBING AND MATERIALS

VI. FEEDS AND NUTRITION

VII. AQUACULTURE PONDS

VIII. CHEMICALS AND TREATMENTS

PREFACE

As the aquaculture industry has expanded throughout the world, it has embraced the experiences of many fields of study to meet increasing technological challenges. The complexities of modern hatchery methodology, more intensive growout systems, and the application of diverse biological and physical sciences to aquatic animal husbandry have reached beyond the ability of most aquaculturists to enjoy an in-depth knowledge of all phases of the aquaculture process. More importantly, in order for the culturist to have at hand the information necessary to make basic decisions, it requires an extensive library of textbooks and scientific literature.

The *Aquaculture Desk Reference* serves as a concise compilation of tables, graphs, conversions, formulas and design specifications useful to the aquaculture industry. It also provides examples, in a straight forward manner, of how information in tabulature can be used to derive values for specific system design and process strategies. Tables and graphs in this volume also provide background documentation and authority for further reference.

The *Aquaculture Desk Reference* is a convenient source book that will alleviate the need for an extensive personal library to access basic information useful for practicing aquaculturists.

Many thanks to Mrs. Ruth Aldrich for her assistance in the preparation of this book. My family, friends and associates also deserve my special appreciation for their encouragement and support.

I. Conversions and Equivalents

TABLE 1-1: TEMPERATURE EQUIVALENTS - CENTIGRADE TO FARENHEIT

Temperature in °C. is expressed in the left column and top row with the corresponding temperature in °F. in the body of the table.

°C.	0	1	2	3	4	5	6	7	8	9
0	32.0	33.8	35.6	37.4	39.2	41.0	42.8	44.6	46.4	48.2
10	50.0	51.8	53.6	55.4	57.2	59.0	60.8	62.6	64.4	66.2
20	68.0	69.8	71.6	73.4	75.2	77.0	78.8	80.6	82.4	84.2
30	86.0	87.8	89.6	91.4	93.2	95.0	96.8	98.6	100.4	102.2
40	104.0	105.8	107.6	109.4	111.2	113.0	114.8	116.6	118.4	120.2
50	122.0	123.8	125.6	127.4	129.2	131.0	132.8	134.6	136.4	138.2

TABLE 1-2: TEMPERATURE EQUIVALENTS - FARENHEIT TO CENTIGRADE

Temperature in °F. is expressed in the left column and top row with the corresponding temperature in °C. in the body of the table.

°F.	0	1	2	3	4	5	6	7	8	9
30	-1.1	-0.6	0.0	0.6	1.1	1.7	2.2	2.8	3.3	3.9
40	4.4	5.0	5.6	6.1	6.7	7.7	7.8	8.3	8.9	9.4
50	10.0	10.6	11.1	11.7	12.2	12.8	13.3	13.9	14.4	15.0
60	15.6	16.1	16.7	17.2	17.8	18.3	18.9	19.4	20.0	20.6
70	21.1	21.7	22.2	22.8	23.3	23.9	24.4	25.0	25.6	26.1
80	26.7	27.2	27.8	28.3	28.9	29.4	30.0	30.6	31.1	31.7
90	32.2	32.8	33.3	33.9	34.4	35.0	35.6	36.1	36.7	37.2
100	37.7	38.3	38.9	39.4	40.0	40.6	41.1	41.7	42.2	42.8

For intermediate temperatures or those exceeding the range of the tables, the following formulas may be used:

$$°F = 1.8 \times °C + 32$$

$$C = \frac{F - 32}{1.8}$$

TABLE 1-3: CONVERSIONS FOR UNITS OF WEIGHT

FROM	TO				
	GRAM	KILOGRAM	GRAIN	OUNCE	POUND
GRAM	1	.001	15.43	0.0353	0.0022
KILOGRAM	1000	1	1.54×10^4	35.027	2.205
GRAIN	0.0648	6.48×10^{-5}	1	0.0023	1.43×10^{-4}
OUNCE	28.35	0.0284	437.5	1	0.0625
POUND	453.6	0.4536	7000	16	1

1

TABLE 1-4: CONVERSIONS FOR UNITS OF LENGTH

FROM	TO				
	CENTIMETER	METER	INCHES	FEET	YARDS
CENTIMETER	1.0	0.01	0.3937	0.0328	0.0109
METER	100	1.0	39.37	32.81	1.0936
INCHES	2.540	0.0254	1.0	0.0833	0.0278
FEET	30.48	0.3048	12.0	1.0	0.3333
YARDS	91.44	0.9144	36.0	3.0	1.0

TABLE 1-5: CONVERSIONS FOR UNITS OF VOLUME

FROM	TO								
	CM^3	LITER	$METER^3$	$INCHES^3$	$FEET^3$	FL. OZ.	FL. PT.	FL. QT.	GAL.
CM^3	1.0	0.001	1×10^{-6}	0.061	3.53×10^{-5}	0.0338	0.00211	0.00106	2.64×10^{-4}
LITER	1000	1.0	0.001	60.98	0.0353	33.81	2.113	1.057	0.2624
$METER^3$	1×10^6	1000	1.0	6.1×10^4	5.31	3.38×10^4	2113	1057	264.2
$INCHES^3$	16.39	0.0164	1.64×10^{-5}	1.0	5.79×10^{-4}	0.5541	0.0346	0.0173	0.0043
$FEET^3$	2.83×10^4	28.32	0.0283	1728	1.0	957.5	59.84	29.92	7.481
FL. OZ.	29.75	0.0296	2.96×10^{-5}	1.805	0.00104	1.0	0.0625	0.0313	0.0078
FL. PT.	473.2	0.4732	4.73×10^{-4}	28.88	0.0167	16.0	1.0	0.5000	0.1250
FL. QT.	946.4	0.9436	9.46×10^{-4}	57.75	0.0334	32.0	2.0	1.0	0.2500
GALLON	3785	3.785	0.0038	231.0	0.1337	128.0	8.0	4.0	1.0

TABLE 1-6: CONVERSIONS FOR UNITS OF VELOCITY

FROM	TO					
	FT/MIN	M/S	M/MIN	M/HR	MPH	KNOTS
FEET/MIN	1.0	0.00508	0.30480	180288	0.01130	0.00987
METER/SEC	196.85	1.0	60.00	3600.0	2.2369	1.9425
METER/MIN	3.2808	0.01667	1.0	60.00	0.03728	0.03238
METER/HR	0.05468	0.00028	0.01667	1.0	0.00062	0.00054
MPH	88.0	0.44704	26.822	1609.4	1.0	0.86839
KNOTS	101.34	0.51479	30.887	1853.2	1.1516	1.0

TABLE 1-7: CONVERSIONS FOR UNITS OF ENERGY

FROM	TO		
	BTU	JOULE (J)	FOOT POUND
BTU	1	1055	778
JOULE (J)	0.0009478	1	0.7376
FOOT POUND	0.001285	1.3558	1

(continued on page 3)

(continued from page 2)

BTU = British Thermal Unit, a unit of heat equal to 252 calories, or the quantity of heat required to raise the temperature of one pound of water from 62 °F to 63 °F

JOULE = a unit of electrical energy or work equivalent to the work done to raise one coulomb of electricity one volt, or in maintaining for one second a current of one ampere against a resistance of one ohm

FOOT POUND = a unit of work equal to the amount of energy required to raise a weight of one pound a distance of one foot

TABLE 1-8: CONVERSIONS FOR UNITS OF POWER

FROM	TO			
	HORSEPOWER	WATT (W)	FT. LB/S	BTU/S
HORSEPOWER	1	746	550	0.7068
WATT (W)	0.001341	1	0.7376	0.00095
FT. LB./S	0.00182	1.356	1	0.001285
BTU/S	1.415	1055	778	1

HORSEPOWER = a unit of power equal to a rate of 33,000 foot-pounds per minute (the force required to raise 33,000 pounds at the rate of one foot per minute)

WATT = a unit of electrical power equal to one ampere under one volt of pressure, or one joule per second

TABLE 1-9: CONVERSION FACTORS OF RADIANT ENERGY, POWER, AND INTENSITY UNITS
(from Hollaender, 1956)

FROM	TO				
	ENERGY				
	ERG	JOULE	G-CAL	WATT HOUR	KG-CAL
ERG (DYNE-CM)	1	10^{-7}	0.239×10^{-7}	0.278×10^{-10}	0.239×10^{-10}
JOULE (WATT-SEC)	10^7	1	0.239	0.278×10^{-3}	0.239×10^{-3}
GRAM-CALORIE	4.19×10^7	4.19	1	1.163×10^{-3}	$10-3$
WATT HOUR	3.60×10^{10}	3,600	860	1	0.860
KILOGRAM-CALORIE	4.19×10^{10}	4,190	1,000	1.16	1
	POWER				
	ERG/SEC	µWATT	CAL/MIN	WATTS	CAL/SEC
ERG/SEC	1	0.1	1.43×10^{-6}	10^{-7}	0.239×10^{-7}
µWATTS	10	1	1.43×10^{-5}	10^{-6}	0.239×10^{-6}
CALORIE/MIN	6.98×10^5	6.98×10^4	1	0.0698	0.0166
WATT	10^7	10^6	14.3	1	0.239
CALORIE/SEC	4.19×107	4.19×10^6	60	4.19	1
	INTENSITY				
	ERG/SEC/CM^2	µWATT/CM^2	µWATT/MM^2	WATT/M^2	CAL/MIN/CM^2
ERG/SEC/CM^2	1	0.1	0.001	0.001	1.43×10^6
µWATT/CM^2	10	1	0.01	0.01	1.43×10^{-5}
µWATT/MM^2	1000	100	1	1	1.43×10^{-3}
WATT/M^2	1000	100	1	1	1.43×10^{-3}
CAL/MIN/CM^2	6.98×10^5	6.98×10^4	698	698	1

ERG = a unit of work or energy in the metric system equal to the amount of work done by one dyne acting through a distance of one centimeter

DYNE = a unit of force which in one second can alter the velocity by one centimeter per second of a mass of one gram

CALORIE = the amount of heat needed to raise the temperature of one gram of water one degree Centigrade

TABLE 1-10: CONVERSION FACTORS FOR ILLUMINATION

FROM	TO			
	BRIGHTNESS			
	FOOT-LAMBERT	LAMBERT	CANDLES/CM2	CANDLES/MM2
FOOT-LAMBERT	1	1.08×10^{-3}	3.39×10^{-3}	$3.39 \times 10\text{-}5$
LAMBERT	929	1	0.318	0.318×10^{-3}
CANDLES/CM^2	2920	3.14	1	0.01
CANDLES/MM^2	2.92×10^5	314	100	1
	ILLUMINANCE			
	LUX	FOOT-CANDLES	LUMEN/CM^2	
LUX	1	0.093	10^{-4}	
FOOT CANDLES	10.8	1	1.08×10^{-3}	
LUMEN/CM^2	10^4	929	1	

LAMBERT = the centimeter-gram-second unit of brightness, equal to the brightness of a perfectly diffusing surface that radiates or reflects light at the rate of one lumen per square centimeter

LUX = illumination equal to one lumen per square meter or the illumination of a surface uniformly one meter distant from a point source of one foot candle

FOOT-CANDLE = illumination equal to the amount of direct light thrown by one international candle on a surface one foot away

INTERNATIONAL CANDLE = a measure of the intensity of light, equal to the light given off by the flame of a sperm candle 7/8 inch in diameter burining at the rate of 7.776 grams per hour

LUMEN = a measure for the flow of light, equal to the amount of flow through a unit solid angle from a uniform point source of one international candle

(Source: E. Bickford and S. Dunn, *Lighting for Plant Growth*; 1972)

TABLE 1-11: CONVERSIONS FOR PRESSURE EQUIVALENTS

FROM	TO			
	POUNDS/IN.^{2}	FEET WATER	INCHES MERCURY (in. Hg)	ATMOSPHERES (atm)
POUNDS/IN.^{2}	1.0	2.309	2.036	0.068
FEET WATER	0.433	1.0	0.882	0.0295
INCHES MERCURY .	0.491	1.134	1.0	0.033
ATMOSPHERES (atm)	14.7	33.9	29.92	1.0
KILOPASCALS (kPa)	0.145	0.335	0.295	0.010

INCHES MERCURY = a unit of pressure as measured by a manometer equal to the pressure balanced by the weight of a one-inch column of mercury in the instrument

ATMOSPHERE = the weight of the atmosphere per square inch of surface; the pressure of 14.69 pounds per square inch exerted in all directions at sea level by the atmosphere

KILOPASCAL = 1,000 pascals = a unit of force equal to one Newton per square meter (N/m^2); typically used as pascal second (Pa • s), or 10 poise designating absolute viscosity

BAR = a metric unit of measure often used, equal to 100 kilopascal (kPa) or 14.5 lb/in^2

TABLE 1-12: CONVERTING WATER PRESSURE (psi) TO FEET HEAD [1]

POUNDS/IN^2	FEET HEAD	POUNDS/IN^2	FEET HEAD
1	2.31	100	230.90
2	4.62	110	253.9a
3	6,93	120	277.07
4	9.24	130	30016
5	11.54	140	323 25
6	13.85	150	346.34
7	16.16	160	369.43
8	1 8.47	170	392.52
9	20.78	180	415.61
10	23.09	200	461.78
15	34.63	250	577.24
20	46.18	300	692.69
25	57.72	350	808.13
30	69.27	400	922.58
40	92.36	500	1154.48
50	115.45	600	1385.39
60	138.54	700	1616.30
70	161.63	800	1847 20
80	184 72	900	2078 10
90	207 81	1000	2309.00

[1] One pound of pressure per square inch of water equals 2.309 feet ot water at 62° Fahrenheit.

TABLE 1-13: CONVERTING FEET HEAD OF WATER TO PSI [1]

FEET HEAD	POUNDS/IN^2	FEET HEAD	POUNDS/IN^2
1	.43	1 00	43.31
2	.87	110	47.64
3	1.30	120	51 .97
4	1.73	130	56.30
5	2.17	140	60.63
6	2.60	150	64.96
7	3.03	160	69.29
8	3.46	170	73.63
9	3.90	180	77.96
10	4.33	200	86.62
15	6.50	250	108.27
20	8.66	300	129.93
25	10.83	350	151.58
30	12.99	400	173.24
40	17.32	500	216.55
50	21.65	600	259.85
60	25.99	700	303.16
70	30.32	800	346.47
80	34.65	900	389.78
90	38.98	1000	433.00

[1] One foot of water at 62° Fahrenheit equals 0.433 pounds pressure per square inch.

TABLE 1-14: DECIMAL EQUIVALENTS FRACTIONS

FRACTION	DECIMAL	FRACTION	DECIMAL
1/64	0.015625	7/16	0.4375
1/32	0.03125	29/64	0.453125
3/64	0.046875	15/32	0.46875
1/20	0.05	31/64	0.484375
1/16	0.0625	1/2	0.5
1/13	0.0769	33/64	0.515625
5/64	0.078125	17/32	0.53125
1/12	0.0833	35/64	0.546875
1/11	0.0909	9/16	0.5625
3/32	0.09375	37/64	0.578125
1/10	0.10	19/32	0.59375
7/64	0.109375	39/64	0.609375
1/9	0.111	5/8	0.625
1/8	0.125	41/64	0.640625
9/64	0.140625	21/32	0.65625
1/7	0.1429	43/64	0.671875
5/32	0.15625	11/16	0.6875
1/6	0.1667	45/64	0.703125
11/64	0.171875	23/32	0.71875
3/16	0.1875	47/64	0.734375
1/5	0.2	3/4	0.75
13/64	0.203125	49/64	0.765625
7/32	0.21875	25/32	0.78125
15/64	0.234375	51/64	0.796875
1/4	0.25	13/16	0.8125
17/64	0.265625	53/64	0.828125
9/32	0.28125	27/32	0.84375
19/64	0.296875	55/64	0.859375
5/16	0.3125	7/8	0.875
21/64	0.328125	57/64	0.890625
1/3	0.333	29/32	0.90625
11/32	0.34375	59/64	0.921875
23/64	0.359375	15/16	0.9375
3/8	0.375	61/64	0.953125
25/64	0.390625	31/32	0.96875
13/32	0.40625	63/64	0.984375
27/64	0.421875	1	1.0

TABLE 1-15: MULTIPLIERS FOR CONVERSION OF UNITS

TO GET FROM	MULTIPLY BY	TO OBTAIN
LINEAR		
CENTIMETERS	3.2808×10^{-2}	FEET
	3.9370×10^{-1}	INCHES
	1.0000×10^{4}	MICRONS
	1.0000×10^{1}	MILLIMETERS
FEET	1.2000×10^{1}	INCHES
	3.0480×10^{-4}	KILOMETERS
	3.0480×10^{-1}	METERS
	1.8939×10^{-4}	MILES
	3.0480×10^{2}	MILLIMETERS
INCHES	2.5400×10^{-2}	METERS
	2.54	CENTIMETERS
	2.5400×10^{1}	MILLIMETERS
	0.08333	FEET
	0.027778	YARDS
	0.000'015783	MILES
KILOMETERS	1.0000×10^{3}	METERS
	6.2137×10^{-1}	MILES
METERS	6.2137×10^{-4}	MILES
	39.37	INCHES
	3.28	FEET
	1.0936	YARDS
	1.0000×10^{3}	MILLIMETERS
	100	CENTIMETERS
MICRONS	1.0000×10^{-3}	MILLIMETERS
	1.0000×10^{-6}	METERS
MILES	1.60935	KILOMETERS
	1,760	YARDS
	5,280	FEET
	63,360	INCHES
	1.6093	KILOMETERS
	1.6093×10^{3}	METERS
AREA		
ACRES	43,560	SQUARE FEET
	4,840	SQUARE YARDS
	43,560	SQUARE FEET
	4,047	SQUARE METERS
	0.404687	HECTARES
	0.0015625	SQUARE MILES
	4,840	SQUARE YARDS
	4,047	SQUARE METERS
HECTARE	1.0000×10^{4}	SQUARE METERS
	2.47 ACRES	
SQUARE CENTIMETERS	1.0764×10^{-3}	SQUARE FEET
	1.5500×10^{-1}	SQUARE INCHES
	1.0000×10^{-4}	SQUARE METERS

TABLE 1-15 (cont):

TO GET FROM	MULTIPLY BY	TO OBTAIN
SQUARE FEET	1.4400×10^2	SQUARE INCHES
	9.2903×10^{-2}	SQUARE METERS
	0.000000003587	SQUARE MILES
	0.000022957	ACRES
	0.11111	SQUARE YARDS
SQUARE INCHES	0,0000000002491	SQUARE MILES
	6.4516×10^{-4}	SQUARE METERS
	6.45163	SQUARE CENTIMETERS
	0.0000001594	ACRES
	0.0007716	SQUARE YARDS
	0.006944	SQUARE FEET
SQUARE KILOMETERS	3.8610×10^{-1}	SQUARE MILES
SQUARE METERS	10.76	SQUARE FEET
SQUARE MILE	4,014,489,600	SQUARE INCHES
	27,878,400	SQUARE FEET
	3,097,600	SQUARE YARDS
	640	ACRES
	259	HECTARES
SQUARE YARDS	0.0000003228	SQUARE MILES
	0.0002066	ACRES
	9	SQUARE FEET
	1,296	SQUARE INCHES
GRAMS (FL)/GALLON	2.6455×10^2	MILLIGRAMS/LITER
	3.4392×10^{-2}	OUNCES (FL)/GALLON
	2.6455×10^2	PARTS PER MILLION
	2.6455×10^{-1}	PARTS PER THOUSAND
GRAINS PER GALLON	17.12	MILLIGRAMS PER LITER
	142.9	POUNDS PER MILLION GALLONS
MILLIGRAMS PER LITER	1	PARTS PER MILLION
	0.0584	GRAINS PER GALLON
	8.345	POUNDS PER MILLION GALLONS
	1.3000×10^{-4}	OUNCES (FL)/GALLON
	1.0000	PARTS PER MILLION
	1.0000×10^{-3}	PARTS PER THOUSAND
	3.7800×10^{-3}	GRAMS (FL)/GALLON
OUNCES (FL)/GALLON	7.6923×10^3	PARTS PER MILLION
	7.6923	PARTS PER THOUSAND
	2.9077×10^1	GRAMS (FL)/GALLON
	7.6923×10^3	MILLIGRAMS/LITRE
PARTS PER MILLION	1.0000×10^{-3}	PARTS PER THOUSAND
	3.7800×10^{-3}	GRAMS (FL)/GALLON
	1.0000	MILLIGRAMS/LITER
	1.3000×10^{-4}	OUNCES (FL)/GALLON
PARTS PER THOUSAND	1.3000×10^{-1}	OUNCES (FL)/GALLON
	1.0000×10^3	MILLIGRAMS/LITER
	1.0000×10^3	PARTS PER MILLION
	3.7800	MILLIGRAMS/LITER

TABLE 1-15 (cont):

TO GET FROM	MULTIPLY BY	TO OBTAIN
DENSITY		
KILOGRAMS/CUBIC METER	1.0000×10^{3}	MILLIGRAMS/LITER
	6.2422×10^{-2}	POUNDS/CUBIC FOOT
MILLIGRAMS/LITER	6.2422×10^{-5}	POUNDS/CUBIC FOOT
	10000×10^{-3}	KILOGRAMS/CUBIC METER
POUNDS/CUBIC FOOT	1.6020×10^{1}	KILOGRAMS/CUBIC METER
	1.6020×10^{4}	MILLIGRAMS/LITER
ENERGY		
BTU	1.0543×10^{3}	JOULES
	2.5200×10^{-1}	KG-CALORIES
	2.9285×10^{-4}	KILOWATT-HRS
JOULES	2.3903×10^{-4}	KG-CALORIES
	9.4852×10^{-4}	BTU
KG-CALORIES	1.1621×10^{-3}	KILOWATT-HRS
	3.9683	BTU
	4.1836×10^{3}	JOULES
KILOWATT-HRS	3.4147×10^{3}	BTU
	8.6050×10^{2}	KG-CALORIES
FLOW		
CUBIC CENTIMETERS/SEC	3.5315×10^{-5}	CUBIC FEET/SEC
	1.0000×10^{-6}	CUBIC METERS/SEC
	1.5848×10^{-2}	GALLONS/MIN
	2.6413×10^{-4}	GALLONS/SEC
	6.0000×10^{-2}	LITERS/MIN
CUBIC FEET/SEC	2.8317×10^{-2}	CUBIC METERS/SEC
	4.4876×10^{2}	GALLONS/MIN
	7.4794	GALLONS/SEC
	1.6990×10^{3}	LITERS/MIN
	2.8317×10^{4}	CUBIC CENTIMETERS/SEC
CUBIC METERS/SEC	1.5848×10^{4}	GALLONS/MIN
	2.6413×10^{2}	GALLONS/SEC
	1.0000×10^{6}	CUBIC CENTIMETERS/SEC
	6.0000×10^{4}	LITERS/MIN
	3.5315×10^{1}	CUBIC FEET/SEC
GALLONS/MIN	1.6667×10^{4}	GALLONS/SEC
	3.7860	LITERS/MIN
	6.3100×10^{1}	CUBIC CENTIMETERS/SEC
	2.2283×10^{-3}	CUBIC FEET/SEC
	6.3100×10^{-5}	CUBIC METERS/SEC
	8.021	CUBIC FEET PER HOUR
GALLONS/SEC	3.7860×10^{-3}	CUBIC METERS/SEC
	1.3370×10^{-1}	CUBIC FEET/SEC
	2.2716×10^{2}	LITERS/MIN
	3.7860×10^{3}	CUBIC CENTIMETERS/SEC
	6.0000×10^{1}	GALLONS/MIN

TABLE 1-15 (cont):

TO GET FROM	MULTIPLY BY	TO OBTAIN
LITERS/MIN	1.6667×10^{1}	CUBIC CENTIMETERS/SEC
	5.8858×10^{-4}	CUBIC FEET/SEC
	1.6667×10^{-5}	CUBIC METERS/SEC
	2.6413×10^{-1}	GALLONS/MIN
	4.4022×10^{-3}	GALLONS/SEC
LITERS/SEC	15.85	GALLONS PER MINUTE
MASS		
GRAMS	1.0000×10^{-3}	KILOGRAMS
	1.0000×10^{3}	MILLIGRAMS
	3.5327×10^{-2}	OUNCES
	2.2050×10^{-3}	POUNDS
	1.0000×10^{-6}	TONNES
KILOGRAMS	1.0000×10^{6}	MILLIGRAMS
	3.527×10^{-5}	OUNCES
	2.2050	POUNDS
	1.0000×10^{-3}	TONNES
	9.8438×10^{-4}	TONS (LONG)
	1.1025×10^{-3}	TONS (SHORT)
MILLIGRAMS	2.2050×10^{-6}	POUNDS
OUNCES	7.7778×10^{-2}	POUNDS
	4.5351×10^{-4}	TONNES
	4.4643×10^{-4}	TONS (LONG)
	5.0000×10^{-4}	TONS (SHORT)
	7,000	GRAINS
	453.6	GRAMS
TONNES	1.1025	TONS (SHORT)
TONS (LONG)	1.1200	TONS (SHORT)
VELOCITY		
CENTIMETERS/SEC	1.9685	FEET/MIN
	1.0000×10^{-5}	KILOMETERS/SEC
	1.0000×10^{-2}	METERS/SEC
	2.2369×10^{-2}	MILES/HR
FEET/MIN	1.8288×10^{-2}	KILOMETERS/HR
	5.0800×10^{-3}	METERS/SEC
	1.1364×10^{-2}	MILES/HR
KILOMETERS/HR	6.2137×10^{-1}	MILES/HR
	1.0000×10^{3}	METERS/SEC
METERS/SEC	2.2369	MILES/HR
	3.281	FEET PER SECOND
VOLUME		
ACRE-FOOT	4.3573×10^{4}	CUBIC FEET
	1.2338×10^{3}	CUBIC METERS
	1.6138×10^{3}	CUBIC YARDS
	2.7137×10^{5}	GALLONS (IMP)
	3.2590×10^{5}	GALLONS (US)

TABLE 1-15 (cont):

TO GET FROM	MULTIPLY BY	TO OBTAIN
CUBIC CENTIMETER	0.06102	CUBIC INCHES
CUBIC FEET	1 728	CUBIC INCHES
	7.481	GALLONS
	28.32	LITERS
	1 728	CUBIC INCHES
	7.4805	U. S. GALLONS
	28.317	LITERS
	0.037037	CUBIC YARDS
	0.00022957	ACRE-FEET
CUBIC METERS	35.31	CUBIC FEET
	264.2	GALLONS
	1 000	LITERS
CUBIC YARDS	21	CUBIC FEET
	46,656	CUBIC INCHES
	0.00061983	ACRE-FEET
	0.76456	CUBIC METERS
GALLONS	3785	CUBIC CENTIMETERS
	0.1337	CUBIC FEET
	231	CUBIC INCHES
	3.785	LITERS
GALLONS OF WATER	8.345	POUNDS OF WATER
LITERS	1,000	CUBIC CENTIMETERS
	0.03531	CUBIC FEET
	0.2642	GALLONS
PRESSURE		
ATMOSPHERES	14.7	PSI
	33.9	FEET OF WATER
	29.9	INCHES OF MERCURY
	76.0	CENTIMETERS OF MERCURY
BAR	1.0197	KILOGRAM/SQUARE CENTIMETER
	14.504	PSI
FEET OF WATER	0.4335	POUNDS/SQUARE INCH
	0.0295	ATMOSPHERES
	0.433	PSI
	0.883	INCHES OF MERCURY
INCHES OF MERCURY	1.133	FEET OF WATER
	0.49	PSI
	0.0334	ATMOSPHERES
INCHES OF WATER	0.074	INCHES OF MERCURY
	0.0.036	PSI
PSI	2.31	FEET OF WATER
	2.04	INCHES OF MERCURY
	0.068	ATMOSPHERES
	0.0685	BAR

TABLE 1-16: TREATMENT CONVERSION CHART

(Amounts listed are for active ingredients or a trade name preparation, depending on the recommendations)

PARTS PER MILLION (PPM)	DILUTION	% SOLUTION	MG/L	GM/L	MG/GAL	OZ/GAL	OZ/1,000 GAL	GM/FT³	OZ/1,000 FT³	LBS/ACRE FOOT
0.1	1:10,000,000	0.00001	0.1	0.0001	0.38	0.000013	0.013	0.0028	0.1	0.27
1	1:1,000,000	0.0001	1	0.001	3.8	0.00013	0.134	0.0285	1	2.7
2	1:500,000	0.0002	2	0.002	7.6	0.00029	0.268	0.0567	2	5.
3	1:333,333	0.0003	3	0.003	11.3	0.00040	0.402	0.0851	3	8.1
4	1:250,000	0.0004	4	0.004	15.2	0.00053	0.536	0.1134	3.99	10.8
5	1:200,000	0.0005	5	0.005	19.0	0.00067	0.670	0.1418	4.99	13.5
6	1:161,600	0.0006	6	0.006	22.8	0.00080	0.804	0.1701	5.99	16.2
7	1:142,900	0.0007	7	0.007	26.6	0.00093	0.938	0.1985	6.99	18.9
8	1:125,000	0.0008	8	0.008	30,4	0.00117	1.072	0.2268	7.99	21.6
9	1:111,000	0.0009	9	0.009	34.1	0.00120	1.206	0.2552	8.98	24.3
10	1:100,000	0.0010	10	0.010	38.0	0.0013	1.340	0.2835	9.98	27.0
11	1:90,909	0.0011	11	0.011	41.8	0.0014	1.474	0.3118	10.98	29.7
12	1:83,333	0.0012	12	0.012	45.6	0.0016	1.608	0.3301	11.98	32.4
13	1:76,923	0.0013	13	0.013	49.4	0.0017	1.742	0.3684	12.97	35.1
14	1:71,429	0.0014	14	0.014	53.2	0.0018	1.876	0.3367	13.97	37.8
15	1:66,667	0.0015	15	0.015	57.0	0.00195	2.010	0.4250	14.98	40 5
16	1:62,500	0.0016	16	0.016	60.8	0.0021	2.144	0.4533	15.97	432
17	1:59.235	0.0017	17	0.017	64.6	0.0022	2.278	0.4816	16,97	45.9
18	1:55,555	0.0018	18	0.018	68.4	0.0023	2.412	0.5099	17.96	48.6
19	1:52,632	0.0019	19	0.019	72.2	0.0025	2.546	0.5382	18.96	51.3
20	1:50,000	0.0020	20	0.020	76.0	0.0026	2.680	0.5620	19.97	54.0
100	1:10,000	0.0100	100	0.100	380.0	0.013	13.400	2.8350	99.84	270.0
125	1:8,000	0.0125	125	0.125	475.0	0.016	16.750	3.5338	134.80	337.5
250	1:4,000	0.025	250	0.25	950.0	0.03	33.500	7.0875	249.60	675.0
500	1:2,000	0.05	500	0.5	1,900.0	0.07	67.000	14.1750	49920	1,350.0
750	1:333	0.075	750	0.75	2,950.0	0.10	100.500	21.2625	748.80	2,025.0
1,000	1:1000	0.1	1,000	1	3,800.0	0.13	134.000	28.3500	998.4	2,700.0
2,000	1:500	02	2,000	2	7,600.0	027	268.000	56.7000	2,000.0	5,400.0
3,000	1:333	0.3	3,000	3	11,400	0.40	402.000	85.1000	3,000.0	8,100.0
4,000	1:250	0.4	4,000	4	15.200.0	053	536.000	113.4000	3,990.0	10,800.0
5,000	1:200	0.5	5,000	5	19,000.0	0.67	670.000	141.8000	4,990.0	13,500.0
6,000	1:166.6	0.6	6,000	6	22,800.0	0.80	804.000	170.1000	5,990.0	16,200.0
7,000	1:142.9	0.7	7,000	7	26.600.0	0.93	938.000	198.5000	6,990.0	18,900.0
8,000	1:125	0.8	8,000	8	30,400	1.07	1,072	226.8	7,990	21,600
9,000	1:111	0.9	9,000	9	34,100	1 20	1,206	255.2	8,980	24,300
10,000	1:100	I	10,000	10	38,000	1.34	1,340	283.5	9,984	27,000
20,000	1:50	2	20,000	20	76,000	2.67	2,680	567.0	20,000	54,000
25,000	1:40	2.5	25,000	25	95,000	3.34	3,350	709.0	25,000	67,500
30,000	1:33.33	3	30,000	30	114,000	4.01	4,020	851.0	30,000	81,000
40,000	1:25	4	40,000	40	152,000	5.34	5,360	1,134.0	39,900	108,000
50,000	1:20	5	50,000	50	190,000	6.68	6,700	1,418.0	49,900	135,000
60,000	1:16.16	6	60,000	60	228,000	8.01	8,040	1,701.0	59,900	162,000
70,000	1:14.29	7	70,000	70	266,000	9.35	9,380	1,985.0	69,900	189,000
75.000	1:13.33	7.5	75,000	75	295,000	10.01	10,500	2,127.0	74,900	202,500
80,000	1:12.50	8	80,000	80	304,000	10.68	10,720	2,268.0	79,900	216,000
90,000	1:111.1	9	90,000	90	341,000	12.02	12,060	2.552.0	89,800	243,000
100,000	1:10	10	100,000	100	380,000	13.35	13,400	2,835.0	99,840	270,000

(Source: N. Herwig, *Handbook of Drugs and Chemicals used in the Treatment of Fish Diseases*; 1979))

12

TABLE 1.17: GEOMETRIC FORMULAS

WHERE:

A = AREA

C = CIRCUMFERENCE

H = HEIGHT

R = RADIUS (1/2 DIAMETER)

L = LENGTH

A1 = SURFACE AREA OF SOLIDS,

P = PERIMETER

D = DIAMETER

π = 3.142

V = VOLUME

RECTANGLE

$A = W \, x \, L$

TRIANGLE

$A = \dfrac{W \, x \, H}{2}$

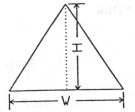

PARALLELOGRAM

$A = H \, x \, L$

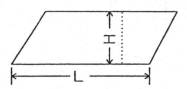

RECTANGULAR SOLID

$A_1 = 2 \,[\,(W \, x \, L) + (L \, x \, H) + (H \, x \, W)\,]$

$V = W \, x \, L \, x \, H$

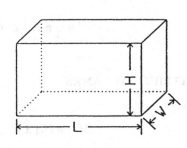

TRAPEZOID

$A = H \, x \, \dfrac{L_1 + L_2}{2}$

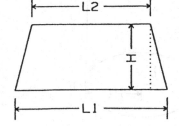

CIRCLE

$A = \pi \, x \, R^2$

$C = 2 \, \pi \, x \, D$

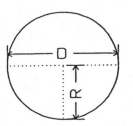

(continued on page 14)

(continued from page 13)

$$R = \frac{D}{2}$$

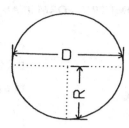

ELLIPSE

$$A = \pi R_1 \times R_2$$

$$C = 2\pi \sqrt{\frac{R_1{}^2 + R_2{}^2}{2}}$$

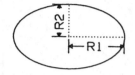

CYLINDER

$$A_1 = 2\pi(R \times H) + 2\pi R^2$$

$$V = \pi R^2 \times H$$

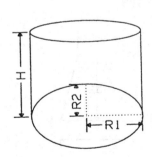

CONE

$$A_1 = \pi(R \times S) + \pi R^2$$

$$V = \frac{1}{3}\pi R^2 \times H$$

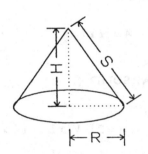

ELLIPTICAL TANKS

$$V = \pi(R_1 \times R_2 \times H)$$

$$A_1 = 2\pi \sqrt{\frac{R_1{}^2 + R_2{}^2}{2}}$$

SPHERE

$$A_1 = 4\pi \times R^2$$

$$V = \frac{4}{3}\pi \times R^3$$

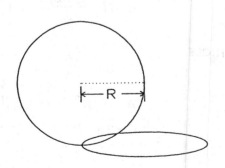

TABLE 1-18: WATER VOLUMES (CUBIC FEET) AND CAPACITIES (US GALLONS) OF CIRCULAR TANKS FILLED TO A 1-FOOT DEPTH

DIAMETER (ft)	VOLUME (ft³)	CAPACITY (gal)	DIAMETER (ft)	VOLUME (ft³)	CAPACITY (gal)
1.00	0.785	5.87	11.0	95.0	711
1.50	1.77	13.2	11.5	104	777
2.00	3.14	23.5	12.0	113	845
2.50	4.91	36.7	12 5	123	918
3.00	7.07	52.9	13.0	133	993
3.50	9.62	72.0	13.5	143	1,070
4.00	12.6	94.0	14.0	154	1,150
4.50	15.9	119	14.5	165	1,240
5.00	19.6	147	15.0	177	1,320
5.50	23.8	178	15.5	189	1,410
6.00	28.3	212	16.0	201	1,500
6.50	33.2	248	16.5	214	1,600
7.00	38.5	288	17.0	227	1,700
7.50	44.2	330	17.5	241	1,800
8.00	50.1	376	18.0	254	1,900
8.50	56.8	424	18.5	269	2,010
9.00	63.6	476	19.0	284	2,120
9.50	70.9	530	19.5	299	2,230
10.0	78.5	588	20.0	314	2,350

For water depths less or greater than 1 foot, multiply the tabulated volumes and capacities by the actual depth in feet

For tanks larger than 20 feet in diameter, multiply the volume and capacity of a tank one-half its diameter by four.

FOR INTERMEDIATE TANK SIZES:

$$\text{VOLUME OF TANK} = \pi R^2 (H) = 3.14 \times (1/2 \text{ DIAMETER})^2 \times \text{WATER DEPTH}$$

EXAMPLE: Calculate volume for the tank below filled to 10 cm below top with 10 cm (4") standpipe.

$$V = [\pi(1/2)(A^2) \times (B-10)] - [\pi(5^2) \times (B-10)] =$$

$$V = [\pi(120^2) \times (120-10)] - [\pi(5^2) \times (120-10)] = 4{,}973{,}760 - 8{,}635 \text{ cm}^3 = 4{,}965{,}125 \text{ cm}^3 = 4{,}965.13 \text{ liters}$$

$$\frac{4{,}965.13 \text{ liters}}{3.784 \text{ liters/gallon}} = 1312.14 \text{ gallons}$$

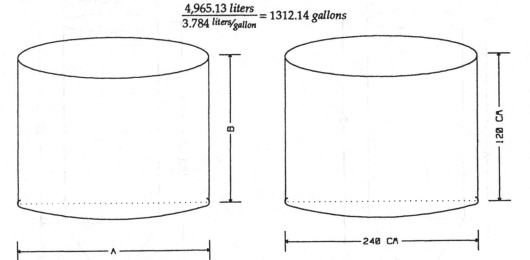

(Source: US Fish and Wildlife Service, *Fish Hatchery Management*, 1979)

15

TABLE 1-19: CALCULATING THE VOLUME OF A CONE BOTTOMED TANK

CALCULATING THE VOLUME OF THE CONE PORTION OF THE TANK:

Using the formula for the volume of a cone:

$$V = \frac{1}{3}\pi R^2 \, x \, H$$

WHERE:

R = 1/2 diameter = 1/2 "B" = 30 cm
H = "C" = 30 cm

$$V = \frac{1}{3}\pi \, 30^2 \, x \, 30$$
$$V = 27,977.4 \text{ cm}^3 = 27.977 \text{ liters}$$

CALCULATING THE VOLUME OF THE CYLINDRICAL PORTION OF THE TANK:

Using the formula for the volume of a cylinder:

$$V = \pi R^2 \, x \, H$$

WHERE:

R = 1/2 diameter = 1/2 "B" = 30 cm
H = "D" = 150 cm

$$V = \pi \, 30^2 \, x \, 150$$
$$V = 423,900 \text{ cm}^3 = 423.9 \text{ liters}$$

TOTAL VOLUME = 423.9 + 27.977 = 451.9 liters

TO CONVERT TO GALLONS, MULTIPLY

$$423,900 \text{ cm}^3 \, (2.6413 \text{ x } 10^{-4)} = 111.95 \text{ gallons}$$

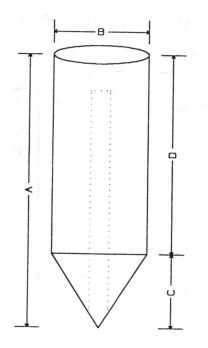

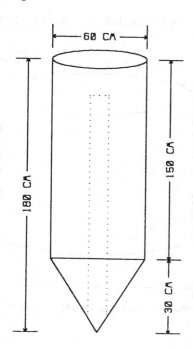

II. Water Chemistry

TABLE 2-1: BASIC INFORMATION ON COMMON ELEMENTS

NAME	SYMBOL	ATOMIC WEIGHT	COMMON VALENCE	EQUIVALENT WEIGHT[1]
ALUMINUM	Al	27.0	3+	9.0
ARSENIC	As	74.9	3+	25.0
BARIUM	Ba	137.3	2+	68.7
BORON	B	10.8	3+	3.6
BROMINE	Br	79.9	1-	79.9
CADMIUM	Cd	112.4	2+	56.2
CALCIUM	Ca	40.1	2+	20.0
CARBON	C	12.0	4-	-
CHLORINE	Cl	35.5	1-	35.5
CHROMIUM	Cr	52.0	3+; 6+	17.3
COPPER	Cu	63.5	2+	31.8
FLUORINE	F	19.0	1-	19.0
HYDROGEN	H	1.0	1+	1.0
IODINE	I	126.9	1+	126.9
IRON	Fe	55.8	2+; 3+	27.9
LEAD	Pb	207.2	2+	103.6
MAGNESIUM	Mg	24.3	2+	12.2
MANGANESE	Mn	54.9	2+; 4+	27.5
MERCURY	Hg	200.6	2+	100.3
NICKEL	Ni	58.7	2+	29.4
NITROGEN	N	14.0	3+	
OXYGEN	O	16.0	2+	8.0
PHOSPHORUS	P	31.0	5+	6.0
POTASSIUM	K	39.1	1+	39.1
SELENIUM	Se	79.0	6+	13.1
SILICON	Si	28.1	4+	6.5
SILVER	Ag	107.9	1+	107.9
SODIUM	Na	23.0	1+	23.0
SULPHUR	S	32.1	2-	16.0
ZINC	Zn	65.4	2+	32.7

[1] Equivalent weight (combining weight) equals atomic weight divided by valence.

(Source: M.J. Hammer, *Waste and Waste Water Technology*; copyright © 1975 - reprinted with permission of John Wiley & Sons, Inc.)

TABLE 2-2: BASIC INFORMATION ON COMMON INORGANIC CHEMICALS USED IN WATER TREATMENT

NAME	FORMULA	COMMON USAGE	MOLECULAR WEIGHT	EQUIVALENT WEIGHT
ACTIVATED CARBON	C	Taste and odor control	12.0	N.A.[1]
ALUMINUM SULFITE (filter alum)	$Al(SO_4)_3 \cdot 14.3 H_2O$	Coagulation	600	100
ALUMINUM HYDROXIDE	$Al(OH)_3$	(Hypothetical combination)	78.0	26.0
AMMONIA	NH_3	Chloramine disinfection	17.0	N.A.
AMMONIUM FLUOSILICATE	$(NH_4)_2SiF_6$	Fluoridation	178	N.A.
AMMONIUM SULFATE	$(NH_4)_2SO_4$	Coagulation	132	66.1
CALCIUM BICARBONATE	$Ca(HCO_3)_2$	(Hypothetical combination)	162	81.0
CALCIUM CARBONATE	$CaCO_3$	Corrosion control	100	50.0
CALCIUM FLUORIDE	CaF_2	Fluoridation	78.1	N.A.
CALCIUM HYDROXIDE	$Ca(OH)_2$	Softening	74.1	37.0
CALCIUM HYPOCHLORITE	$Ca(ClO)_2 \cdot 2H_2O$	Disinfection	179	N.A.
CALCIUM OXIDE (lime)	CaO	Softening	56.1	28.0
CARBON DIOXIDE	CO_2	Recarbonation	44.0	22.0
CHLORINE	Cl_2	Disinfection	71.0	N.A.
CHLORINE DIOXIDE	ClO_2	Taste and odor control	67.0	N.A.
COPPER SULFATE	$CuSO_4$	Algae control	160	79.8
FERRIC CHLORIDE	$FeCl_3$	Coagulation	162	54.1
FERRIC HYDROXIDE	$Fe(OH)_3$	(Hypothetical combination)	107	35.6
FERRIC SULFATE	$Fe_2(SO_4)_3$	Coagulation	400	66.7
FERROUS SULFATE (copperas)	$FeSO_4 \cdot 7H_2O$	Coagulation	278	139
FLUOSILICIC ACID	H_2SiF_6	Fluoridation	144	N.A.
HYDROCHLORIC ACID	HCl	N.A.	36.5	36.5
MAGNESIUM HYDROXIDE	$Mg(OH)_2$	Defluoridation	58.3	29.2
OXYGEN	O	Aeration	32.0	16.0
POTASSIUM PERMANGANATE	$KMnO_4$	Oxidation; disinfection	158	N.A.
SODIUM ALUMINATE	$NaAlO_2$	Coagulation	82.0	N.A.
SODIUM BICARBONATE (baking soda)	$NaHCO_3$	pH adjustment	84.0	84.0
SODIUM CARBONATE (soda ash)	Na_2CO_3	Softening	106	53.0
SODIUM CHLORIDE (common salt)	$NaCl$	Ion-exchanger; regeneration	58.4	58.4
SODIUM FLUORIDE	NaF	Fluoridation	42.0	N.A.
SODIUM HEXAMETAPHOSPHATE	$(NaPO_3)_n$	Corrosion control	N.A.	N.A.
SODIUM HYDROXIDE	$NaOH$	pH adjustment	40.0	40.0
SODIUM HYPOCHLORITE	$NaClO$	Disinfection	74.4	N.A.
SODIUM SILICATE	Na_4SiO_4	Coagulation aid	184	N.A.
SODIUM FLUOSILICATE	Na_2SiF_6	Fluoridation	188	N.A.
SODIUM THIOSULFATE	$Na_2S_2O_3$	Dechlorination	158	N.A.
SULPHUR DIOXIDE	SO_2	Dechlorination	64.1	N.A.
SULFURIC ACID	$H_2SO_4^{2-}$	N.A.	98.1	49.0
WATER	H_2O	N.A.	18.0	N.A.

[1] N.A. = Not Applicable

(Source: M.J. Hammer, *Waste and Waste Water Technology*; copyright © 1975 - reprinted with permission of John Wiley & Sons, Inc.)

TABLE 2-3: SURFACE WATER CRITERIA FOR PUBLIC WATER SUPPLIES

CONSTITUENT OR CHARACTERISTIC	PERMISSIBLE CRITERIA	DESIRABLE CRITERIA
MICROBIOLOGICAL:		
COLIFORM ORGANISMS	10,000/100 ml	<100/100ml
FECAL COLIFORMS	2,000/100ml	<20/100ml
INORGANIC CHEMICALS:	(mg/l)	(mg/l)
ALKALINITY	30 to < 400-500	
AMMONIA	0.5 (as N)	<0.01
ARSENIC[1]	0.05	ABSENT[2]
BARIUM[1]	1.0	ABSENT
BORON[1]	1.0	ABSENT
CADMIUM[1]	0.01	ABSENT
CHLORIDE[1]	250	<25
CHROMIUM (hexavalent[1])	0.05	ABSENT
COPPER[1]	1.0	VIRTUALLY ABSENT[3]
DISSOLVED OXYGEN	> 4 (monthly mean)	NEAR SATURATION
FLUORIDE[1]	0.6	
HARDNESS[1]	VARIABLE	
IRON (filterable)	0.3	VIRTUALLY ABSENT
LEAD[1]	0.05	ABSENT
MANGANESE (filterable)[1]	0.05	ABSENT
NITRATES PLUS NITRITES[1]	10 (as N)	VIRTUALLY ABSENT
pH (range)	6.0-8.5	
PHOSPHORUS[1]	VARIABLE	
SELENIUM	0.01	ABSENT
SILVER[1]	0.05	ABSENT
SULFATE[1]	250	2
TOTAL DISSOLVED SOLIDS (filterable residue)[1]	500	200
URANYL IONS	5	ABSENT
ZINC[1]	5	VIRTUALLY ABSENT
ORGANIC CHEMICALS:		
CARBON CHLOROFORM EXTRACT (CCE)[1]	0.15	
CYANIDE[1]	0.20	ABSENT
METHYLENE BLUE ACTIVE SUBSTANCES[1]	0.5	VIRTUALLY ABSENT
OIL AND GREASE[1]	VIRTUALLY ABSENT	ABSENT
PESTICIDES:		
ALDRIN[1]	0.017	ABSENT
CHLORDANE[1]	0.003	ABSENT
DDT[1]	0.042	ABSENT
DIELDRIN[1]	0.017	ABSENT
ENDRIN[1]	0.001	ABSENT
HEPTACHLOR[1]	0.018	ABSENT
HEPTACHLOR EPOXIDE[1]	0.018	ABSENT
LINDANE[1]	0.056	ABSENT

TABLE 2-3 (cont.):

CONSTITUENT OR CHARACTERISTIC	PERMISSIBLE CRITERIA	DESIRABLE CRITERIA
METHOXYCHLOR[1]	0.035	ABSENT
ORGANIC PHOSPHATES PLUS CARBAMATES[1]	0.1	ABSENT
TOXAPHENE[1]	0.005	ABSENT
HERBICIDES:		
2,4D PLUS 2,4,5-T; PLUS 2,4,5-TP	0.1	ABSENT
PHENOLS[1]	0.001	ABSENT
RADIOACTIVITY:[4]	(pc/l)	(pc/l)
GROSS BETA[1]	1,000	<100
RADIUM 226[1]	3	<1
STRONTIUM[1]	10	<2

[1] Substances that are not significantly affected by the following treatment process; coagulation (less than about 50 mg/l alum, ferric sulfate, or copper as with alkali addition as necessary but without coagulant aids or activated carbon); sedimentation (6 hr or less); rapid sand filtration (3 gpm/sq ft or less); disinfection with chlorine (without consideration to concentration or form of chlorine residual)

[2] Absent = Not detectable by the most sensitive analytical procedure in *Standard Methods*.

[3] Virtually Absent = Detected in *very low* concentrations.

[4] (pc/l) = pico curies per liter

(Source: EPA, *Report of the Committee on Water Quality Criteria*; 1968.)

TABLE 2-4: COMMON RADICALS ENCOUNTERED IN FRESHWATER

NAME	FORMULA	MOLECULAR WEIGHT	ELECTRICAL CHARGE	EQUIVALENT WEIGHT
AMMONIUM	NH_4^+	18.0	1+	18.0
HYDROXYL	OH^-	17.0	1-	17.0
BICARBONATE	HCO_3^-	61.0	1-	61.0
CARBONATE	CO_3^{2-}	60.0	2-	30.0
ORTHOPHOSPHATE	PO_4^{3-}	95.0	3-	31.7
ORTHOPHOSPHATE (mono-hydrogen)	HPO_4^{2-}	96.0	2-	48.0
ORTHOPHOSPHATE (di-hydrogen)	$H_2PO_4^{2-}$	97.0	1-	97.0
BISULFATE	HSO_4^-	97.0	1-	97.0
SULFATE	SO_4^{2-}	96.0	2-	48.0
BISULFITE	HSO_3^{2-}	81.0	1-	81.0
SULFITE	SO_3^{2-}	80.0	2-	40.0
NITRITE	NO_2^-	46.0	1	46.0
NITRATE	NO_3^-	62.0	1-	62.0
HYPOCHLORITE	OCl^-	51.5	1-	51.5

(Source: M.J. Hammer, *Waste and Waste Water Technology*; copyright © 1975 - reprinted by permission of John Wiley & Sons, Inc.)

TABLE 2-5: DISSOLVED OXYGEN (PPM) FOR FRESH WATER IN EQUILIBRIUM WITH AIR AT ALTITUDE
(after Leitritz and Lewis, 1976.)

°F	ELEVATION IN FEET										
	0	1,000	2,000	3,000	4,000	5,000	6,000	7,000	8,000	9,000	10,000
40	13.0	12.5	12.1	11.6	11.2	10.8	10.4	10.0	9.6	9.3	9.0
45	12.1	11.7	11.2	10.8	10.5	10.1	9.7	9.3	9.0	8.7	8.4
46	11.9	11.5	11.1	10.7	10.3	9.9	9.6	9.2	8.9	8.6	8.3
47	11.8	11.3	10.9	10.5	10.2	9.8	9.4	9.1	8.8	8.5	8.2
48	11.6	11.2	10.8	10.4	10.0	9.7	9.3	9.0	8.7	8.3	8.0
49	11.5	11.1	10.6	10.3	9.9	9.5	9.2	8.9	8.6	8.2	7.9
50	11.3	10.9	10.5	10.1	9.8	9.4	9.1	8.7	8.4	8.1	7.8
51	11.2	10.8	10.4	10.0	9.7	9.3	9.0	8.6	8.3	8.0	7.7
52	11.0	10.6	10.2	9.9	9.5	9.2	8.9	8.5	8.2	7.9	7.6
53	10.9	10.5	10.1	9.8	9.4	9.1	8.7	8.4	8.1	7.8	7.5
54	10.8	10.4	10.0	9.6	9.3	9.0	8.6	8.3	8.0	7.7	7.4
55	10.6	10.3	9.9	9.5	9.2	8.9	8.5	8.2	7.9	7.6	7.3
60	10.0	9.6	9.3	8.9	8.6	8.3	8.0	7.7	7.4	7.1	6.8
65	9.4	9.1	8.8	8.4	8.1	7.8	7.5	7.2	7.0	6.7	6.4
70	9.0	8.7	8.4	8.0	7.8	7.4	7.2	6.9	6.7	6.4	6.1
75	8.6	8.3	8.0	7.7	7.4	7.1	6.8	6.5	6.3	6.1	5.8

(Source: US Fish and Wildlife Service, *Fish Hatchery Management; 1982*)

TABLE 2-6: DENSITY OF FRESHWATER (g/cm^3) AT DIFFERENT TEMPERATURES

°C	g/cm^3	°C	g/cm^3
0	0.9998679	16	0.9989701
1	0.9999267	17	0.9988022
2	0.9999679	18	0.9986232
3	0.9999922	19	0.9984331
4	1.0000000	20	0.9982323
5	0.9999919	21	0.9980210
6	0.9999681	22	0.9977993
7	0.9999295	23	0.9975674
8	0.9998762	24	0.9973256
9	0.9998088	25	0.9970739
10	0.9997277	26	0.9968128
11	0.9996328	27	0.9965421
12	0.9995247	28	0.9962623
13	0.9994040	29	0.9959735
14	0.9992712	30	0.9956756
15	0.9991265		

(Source: C.E. Boyd, *Water Quality in Ponds for Aquaculture; 1990*)

TABLE 2-7: CORRESPONDING DENSITIES AND SALINITIES (Zerbe and Taylor, 1953)

DENSITY (g/cm³)	SALINITY (ppt)	DENSITY (g/cm³)	SALINITY (ppt)	DENSITY (g/cm³)	SALINITY (ppt)	DENSITY (g/cm³)	SALINITY (ppt)
0.9991	0.0	1.0036	5.8	1.0081	11.6	1.0126	17.5
0.9992	0.0	1.0037	5.9	1.0082	11.8	1.0127	17.7
0.9993	0.2	1.0038	6.0	1.0083	11.9	1.0128	17.8
0.9994	0.3	1.0039	6.2	1.0084	12.0	1.0129	17.9
0.9995	0.4	1.0040	6.3	1.0085	12.2	1.0130	18.0
0.9996	0.6	1.0041	6.4	1.0086	12.3	1.0131	18.2
0.9997	0.7	1.0042	6.6	1.0087	12.4	1.0132	18.3
0.9998	0.8	1.0043	6.7	1.0088	12.6	1.0133	18.4
0.9999	0.6	1.0044	6.8	1.0089	12.7	1.0134	18.6
1.0000	1.1	1.0045	6.9	1.0090	12.8	1.0135	18.7
1.0001	1.2	1.0046	7.1	1.0091	12.9	1.0136	18.8
1.0002	1.3	1.0047	7.2	1.0092	13.1	1.0137	19.0
1.0003	1.5	1.0048	7.3	1.0093	13.2	1.0138	19.1
1.0004	1.6	1.0049	7.5	1.0094	13.3	1.0139	19.2
1.0005	1.7	1.0050	7.6	1.0095	13.5	1.0140	19.3
1.0006	1.9	1.0051	7.7	1.0096	13.6	1.0141	19.5
1.0007	2.0	1.0052	7.9	1.0097	13.7	1.0142	19.6
1.0008	2.1	1.0053	8.0	1.0098	13.9	1.0143	19.7
1.0009	2.2	1.0054	8.1	1.0099	14.0	1.0144	19.9
1.0010	2.4	1.0055	8.2	1.0100	14.1	1.0145	20.0
1.0011	2.5	1.0056	8.4	1.0101	14.2	1.0146	20.1
1.0012	2.6	1.0057	8.5	1.0102	14.4	1.0147	20.3
1.0013	2.8	1.0058	8.6	1.0103	14.5	1.0148	20.4
1.0014	2.9	1.0059	8.8	1.0104	14.6	1.0149	20.5
1.0015	3.0	1.0060	8.9	1.0105	14.8	1.0150	20.6
1.0016	3.2	1.0061	9.0	1.0106	14.9	1.0151	20.8
1.0017	3.3	1.0062	9.2	1.0107	15.0	1.0152	20.9
1.0018	3.4	1.0063	9.3	1.0108	15.2	1.0153	21.0
1.0019	3.5	1.0064	9.4	1.0109	15.3	1.0154	21.2
1.0020	3.7	1.0065	9.6	1.0110	15.4	1.0155	21.3
1.0021	3.8	1.0066	9.7	1.0111	15.6	1.0156	21.4
1.0022	3.9	1.0067	9.8	1.0112	15.7	1.0157	21.6
1.0023	4.1	1.0068	9.9	1.0113	15.8	1.0158	21.7
1.0024	4.2	1.0069	10.1	1.0114	16.0	1.0159	21.8
1.0025	4.3	1.0070	10.2	1.0115	16.1	1.0160	22.0
1.0026	4.5	1.0071	10.3	1.0116	16.2	1.0161	22.1
1.0027	4.6	1.0072	10.5	1.0117	16.3	1.0162	22.2
1.0028	4.7	1.0073	10.6	1.0118	16.5	1.0163	22.4
1.0029	4.8	1.0074	10.7	1.0119	16.6	1.0164	22.5
1.0030	5.0	1.0075	10.8	1.0120	16.7	1.0165	22.6
1.0031	5.1	1.0076	11.0	1.0121	16.9	1.0166	22.7
1.0032	5.2	1.0077	11.1	1.0122	17.0	1.0167	22.9
1.0033	5.4	1.0078	11.2	1.0123	17.1	1.0168	23.0
1.0034	5.5	1.0079	11.4	1.0124	17.3	1.0169	23.1
1.0035	5.6	1.0080	11.5	1.0125	17.4	1.0170	23.3

TABLE 2-7(cont.):

DENSITY (g/cm^3)	SALINITY (ppt)	DENSITY (g/cm^3)	SALINITY (ppt)	DENSITY (g/cm^3)	SALINITY (ppt)	DENSITY (g/cm^3)	SALINITY (ppt)
1.0171	23.4	1.0211	28.6	1.0251	33.8	1.0291	39.0
1.0172	23.5	1.0212	28.8	1.0252	34.0	1.0292	39.2
1.0173	23.7	1.0213	28.9	1.0253	34.1	1.0293	39.3
1.0174	23.8	1.0214	29.0	1.0254	34.2	1.0294	39.4
1.0175	23.9	1.0215	29.1	1.0255	34.4	1.0295	39.6
1.0176	24.1	1.0216	29.3	1.0256	34.5	1.0296	39.7
1.0177	24.2	1.0217	29.4	1.0257	34.6	1.0297	39.8
1.0178	24.3	1.0218	29.5	1.0258	34.8	1.0298	39.9
1.0179	24.4	1.0219	29.7	1.0259	34.9	1.0299	40.1
1.0180	24.6	1.0220	29.8	1.0260	35.0	1.0300	40.2
1.0181	24.7	1.0221	29.9	1.0261	35.1	1.0301	40.3
1.0182	24.8	1.0222	30.1	1.0262	35.3	1.0302	40.4
1.0183	25.0	1.0223	30.2	1.0263	35.4	1.0303	40.6
1.0184	25.1	1.0224	30.3	1.0264	35.5	1.0304	40.7
1.0185	25.2	1.0225	30.4	1.0265	35.7	1.0305	40.8
1.0186	25.4	1.0226	30.6	1.0266	35.8	1.0306	41.0
1.0187	25.5	1.0227	30.7	1.0267	35.9	1.0307	41.1
1.0188	25.6	1.0228	30.8	1.0268	36.0	1.0308	41.2
1.0189	25.8	1.0229	31.0	1.0269	36.2	1.0309	41.4
1.0190	25.9	1.0230	31.1	1.0270	36.3	1.0310	41.5
1.0191	26.0	1.0231	31.2	1.0271	36.4	1.0311	41.6
1.0192	26.1	1.0232	31.4	1.0272	36.6	1.0312	41.7
1.0193	26.3	1.0233	31.5	1.0273	36.7	1.0313	41.9
1.0194	26.4	1.0234	31.6	1.0274	36.8	1.0314	42.0
1.0195	26.5	1.0235	31.8	1.0275	37.0	1.0315	42.1
1.0196	26.7	1.0236	31.9	1.0276	37.1	1.0316	42.3
1.0197	26.8	1.0237	32.0	1.0277	37.2	1.0317	42.4
1.0198	26.9	1.0238	32.1	1.0278	37.3	1.0318	42.5
1.0199	27.1	1.0239	32.3	1.0279	37.5	1.0319	42.7
1.0200	27.2	1.0240	32.4	1.0280	37.6	1.0320	42.8
1.0201	27.3	1.0241	32.5	1.0281	37.7		
1.0202	27.5	1.0242	32.7	1.0282	37.9		
1.0203	27.6	1.0243	32.8	1.0283	38.0		
1.0204	27.7	1.0244	32.9	1.0284	38.1		
1.0205	27.8	1.0245	33.1	1.0285	38.2		
1.0206	28.0	1.0246	33.2	1.0286	38.4		
1.0207	28.1	1.0247	33.3	1.0287	38.5		
1.0208	28.2	1.0248	33.5	1.0288	38.6		
1.0209	28.4	1.0249	33.6	1.0289	38.8		
1.0210	28.5	1.0250	33.7	1.0290	38.9		

5 °C. Salinity in parts per thousand. (ppt, ‰)

(Source: S. Spotte, *Fish and Invertebrate Culture*; 1979)

TABLE 2-8: DIFFERENCES TO CONVERT HYDROMETER READINGS AT ANY TEMPERATURE TO DENSITY
(Zerbe and Taylor, 1953)

OBSERVED READING	TEMPERATURE OF WATER (°C)												
	2.0	1.0	0.0	1.0	2.0	3.0	4.0	5.0	6.0	7.0	8.0	9.0	10.0
0.9960													
0.9970													
0.9980													
0.9990	-1	-2	-3	-4	-5	-5	-6	-6	-6	-6	-6	-5	-5
1.0000	-2	-3	-4	-5	-5	-6	-6	-6	-6	-6	-6	-5	-5
1.0010	-3	-4	-4	-5	-6	-6	-6	-7	-7	-6	-6	-6	-5
1.0020	-3	-4	-5	-6	-6	-7	-7	-7	-7	-7	-6	-6	-5
1.0030	-4	-5	-6	-6	-7	-7	-7	-7	-7	-7	-6	-6	-5
1.0040	-4	-5	-6	-7	-7	-7	-8	-8	-7	-7	-7	-6	-6
1.0050	-5	-6	-6	-7	-8	-8	-8	-8	-8	-7	-7	-6	-6
1.0060	-6	-6	-7	-8	-8	-8	-8	-8	-8	-8	-7	-6	-6
1.0070	-6	-7	-8	-8	-8	-8	-8	-8	-8	-8	-7	-7	-6
1.0080	-7	-8	-8	-9	-9	-9	-9	-9	-8	-8	-7	-7	-6
1.0100	-8	-9	-9	-10	-10	-10	-10	-9	-9	-8	-8	-7	-6
1.0110	-9	-9	-10	-10	-10	-10	-10	-10	-9	-9	-8	-7	-6
1.0120	-9	-10	-10	-10	-10	-10	-10	-10	-10	-9	-8	-7	-7
1.0130	-10	-10	-11	-11	-11	-11	-11	-10	-10	-9	-8	-8	-7
1.0140	-10	-11	-11	-11	-11	-11	-11	-11	-10	-10	-9	-8	-7
1.0150	-11	-11	-12	-12	-12	-12	-11	-11	-10	-10	-9	-8	-7
1.0160	-12	-12	-12	-12	-12	-12	-12	-11	-11	-10	-9	-8	-7
1.0170	-12	-12	-12	-13	-13	-12	-12	-12	-11	-10	-9	-8	-7
1.0180	-13	-13	-13	-13	-13	-13	-12	-12	-11	-10	-9	-8	-7
1.0190	-13	-13	-14	-14	-13	-13	-13	-12	-12	-11	-10	-9	-8
1.0200	-14	-14	-14	-14	-14	-13	-13	-12	-12	-11	-10	-9	-8
1.0210	-14	-14	-14	-14	-14	-14	-13	-13	-12	-11	-10	-9	-8
1.0220	-15	-15	-15	-15	-15	-14	-14	-13	-12	-11	-10	-9	-8
1.0230	-15	-15	-15	-15	-15	-15	-14	-13	-12	-12	-10	-9	-8
1.0240	-16	-16	-16	-16	-15	-15	-14	-14	-13	-12	-11	-10	-8
1.0250	-16	-16	-16	-16	-16	-15	-15	-14	-13	-12	-11	-10	-8
1.0260	-17	-17	-17	-16	-16	-16	-15	-14	-13	-12	-11	-10	-8
1.0270	-18	-17	-17	-17	-17	-16	-15	-14	-14	-12	-11	-10	-9
1.0280	-18	-18	-18	-17	-17	-16	-16	-15	-14	-13	-11	-10	-9
1.0290	-19	-18	-18	-18	-17	-17	-16	-15	-14	-13	-12	-10	-9
1.0300	-19	-19	-19	-18	-18	-17	-16	-15	-14	-13	-12	-10	-9
1.0310	-20	-19	-19	-19	-18	-17	-16	-16	-15	-13	-12	-10	-9

24

TABLE 2-8 (cont.):

OBSERVED READING	TEMPERATURE OF WATER (°C)											
	11.0	12.0	13.0	14.0	15.0	16.0	17.0	18.0	18.5	19.0	19.5	20.0
0.9960												
0.9970												
0.9980						3	4	5	6	7	8	
0.9990	-4	-3	-2	-1	0	1	3	4	5	6	7	8
1.0000	-4	-3	-2	-1	0	1	3	4	5	6	7	8
1.0010	-4	-3	-2	-1	0	1	3	4	5	6	7	8
1.0020	-4	-3	-2	-1	0	1	3	4	5	6	7	8
1.0030	-4	-3	-2	-1	0	1	3	4	5	6	7	8
1.0040	-5	-4	-3	-1	0	2	3	5	6	6	7	8
1.0050	-5	-4	-3	-1	0	2	3	5	6	7	8	9
1.0060	-5	-4	-3	-1	0	2	3	5	6	7	8	9
1.0070	-5	-4	-3	-2	0	2	3	5	6	7	8	9
1.0080	-5	-4	-3	-2	0	2	3	5	6	7	8	9
1.0090	-5	-4	-3	-2	0	2	3	5	6	7	8	9
1.0100	-5	-4	-3	-2	0	2	3	5	6	7	8	9
1.0110	-5	-4	-3	-2	0	2	3	5	6	7	8	9
1.0120	-6	-4	-3	-2	0	2	3	5	6	7	8	9
1.0130	-6	-4	-3	-2	0	2	3	5	6	7	8	10
1.0140	-6	-4	-3	-2	0	2	4	5	6	8	9	10
1.0150	-6	-4	-3	-2	0	2	4	5	6	8	9	10
1.0160	-6	-5	-3	-2	0	2	4	6	7	8	9	10
1.0170	-6	-5	-3	-2	0	2	4	6	7	8	9	10
1.0180	-6	-5	-3	-2	0	2	4	6	7	8	9	10
1.0190	-6	-5	-3	-2	0	2	4	6	7	8	9	10
1.0200	-6	-5	-3	-2	0	2	4	6	7	8	9	10
1.0210	-6	-5	-3	-2	0	2	4	6	7	8	9	10
1.0220	-7	-5	-3	-2	0	2	4	6	7	8	9	11
1.0230	-7	-5	-4	-2	0	2	4	6	7	8	9	11
1.0240	-7	-5	-4	-2	0	2	4	6	7	8	10	11
1.0250	-7	-5	-4	-2	0	2	4	6	7	8	10	11
1.0260	-7	-5	-4	-2	0	2	4	6	7	9	10	11
1.0270	-7	-5	-4	-2	0	2	4	6	7	9	10	11
1.0280	-7	-6	-4	-2	0	2	4	6	8	9	10	11
1.0290	-7	-6	-4	-2	0	2	4	6	8	9	10	11
1.0300	-7	-6	-4	-2	0	2	4	6	8	9	10	12
1.0310	-8	-6	-4	-2	0	2	4					

TABLE 2-8(cont.):

OBSERVED READING	TEMPERATURE OF WATER (°C)												
	20.5	21.0	21.5	22.0	22.5	23.0	23.5	24.0	24.5	25.0	25.5	26.0	26.5
0.9960						19	20	21					
0.9970			10	11	12	14	15	16	17	18	19	20	22
0.9980	9	10	11	12	13	14	15	16	17	18	19	21	22
0.9990	9	10	11	12	13	14	15	16	17	18	20	21	22
1.0000	9	10	11	12	13	14	15	16	17	19	20	21	22
1 0010	9	10	11	12	13	14	15	17	18	19	20	21	23
1 0020	9	10	11	12	13	14	16	17	18	19	20	22	23
1.0030	9	10	11	12	13	15	16	17	18	19	21	22	23
1.0040	9	10	11	12	14	15	16	17	18	20	21	22	23
1 0050	10	11	12	13	14	15	16	17	19	20	21	22	24
1 0060	10	11	12	13	14	15	16	18	19	20	21	23	24
1.0070	10	11	12	13	14	15	17	18	19	20	21	23	24
1.0080	10	11	12	13	14	16	17	18	19	20	22	23	24
1.0090	10	11	12	13	1S	16	17	18	19	21	22	23	25
1.0100	10	11	12	14	15	16	17	18	20	21	22	24	25
1.0110	10	12	13	14	15	16	17	19	20	21	22	24	25
1.0120	10	12	13	14	15	16	18	19	20	21	23	24	25
1.0130	11	12	13	14	15	16	18	19	20	22	23	24	26
1.0140	11	12	13	14	15	17	18	19	20	22	23	24	26
1.0150	11	12	13	14	16	17	18	20	21	22	23	25	26
1.0160	11	12	13	14	16	17	18	20	21	22	24	25	26
1.0170	11	12	13	15	16	17	18	20	21	22	24	25	27
1.0180	11	12	14	15	16	17	19	20	21	23	24	25	27
1.0190	11	12	14	15	16	18	19	20	21	23	24	26	27
1.0200	11	13	14	15	16	18	19	20	22	23	24	26	27
1.0210	12	13	14	15	17	18	19	21	22	23	25	26	27
1.0220	12	13	14	15	17	18	19	21	22	23	25	26	28
1.0230	12	13	14	16	17	18	20	21	22	24	25	26	28
1.0240	12	13	14	16	17	18	20	21	22	24	25	27	28
1.0250	12	13	15	16	17	18	20	21	23	24	25	27	28
1.0260	12	13	15	16	17	19	20	22	23	24	26	27	29
1.0270	12	14	15	16	17	19	20	22	23	24	26	27	29
1.0280	12	14	15	16	18	19	20	22	23	25	26	28	29
1.0290	13	14	15	16	18	19	21	22	23				
1.0300	13	14	15	16	18								
1.0310													

TABLE 2-8 (cont.):

TEMPERATURE OF WATER (°C)

OBSERVED READING	27.0	27.5	28.0	28.5	29.0	29.5	30.0	30.5	31.0	31.5	32.0	32.5	33.0
0.9960	23	24	25	27	28	29	31	32	34	35	37	38	40
0.9970	23	24	26	27	28	30	31	33	34	36	37	39	40
0.9980	23	25	26	27	29	30	31	33	34	36	38	39	41
0.9990	24	25	26	28	29	30	32	33	35	36	38	39	41
1.0000	24	25	26	28	29	31	32	34	35	37	38	40	41
1.0010	24	25	27	28	30	31	32	34	35	37	39	40	42
1.0020	24	26	27	28	30	31	33	34	36	37	39	41	42
1.0030	25	26	27	29	30	32	33	35	36	38	39	41	42
1.0040	25	26	28	29	30	32	33	35	36	38	40	41	43
1.0050	25	26	28	29	31	32	34	35	37	38	40	42	43
1.0060	25	27	28	30	31	32	34	36	37	39	40	42	44
1.0070	26	27	28	30	31	33	34	36	38	39	41	42	44
1.0080	26	27	29	30	32	33	35	36	38	39	41	43	44
1.0090	26	28	29	30	32	33	35	36	38	40	41	43	45
1.0100	26	28	29	31	32	34	35	37	38	40	42	43	45
1.0110	27	28	30	31	32	34	36	37	39	40	42	44	45
1.0120	27	28	30	31	33	34	36	37	39	41	42	44	46
1.0130	27	29	30	32	33	35	36	38	39	41	43	44	46
1.0140	27	29	30	32	33	35	36	38	40	41	43	45	46
1.0150	28	29	31	32	34	35	37	38	40	42	43	45	47
1.0160	28	29	31	32	34	35	37	39	40	42	44	45	47
1.0170	28	30	31	33	34	36	37	39	40	42	44	46	47
1.0180	28	30	31	33	34	36	38	39	41	42	44	46	48
1.0190	29	30	32	33	35	36	38	39	41	43	44	46	48
1.0200	29	30	32	33	35	37	38	40	41	43	45	47	48
1.0210	29	31	32	34	35	37	38	40	42	43	45	47	49
1.0220	29	31	32	34	36	37	39	40	42	44	45	47	49
1.0230	30	31	33	34	36	37	39	41	42	44	46	47	49
1.0240	30	31	33	34	36	37	39	41	42	44	46	48	49
1.0250	30	31	33	35	36	38	39	41	43	44	46	48	50
1.0260	30	32	33	35	37	38	40	41	43	45	46	48	50
1.0270	30	32	34	35	37	38	40						
1.0280	31	32											
1.0290													
1.0300													
1.0310													

Measure the actual specific gravity and temperature simultaneously. Estimate specific gravity to the fourth decimal place (ten thousandths). Correct the temperature by reading from the table values and adding to the observed reading

EXAMPLE: Actual specific gravity of observed reading of 1.022 at a temperature of 20° C, from the table 1.022 + 11 = 1.0231.

(Source: S. Spotte, *Fish and Invertebrate Culture*; 1979)

TABLE 2-9: AIR SOLUBILITY OF OXYGEN (mg/l) IN SEAWATER (0-40 g/kg, ‰)

TEMPERATURE (°C)	SALINITY (g/kg)								
	0	5	10	15	20	25	30	35	40
0	14.621	14.120	13.636	13.167	12.714	12.277	11.854	11.445	11.051
1	14.216	13.733	13.266	12.815	12.378	11.956	11.548	11.154	10.773
2	13.829	13.364	12.914	12.478	12.057	11.650	11.256	10.875	10.507
3	13.460	13.011	12.577	12.156	11.750	11.356	10.976	10.608	10.252
4	13.107	12.674	12.255	11.849	11.456	11.076	10.708	10.352	10.008
5	12.770	12.352	11.947	11.554	11.175	10.807	10.451	10.107	9.774
6	12.447	12.043	11.652	11.272	10.905	10.550	10.206	9.872	9.550
7	12.139	11.748	11.369	11.002	10.647	10.303	9.970	9.647	9.335
8	11.843	11.465	11.098	10.743	10.399	10.066	9.744	9.431	9.128
9	11.559	11.194	10.839	10.495	10.162	9.839	9.526	9.223	8.930
10	11.288	10.933	10.590	10.257	9.934	9.621	9.318	9.024	8.739
11	11.027	10.684	10.351	10.028	9.715	9.412	9.117	8.832	8.556
12	10.777	10.444	10.121	9.808	9.505	9.210	8.925	8.648	8.379
13	10.537	10.214	9.901	9.597	9.302	9.017	8.739	8.470	8.210
14	10.306	9.993	9.689	9.394	9.108	8.830	8.561	8.300	8.046
15	10.084	9.780	9.485	9.198	8.921	8.651	8.389	8.135	7.888
16	9.870	9.575	9.289	9.010	8.740	8.478	8.223	7.976	7.737
17	9.665	9.378	9.099	8.829	8.566	8.311	8.064	7.823	7.590
18	9.467	9.188	8.917	8.654	8.399	8.151	7.910	7.676	7.449
19	9.276	9.005	8.742	8.486	8.237	7.995	7.761	7.533	7.312
20	9.092	8.828	8.572	8.323	8.081	7.846	7.617	7.395	7.180
21	8.914	8.658	8.408	8.166	7.930	7.701	7.479	7.262	7.052
22	8.743	8.493	8.250	8.014	7.785	7.561	7.344	7.134	6.929
23	8.578	8.334	8.098	7.867	7.644	7.426	7.214	7.009	6.809
24	8.418	8.181	7.950	7.725	7.507	7.295	7.089	6.888	6.693
25	8.263	8.032	7.807	7.588	7.375	7.168	6.967	6.771	6.581
26	8.113	7.888	7.668	7.455	7.247	7.045	6.849	6.658	6.472
27	7.968	7.748	7.534	7.326	7.123	6.926	6.734	6.548	6.366
28	7.827	7.613	7.404	7.201	7.003	6.810	6.623	6.441	6.263
29	7.691	7.482	7.278	7.079	6.886	6.698	6.515	6.337	6.164
30	7.558	7.354	7.155	6.961	6.772	6.589	6.410	6.236	6.066
31	7.430	7.230	7.036	6.846	6.662	6.483	6.308	6.137	5.972
32	7.305	7.110	6.920	6.735	6.555	6.379	6.208	6.042	5.880
33	7.183	6.993	6.807	6.626	6.450	6.278	6.111	5.948	5.790
34	7.065	6.879	6.697	6.520	6.348	6.180	6.017	5.857	5.702
35	6.949	6.767	6.590	6.417	6.248	6.084	5.924	5.768	5.617
36	6.837	6.659	6.485	6.316	6.151	5.991	5.834	5.681	5.533
37	6.727	6.553	6.383	6.218	6.056	5.899	5.746	5.597	5.451
38	6.619	6.449	6.283	6.121	5.963	5.810	5.660	5.513	5.371
39	6.514	6.348	6.186	6.027	5.873	5.722	5.575	5.432	5.292
40	6.412	6.249	6.090	5.935	5.783	5.636	5.492	5.352	5.215

Based on Benson and Krause. 1984.

(Source: J. Huguenin and J. Colt, *Design and Operating Guide for Aquaculture Seawater Systems*; 1989)

TABLE 2-10: SEAWATER PROPERTIES AS A FUNCTION OF TEMPERATURE AND SALINITY

See key for parameters and units The kinematic viscosity equals the table values x 10^{-6}. Also provided is an equation for calculating the dynamic (absolute) viscosity from given data and some relevant units and conversions. All measurements are at one atmosphere of pressure. Values will be much more precise than required for most calculations.

$$\text{MASS DENSITY (MD)} = \text{slugs/ft}^3 \text{ (kg/m}^3)$$

$$\text{SPECIFIC WEIGHT (SW)} = \text{lb/ft}^3 \text{ (kn/m}^3)$$

$$\text{KINEMATIC VISCOSITY (KV)} = \text{X } 10^{-6} \text{ ft}^2\text{/s (X } 10^{-6} \text{ m}^2\text{/s)}$$

TEMPERATURE	SALINITY (g/kg, ‰)				
	10	20	25	30	35
5 °C 41 °F	MD = 1.9556 (1007.9)	MD = 1.9709 (1015.8)	MD = 1.9787 (1019.8)	MD = 1.9862 (1023.7)	MD = 1.9939 (1027.7)
	SW = 62.915 (9.884)	SW = 63.412 (9.962)	SW = 63.654 (10.000)	SW = 63.902 (10.039)	SW = 64.150 (10.078)
	KV = 16.485 (1.5315)	KV = 16.614 (1.5435)	KV = 16.679 (1.5495)	KV = 16.743 (1.5555)	KV = 16.809 (1.5616)
15 °C 59 °C	MD = 1.9534 (1006.8)	MD = 1.9682 (1014.4)	MD = 1.9757 (1018.3)	MD = 1.9831 (1022.1)	MD = 1.9906 (1026.0)
	SW = 62.845 (9.873)	SW = 63.323 (9.948)	SW = 63.565 (9.986)	SW = 63.806 (10.024)	SW = 64.042 (10_061)
	KV = 12.403 (1.1523)	KV = 12.531 (1.1642)	KV = 12.595 (1.1701)	KV = 12.658 (1.1760)	KV = 12.721 (1.1818)
20 °C 68 °F	MD = 1.9515 (1005.8)	M D = 1.9662 (1013.4)	MD = 1.9736 (1017.2)	M D = 1.9810 (1021.0)	M D = 1.9884 (1024.8)
	SW = 62.782 (9.863)	SW = 63.259 (9.938)	SW = 63.495 (9.975)	SW = 63.730 (10.012)	SW = 63.966 (10.049)
	KV = 10.946 (1.0169)	KV = 11.072 (1.0286)	KV = 11.134 (1.0344)	KV = 11.196 (1.0401)	KV = 11.258 (1.0459)
25 °C 77 °F	MD = 1.9492 (1004.6)	MD = 1.9637 (1012.1)	MD = 1.9709 (1015.8)	MD = 1.9783 (1019.6)	MD = 1.9854 (1023.3)
	SW = 62.705 (9.851)	SW = 63.176 (9.925)	SW= 63.412 (9.962)	SW = 63.647 (9.999)	SW = 63.883 (10.036)
	KV = 9.752 (0.9060)	KV = 9.876 (0.9175)	KV= 9.936 (0.9231)	KV = 9.996 (0.9287)	KV = 10.057 (0.9343)
30 °C 86 °F	MD = 1.9463 (1003.1)	MD = 1.9606 (1010.5)	MD = 1.9680 (1014.3)	MD = 1.9752 (1OIX.O)	MD = 1.9823 (1021.7)
	SW = 62.616 (9.837)	SW = 63.081 (9.910)	SW = 63.310 (9.946)	SW = 63.546 (9.983)	SW = 63.781 (10.020)
	KV = 8.761 (0.8139)	KV = 8.881 (0.8251)	KV = 8.942 (0.8307)	KV = 9.000 (0.8361)	KV = 9.US'J (0.8416)
40 °C 104 °F	MD = 1.9395 (999.6)	MD = 1.9536 (1006.9)	MD = 1.9608 (1010.6)	MD = 1.9680 (1014.3)	MD = 1.9752 (1018.0)
	SW = 62.393 (9.802)	SW = 62.852 (9.874)	SW = 63.087 (9.911)	SW = 63.316 (9.947)	SW = 63.546 (9.983)
	KV = 7.221 (0.6709)	KV= 7.337 (0.6816)	KV= 7.394 (0.6869)	KV = 7.450 (0.6921)	KV = 7.506 (0.6973)

MASS IN SLUGS = weight in lb/gravitational constant in ft/s^2; g = 32.2 ft/s^2 = 9.81 m/s^2

SW= (MD) g; Absolute (dynamic) viscosity = (KV)(MD)

(Source: J. Huegenin and J. Colt, *Design and Operating Guide for Aquaculture Seawater Systems*; 1989)

TABLE 2-11: MOLE FRACTION OF UN-IONIZED AMMONIA: 0-5 g/kg SALINITY (Emerson et al., 1975)

TEMPERATURE (°C)	pH							
	7.0	7.8	7.9	8.0	8.1	8.2	8.3	9.0
5	0.0012	0.0078	0.0098	0.0123	0.0154	0.0193	0.0242	0.1107
10	0.0019	0.0116	0.0145	0.0182	0.0229	0.0286	0.0357	0.1567
15	0.0027	0.0169	0.0212	0.0266	0.0332	0.0415	0.0516	0.2144
20	0.0039	0.0243	0.0304	0.0380	0.0474	0.0590	0.0731	0.2833
25	0.0056	0.0346	0.0431	0.0537	0.0667	0.0825	0.1017	0.3621
30	0.0080	0.0483	0.0600	0.0744	0.0919	0.1130	0.1382	0.4455
35	0.0111	0.0663	0.0820	0.1011	0.1240	0.1513	0.1833	0.5293
40	0.0153	0.0894	0.1100	0.1345	0.1638	0.1978	0.2367	0.6088

Based on freshwater equilibrium constants (Emerson et al., 1975).

TABLE 2-12: MOLE FRACTION OF UN-IONIZED AMMONIA: 5-40 g/kg SALINITY (Emerson et al., 1975)

TEMPERATURE (°C)	pH							
	7.0	7.8	7.9	8.0	8.1	8.2	8.3	9.0
5	0.0007	0.0043	0.0054	0.0068	0.0085	0.0107	0.0135	0.06410
10	0.0010	0.0064	0.0081	0.0101	0.0127	0.0160	0.0200	0.0928
20	0.0022	0.0136	0.0171	0.0215	0.0269	0.0336	0.0419	0.1798
25	0.0031	0.0195	0.0244	0.0305	0.0381	0.0475	0.0591	0.2394
30	0.0044	0.0274	0.0343	0.0428	0.0532	0.0661	0.0818	0.3088
35	0.0062	0.0381	0.0475	0.0591	0.0733	0.0905	0.1114	0.3858
40	0.0086	0.0521	0.0647	0.0801	0.0988	0.1213	0.1481	0.4665

Saltwater data from Khoo et al.(1977), salinity and the equation for the computation of ionic strength (Whitfield, 1974). Converted to the NBS pH scale by addition of 0.149 to freshwater negative logarithm of the equilibrium constants (Bates, 1975).

DETERMINATION OF NH₃-N CONCENTRATION FROM TOTAL AMMONIA NITROGEN (TAN)

$$NH3\text{-}N = (a)\,(TAN)$$

WHERE:,

a = Mole Fraction of Un-ionized Ammonia

TAN = Total Ammonia Nitrogen

EXAMPLE:

A tank of seawater (35ppt) at a temperature of 25°C has a pH of 8.2 and a Total Ammonia Nitrogen level of 0.4 mg/l. Calculate the concentration of un-ionized ammonia in the tank.

NH_3-N = (a) TAN where a = 0.475 (taken from the table above at 25°C, pH = 8.2)

NH_3-N = (0.0475) (0.4) = 0.0190 mg/l = 19 µg/l or ppb

(Source: J. Huegenin and J. Colt, *Design and Operating Guide for Seawater Aquaculture Systems*; 1989)

TABLE 2-13: SOLUBILITY OF NITROGEN (mg/liter) IN WATER AT DIFFERENT TEMPERATURES AND SALINITIES FROM MOIST AIR WITH PRESSURE OF 760 MM HG. (Colt, 1984)

TEMPERATURE (°C)	SALINITY (ppt)								
	0	5	10	15	20	25	30	35	40
0	23.04	22.19	21.38	20.60	19.85	19.12	18.42	17.75	170
5	20.33	19.61	18.92	18.26	17.61	16.99	16.40	15.82	15.26
0	18.14	17.53	16.93	16.36	15.81	15.27	14.75	14.25	13.77
15	16.36	15.82	15.31	14.81	14.32	13.86	13.40	12.97	12.54
20	14.88	14.41	13.96	13.52	13.09	12.68	12.28	11.89	11.52
25	13.64	13.22	12.82	12.43	12.05	11.69	11.33	10.99	10.65
30	12.58	12.21	11.85	11.50	11.17	10.84	10.52	10.21	9.91
35	11.68	11.34	11.02	10.71	10.40	10.10	9.82	9.54	9.26
40	10.89	10.59	10.29	11.01	9.73	9.46	9.20	8.94	8.70

(Source: C. E. Boyd, *Water Quality in Ponds for Aquaculture*; 1990)

TABLE 2-14: VAPOR PRESSURE OF PURE WATER AT DIFFERENT TEMPERATURES

°C	mm Hg	°C	mm Hg	°C	mm Hg
0	4.579	12	10.518	24	22.377
1	4.926	13	11.231	25	23.756
2	5.294	14	11.987	26	25.209
3	5.685	15	12.788	27	26.739
4	6.101	16	13.634	28	28.349
5	6.543	17	14.530	29	30.043
6	7.013	18	15.477	30	31.824
7	7.513	19	16.477	31	31.695
8	8.045	20	17.535	32	35.663
9	8.609	21	18.650	33	37.729
10	9.209	22	19.827	34	39.898
11	9.844	23	21.068	35	42.175

(Source: C. E. Boyd, *Water Quality in Ponds for Aquaculture*; 1990)

TABLE 2-15: PERCENTAGE UN-IONIZED HYDROGEN SULFIDE IN AQUEOUS SOLUTION AT DIFFERENT pH VALUES AND TEMPERATURES

pH	TEMPERATURE (°C)								
	16	18	20	22	24	26	28	30	32
5.0	99.3	99.2	99.2	99.1	99.1	99.0	98.9	98.9	98.9
5.5	97.7	97.6	97.4	97.3	97.1	96.9	96.7	96.5	96.3
6.0	93.2	92.8	92.3	92.0	91.4	90.8	90.3	89.7	89.1
6.5	81.2	80.2	79.2	78.1	77.0	75.8	74.6	73.4	72.1
7.0	57.7	56.2	54.6	53.0	51.4	49.7	48.2	46.6	45.0
7.5	30.1	28.9	27.5	26.3	25.0	23.8	22.7	21.6	20.6
8.0	12.0	11.4	10.7	10.1	9.6	9.0	8.5	8.0	7.6
8.5	4.1	3.9	3.7	3.4	3.2	3.0	2.9	2.7	2.5
9.0	1.3	1.3	1.2	1.1	1.0	1.0	0.9	0.9	0.8

(Source: C. E. Boyd, *Water Quality in Ponds for Aquaculture*; 1990)

TABLE 2-16: FACTORS FOR CONVERTING TOTAL ALKALINITY TO MILLIGRAMS OF AVAILABLE CARBON PER LITER.

MULTIPLY FACTORS BY TOTAL ALKALINITY.(after Backmann, 1962)

pH	TEMPERATURE (^{o}C)					
	5	10	15	20	25	30[1]
5.0	8.19	7.16	6.55	6.00	5.61	5.20
5.5	2.75	2.43	2.24	2.06	1.94	1.84
6.0	1.03	0.93	0.87	0.82	0.78	0.73
6.5	0.49	0.46	0.44	0.42	0.41	0.40
7.0	0.32	0.31	0.30	0.30	0.29	0.29
7.5	0.26	0.26	0.26	0.26	0.26	0.26
8.0	0.25	0.25	0.25	0.24	0.24	0.24
8.5	0.24	0.24	0.24	0.24	0.24	0.24
9.0	0.23	0.23	0.23	0.23	0.23	0.23

[1] estimated by extrapolation.

The amount of carbon dioxide available to plants for use in phtosynthesis is a function of pH, temperature, and total alkalinity. To calculate total carbon availability from the above table:

EXAMPLE:

What is the total available carbon (mg/l) for water of pH 8 at 15° C with a total alkalinity of 80 mg/l?

TOTAL CARBON = 80 X 0.25 = 20 mg/l

(Source: C. E. Boyd, *Water Quality in Ponds for Aquaculture*; 1990)

TABLE 2-17: SOLUBILITY OF CARBON DIOXIDE (mg/liter) IN WATER AT DIFFERENT TEMPERATURES AND SALINITIES FROM MOIST AIR WITH PRESSURE OF 760 MM Hg (Colt, 1984)

TEMPERATURE (^{o}C)	SALINITY (ppt)								
	0	5	10	15	20	25	30	35	40
0	1.09	1.06	1.03	1.00	0.98	0.95	0.93	0.90	0.88
5	0.89	0.87	0.85	0.83	0.81	0.79	0.77	0.75	0.73
10	0.75	0.73	0.71	0.69	0.68	0.66	0.64	0.63	0.61
15	0.63	0.62	0.60	0.59	0.57	0.56	0.54	0.53	0.52
20	0.54	0.53	0.51	0.50	0.49	0.48	0.47	0.46	0.45
25	0.46	0.45	0.44	0.43	0.42	0.41	0.41	0.40	0.39
30	0.40	0.39	0.39	0.38	0.37	0.36	0.35	0.35	0.34
35	0.35	0.35	0.34	0.33	0.33	0.32	0.31	0.31	0.30
40	0.31	0.30	0.30	0.29	0.29	0.28	0.28	0.27	0.27

(Source: C. E. Boyd, *Water Quality in Ponds for Aquaculture*; 1990)

TABLE 2-18: ELEMENTAL COMPOSITION OF SEAWATER (From Bowen, 1966)

ELEMENT	CHEMICAL FORM	CONCENTRATION (mg/l)
Ag	$AgCl_2-$	0.0003
Al	-	0.01
Ar	Ar	0.6
As	AsO_4H^{2-}	0.003
Au	$AuCl_4-$	0.000011
B	$B(OH)_3$	4.6
Ba	Ba^{2+}	0.03
Be	-	0.0000006
Bi	-	0.000017
Br	Br^-	65
C	CO_3H^- (organic C)	28
Ca	Ca^{2+}	400
Cd	Cd^{2+}	0.00011
Ce	-	0.0004
Cl	Cl^-	19,000
Co	Co^{2+}	0.00027
Cr	-	0.00005
Cs	Cs^+	0.0005
Cu	Cu^{2+}	0.003
F	F^-	1.3
Fe	$Fe(OH)_3$	0.01
Ga	-	0.00003
Ge	$Ge(OH)_4$	0.00007
H	H_2O	108,000
He	He	0.0000069
Hf	-	<0.000008
Hg	$HgCl_4^{2-}$	0.00003
I	$I^-;IO_3-$	0.06
In	-	<0.02
K	K^+	380
Kr	Kr	0.0025
La	-	0.000012
Li	Li^+	0.18
Mg	Mg^{2+}	1350
Mn	Mn^{2+}	0.002
Mo	$MoO_4 2-$	0.01
N	Organic N; NO_3-; NH^{4+}	0.5
Na	Na^+	10,500
Nb	-	0.00001
Ne	Ne	0.00014
Ni	Ni^{2+}	0.0054
O	$H_2O; O_2; SO_4^{2-}$	857,000
P	PO_4H^{2-}	0.07
Pa	-	2×10^{-9}
Pb	Pb^{2+}	0.00003
Ra	-	6×10^{-11}
Rb	Rb^+	0.12
Rn	Rn	6×10^{-16}
S	$SO_4 2-$	885
Sb	-	0.00033
Sc	-	<0.000004
Se	-	0.0009
Si	$Si(OH)_4$	3
Sn	-	0.003
Sr	Sr^{9+}	8.1
Ta	-	<0.0000025
Th	-	0.0005
Ti	-	0.0001
Tl	Tl^+	<0.00001
U	$UO_2(CO_3)_3^{4-}$	0.003
V	$VO_5H_3^{2-}$	0.002
W	WO_4^{2-}	0.0001
Xe	Xe	0.000052
Y	-	0.0003
Zn	Zn^{2+}	0.01
Zr	-	0.000022

(Source: S. Spotte, *Fish and Invertebrate Culture*; 1979)

III. Enrichment Formulations

TABLE 3-1: SALT CONCENTRATIONS FOR THE MODIFIED SEGEDI-KELLEY MEDIUM FORMULA (S = 35.3 ‰)
(Segedi and Kelley, 1964)

ADDITIVE	CONCENTRATION
NaCl	27.60 g/l
$MgSO_4 \cdot 7H_2O$	6.89 g/l
$MgCl_2 \cdot 6H_2O$	5.40 g/l
$CaCl_2 \cdot 2H_2O$	1.38 g/l
KCl	0.60 g/l
$NaHCO_3$	0.21 g/l
KBr	26.90 mg/l
$SrCl_2 \cdot 6H_2O$	19.84 mg/l
$MnSO_4 \cdot H_2O$	3.97 mg/l
$NaH_2PO_4 \cdot 7H_2O$	3.97 mg/l
LiCl	0.99 mg/l
$Na_2MoO_4 \cdot 2H_2O$	0.99 mg/l
$Na_2S_2O_3 \cdot 5H_2O$	0.99 mg/l
$Al_2(SO_4)_3 \cdot 18H_2O$	0.85 mg/l
RbCl	149.00 µg/l
$ZnSO_4 \cdot 7H_2O$	95.90 µg/l
$CoSO_4 \cdot 7H_2O$	89.30 µg/l
KI	89.30 µg/l
$CuSO_4 \cdot 5H_2O$	9.90 µg/l

(Source: S. Spotte, *Seawater Aquaria: The Captive Environment*; 1979)

TABLE 3-2: OTT'S (1965) ARTIFICIAL SEAWATER

ADDITIVE	CONCENTRATION
SALTS	**MAKE UP TO 1 LITER DISTILLED H₂O**
NaCl	21 g/l
$MgSO_4 \cdot 7H_2O$	6 g/l
$MgCl_2 \cdot 6H_2O$	5 g/l
$CaCl_2 \cdot 2H_2O$	1 g/l
KCl	0.8 g/l
NaBr	0.1 g/l
$NaNO_3$	0.2 g/l
$NaHCO_3$	0.2 g/l
H_3BO_3	0.06 g/l
$Na_2SiO_3 \cdot 9H_2O$	0.01 g/l
$Sr(NO_3)_2$	0.03 g/l
Na_2HPO_4	0.02 g/l

To the above, add 1 ml each of the micronutrients listed under the formula for BOLD'S BASAL MEDIA (Table 3-7). This artificial seawater may be used in preparing Erdschreiber or von Stosch's enrichment media.

(Source: H. C. Bold and M. J. Wynne, *Introduction to the Algae*; 1978)

34

TABLE 3-3: INSTANT OCEAN[TM] ARTIFICIAL SEAWATER MIXTURE [1]

ADDITIVE	CONCENTRATION (μm/l)	ADDITIVE	CONCENTRATION (μm/l)
Cl	5.19×10^5	MoO_4	4.40
Na	4.44×10^5	S_2O_3	3.60
SO_4	2.60×10^4	Li	28.80
Mg	4.94×10^4	Rb	1.20
K	9.50×10^3	I	0.55
Ca	9.20×10^3	EDTA	0.13
HCO_3	2.30×10^3	Al	1.50
H_3BO_3	4.04×10^2	Zn	0.31
Br	2.50×10^2	V	0.39
Sr	91.30	Co	0.17
PO_4	10.50	Fe	0.18
Mn	18.00	Cu	0.05

[1] Aquarium Systems, Inc. Twinbrook, Mentor, Ohio 44060

(Source: J.P. McVey, CRC *Handbook of Mariculture: Volume I - Crustacean Aquaculture*; copyright ©1983 - reprinted with permission of CRC Press, Boca Raton, FL)

TABLE 3-4: SALT CONCENTRATIONS FOR THE GP MEDIUM FORMULA (S = 33.1 ‰, ppt)

SALT	SOLUTION	CONCENTRATION
NaCl	A	26.00 g/l
$MgSO_4 \cdot 7H_2O$	A	6.58 g/l
$MgCl_2 \cdot 6H_2O$	A	4.88 g/l
$CaCl_2 \cdot 2H_2O$	A	1.46 g/l
KCl	A	0.675 g/l
$NaHCO_3$	A	0.184 g/l
KBr	B	95.3 mg/l
$SrCl_2 \cdot 6H_2O$	B	24.2 mg/l
$NaH_2PO_4 \cdot 7H_2O$	B	4.0 mg/l
LiCl	B	1.04 mg/l
$Al_2(SO_4)_3 \cdot 18H_2O$	B	0.0235 mg/l
H_3BO_3	C	24,2000.0 μg/l
Na_2EDTA	C	9,440.0 μg/l
Fe citrate H_2O	C	3,830.0 μg/l
$Na_2MoO_4 \cdot 2H_2O$	C	2,220.0 μγ/l
$MnSO_4 \cdot H_2O$	C	1,610.0 μg/l
$ZnSO_4 \cdot 7H_2O$	C	1,425.0 μg/l
$CuSO_4 \cdot 5H_2O$	C	97.7 μg/l
KI	C	79.1 μg/l
$CoSO_4 \cdot 7H_2O$	C	13.4 μg/l
$Na_2VO_4 \cdot 4H_2O$	C	9.24 μg/l
Thiamine HCl	D	1,953.0 μg/l
Cyanocobalamin	D	0.977 μg/l

SOLUTION A: Dissolve salts separately; dilute to approximately 75% by volume; cover and aerate.

SOLUTION B: Dissolve in distilled water; add to Solution A on second day

SOLUTION C: Dissolve each salt in distilled water with 2 molar equiv. of Na_2EDTA; boil and dilute to volume; add to solutions A and B on third day.

SOLUTION D: Requires no special preparation; dissolve in distilled water

(Source: S. Spotte, *Seawater Aquaria: The Captive Environment*; copyright ©1979 - reprinted by permission of John Wiley &Sons, Inc.)

TABLE 3-5: GATES AND WILSON'S NH ARTIFICIAL SEAWATER MEDIUM

ADDITIVE	CONCENTRATION
NaCl	24.0 g/l
KCl	0.6 g/l
$MgCl_2 \cdot 6H_2O$	4.5 g/l
$MgSO_4 \cdot 7H_2O$	6.0 g/l
$CaCl_2$	0.7 g/l
K_2HPO_4	10.0 mg/l
Vitamin B_{12}	1.0 µg/l
Thiamin HCl	10.0 mg/l
Biotin	0.5 ug/l
SULPHIDES[1]	1.0 ml/l
VITAMIN MIX [2]	0.1 ml/l
TRACE METALS MIX[3]	5.0 ml/l
Adenine sulphate	1.0 mg/l
Tris buffer	0.1 g/l
NaEDTA	10.0 mg/l
NUTRIENT MIXES	
[1] SULPHIDES	**MAKE UP TO ONE LITER DISTILLED WATER**
NH_4Cl	0.2 g
KH_2PO_4	0.1 g
$MgCl_2 \cdot 6H_2O$	0.04 g
$NaHCO_3$	0.2 g
$Na_2SiO_3 \cdot 9H_2O$	0.15 g
[2] VITAMIN MIX	**MAKE UP TO 100 ML DISTILLED WATER**
Thiamin-HCl	20 mg
Biotin	50 µg
Vitamin B_{12}	5 µg
Folic acid	0.25 mg
PABA	1.0 mg
Nicotine acid	10 mg
Thymine	80 mg
Choline	50 mg
Inositol	100 mg
Patrescine	0.8 mg
Riboflavin	0.5 mg
Pyridoxine	4.0 mg
Orotic acid	26 mg
Fe Tartrate	2.5 ml (5 mg Fe)
[3] TRACE METAL MIX (1 % SOLUTION)	**MAKE UP TO 100 ML DISTILLED WATER**
H_3BO_3	3.0 ml (5.1 mg B)
H_2SeO_3	0.1 ml (1.0 mg Se)
NH_4VO_3	0.12 ml (0.5 mg V)
K_2CrO_4	0.11 ml (0.2 mg Cr)
$MnCl_2$	0.37 ml (1.0 mg Mn)
TiO_2	0.11 ml (5.0 mg Ti)
Na_2SiO_3	5.0 ml (5.0 mg Si)
$ZrOCl_2$	0.4 ml (2.0 mg Zr)
$BaCl_2$	0.15 ml (1.0 mg Ba)

(Source: T.V.R. Pillay, *Aquaculture Principles and Practices*; 1990)

TABLE 3-6: BOLDS BASAL MEDIUM (BBM) (Bischoff and Bold, 1963)

BBM is a medium useful for culturing Chlorophyceae, Chrysophyceae, Cyanophyceae and Rhodophyceae.

Six macronutrient and four trace metal stock solutions are prepared.

ADDITIVE	CONCENTRATION
MACRONUTRIENT (STOCK)	**ADD 10 ML / 940 ML DISTILLED H_2O**
$NaNO_3$	10.0 g/400 ml
$CaCl_2 \cdot 2H_2O$	1.0 g/400 ml
$MgSO_4 \cdot 7H_2O$	3.0 g/400 ml
K_2HPO_4	3.0 g/400 ml
KH_2PO_4	7.0 g/400 ml
$NaCl$	1.0 g/400 ml

To 940 ml distilled water, add 1.0 ml of each of the stock trace–element solutions prepared as follows:

1. 50 g EDTA and 31 g KOH dissolved in 1 liter distilled H_2O (or 50 g $Na_2 \cdot$ EDTA)

2. 4.98 g $FeSO_4 \cdot 7H_2O$ dissolved in 1 liter of acidified water (acidified H_2O :1.0 ml H_2SO_4 dissolved in 1 liter distilled H_2O).

3. 11.42 g H_3BO_3 dissolved in 1 liter distilled H_2O.

4. Trace Element Stock Solution (below)

TABLE 3-7: BBM TRACE METAL STOCK SOLUTION

ADDITIVE	CONCENTRATION
TRACE ELEMENT (STOCK)	**MAKE UP TO ONE LITER H_2O**
$ZnSO_4 \cdot 7H_2O$	8.82 g/l
$MnCl_2 \cdot 4H_2O$	1.44 g/l
MoO_3	0.71 g/l
$CuSO_4 \cdot 5H_2O$	1.57 g/l
$Co(NO_3)_2 \cdot 6H_2O$	0.49 g/l

This may be enriched by substituting 30 ml of stock $NaNO_3$ per liter to the definitive solution (3 x Nitrogen BBM). Alternately, many algae thrive when urea is substituted as the nitrogen source; it may be provided at the level of 3 x or 6 x the level of nitrogen in BBM.

Vitamins, most frequently B_1, B_6, and B_{12}, may enhance the growth of algae in BBM. These may be added to a liter of BBM as 5 ml of Eagle's mixture [1] and B_2 (cyanocobalamine) at concentrations of 0.1 ml of a 1.0 mg/ml solution (equivalent to 100 mg/liter).

[1] "TC-Vitamins Minimal Eagle, 100 x " (Difco Laboratories, Detroit, Mich.).

(Source: H. C. Bold and M. J. Wynne, *Introduction to the Algae*; 1978)

TABLE 3-8: CHU 10 FRESHWATER MEDIUM pH 6.5-7.0 (Chu, 1942)

Chu's medium is useful for culturing Bacillariophyceae, Chlorophyceae, Chrysophyceae and Cyanophyceae

ADDITIVE	CONCENTRATION
MACRONUTRIENTS	
$Ca(NO_3)_2$	0.04 g/l
K_2HPO_4	0.01 or 0.005 g/l
$MgSO_4 \cdot 7H_2O$	0.025 g/l
Na_2CO_3	0.02 g/l
Na_2SiO_3	0.025 g/l
$FeCl_3$	0.0008 g/l

TABLE 3-9: MODIFIED CHU NO. 10 SOLUTION (Modified by Wright and Guillard; and by Van Dover)

1. Make stock solutions by dissolving the salts listed in the amounts indicated (in grams) each in 100 ml of distilled or deionized water; autoclave them and keep sterile.

ADDITIVE	CONCENTRATION
MACRONUTRIENTS (STOCK)	ADD 1ML/L
$CaCl_2 \cdot 2H_2O$	3.67 g/100 ml distilled H_2O
$MgSO_4 \cdot 7H_2O$	3.69 g/100 ml
$NaHCO_3$	1.26 g/100 ml
K_2HPO_4	0.87 g/100 ml
$NaNO_3$	8.5 g/100 ml
$Na_2SiO_3 \cdot 9H_2O$ (metasilicate)	2.84 g/100 ml

2. Prepare an iron solution by dissolving 3.35 g citric acid ($C_6H_8O_7 \cdot H_2O$) in 100 ml distilled water; then add 3.35 g ferric citrate ($FeC_6H_5O_3 \cdot 5H_2O$), autoclave to dissolve, dispense in sterile tubes, and keep sterile. Refrigerate in darkness (wrapped in aluminum foil).

3. Prepare a trace-elements solution by dissolving the salts in the amounts (milligram) indicated together in 1 liter of distilled water. Autoclave and keep sterile.

TABLE 3-10: MODIFIED CHU NO. 10 TRACE ELEMENT SOLUTION

ADDITIVE	CONCENTRATION
$CuSO_4 \cdot 5H_2O$	19.6 mg/l
$ZnSO_4 \cdot 7H_2O$	44.0 mg/l
$CoCl_2 \cdot 6H_2O$	20.0 mg/l
$MnCl_2 \cdot 4H_2O$	36.0 mg/l
$Na_2MoO_4 \cdot 2H_2O$	12.6 mg/l
H_3BO_3	618.4 mg/l

To prepare the definitive solution, add aseptically 1 ml of each of the six stock solutions in step 1 to a liter of sterile double distilled or deionized water. Then add aseptically 1 ml of stock solution in step 2, and 1 ml of the Trace Elements Solution in Table 3-10. Dispense aseptically into sterile containers.

(Source: H.C. Bold and M.J. Wynne, *Introduction to the Algae*; 1978)

TABLE 3-11: HUGHES, GORHAM, AND ZEHNDER'S (1958) MEDIUM

This media is suitable for culturing a variety of freshwater algae.

ADDITIVE	CONCENTRATION
$NaNO_3$	0.496 g/l
K_2HPO_4	0.0399 g/l
$MgSO_4 \cdot 7H_2O$	0.75 g/l
$CuCl_2 \cdot 2H_2O$	0.036 g/l
Na_2CO_3	0.020 g/l
$Na_2SiO_3 \cdot 9H_2O$	0.058 g/l
Ferric Citrate	0.006 g/l
Citric acid	0.006 g/l
EDTA	0.001 g/l
GAFFRON'S TRACE-ELEMENT SOLUTION	0.08 ml
DISTILLED OR DEIONIZED WATER	TO 1 LITER

(Source: E.O Hughes, P. R. Gorham, and A. ZehnderH, *Canadian Journal of Microbiology*; 1958)

TABLE 3-12: GAFFRON'S TRACE-ELEMENT SOLUTION

ADDITIVE	CONCENTRATION
H_3BO_3	3.100 g/l
$MnSO_4 \cdot 4H_2O$	2.230 g/l
$ZnSO_4 \cdot 7H2O$	.0.287 g/l
$(NH_4)_6Mo_7O_{24} \cdot 4H_2O$	0.088 g/l
$CO(NO_3)_2 \cdot 4H_2O$	0.146 g/l
$Na_2WO_4 \cdot 2H_2O$	0.033 g/l
KBr	0.119 g/l
KI	0.083 g/l
$Cd(NO_3)_2 \cdot 4H_2O$	0.154 g/l
$NiSO_4(NH_4)_2SO_4 \cdot 6H_2O$	0.198 g/l
$VOSO_4 \cdot 2H_2O$	0.020 g/l
$Al_2(SO_4)_3 \cdot K_2SO_4 \cdot 24H_2O$	0.474 g/l

(Source: H. C. Bold and M. J. Wynne, *Introduction to the Algae*; 1978)

TABLE 3-13: ALLEN'S (1968) MODIFICATION OF HUGHES, GORHAM, AND ZEHNDER'S (1958) MEDIUM

ADDITIVE	CONCENTRATION
$NaNO_3$	1.5 g/l
K_2HPO_4	0.039 g/l
$MgSO_4 \cdot 7H_2O$	0.075 g/l
$CaCl_2 \cdot 2H_2O$	0.027 g/l
Na_2CO_3	0.020 g/l
$Na_2SiO_3 \cdot 9H_2O$	0.058 g/l
Ferric citrate	0.006 g/l
Citric acid	0.006 g/l
EDTA	0.001 g/l
ALLEN'S TRACE-ELEMENT SOLUTION	1.0 ml
DISTILLED OR DEIONIZED WATER	TO 1 LITER
pH of medium	7.8

(continued on page 40)

(continued from page 39)

Autoclave ferric citrate separately and add aseptically after medium has stood for 24 hours. To prepare solid media, equal volumes of double-strength mineral base and double-strength agar are separately sterilized and combined after cooling to 48°C.

(Source: H. C. Bold and M. J. Wynne, *Introduction to the Algae*; 1978)

TABLE 3-14: ALLEN'S (1968) TRACE-ELEMENT SOLUTION (Modified)

Dissolve in 1 liter of distilled water

ADDITIVE	CONCENTRATION
H_3BO_3	2.86 g/l
$MnCl_2 \cdot 4H_2O$	1.81 g/l
$ZnSO_4 \cdot 7H_2O$	0.222 g/l
$Na_2MoO_4 \cdot 2H_2O$	0.391 g/l
$CuSO_4 \cdot 5H_2O$	0.079 g/l
$Co(NO_3)_2 \cdot 6H_2O$	0.0494 g/l

Media may be solidified by adding 15 g/liter of agar.

(Source: H. C. Bold and M. J. Wynne, *Introduction to the Algae*; 1978)

TABLE 3-15: RODHE VIII pH 7.0-7.5 FRESHWATER MEDIUM (Rodhe 1948).

Useful for the culture of Bacillariophyceae, Chlorophyceae, Chrysophyceae and Cyanophyceae

ADDITIVE	CONCENTRATION
$Ca(NO_3)_2$	60 mg/l
$MgSO_4$	5 mg/l
Na_2SiO_3	20 mg/l
K_2HPO_4	5 mg/l
Ferric citrate	1.0 mg/l
Citric acid	1.0 mg/l
$MnSO_4$	0.03 mg/l

Prior mixing of stock solutions is suggested to avoid repetitive weighing.

(Source: H.W. Nichols, In: J.R. Stein, *Phycological Methods: Culture Methods & Growth Measurements*; copyright ©1973, - with permission Cambridge University Press)

TABLE 3-16: WARIS pH 6.0 FRESHWATER MEDIUM (Waris, 1953)

Useful for the culture of Chlorophyceae

ADDITIVE	CONCENTRATION
MACRONUTRIENTS	
KNO_3	0.1 g/l
$MgSO_4 \cdot 7H_2O$	0.02 g/l
$(NH4) \cdot 2HPO4$	0.02 g/l
$CaSO_4$	0.05 g/l
IRON SEQUESTRINE	
EDTA	1.30 g/l
$FeSO_4 \cdot 7H_2O$	1.25 g/l in 13.5 ml/l KOH

(continued from page 40)

pH is adjusted to 6.0. FeNaEDTA (ferric sequestrine) may be substituted for the solution (use 2.49 g FeNaEDTA/liter)

Separate stocks of each macronutrient and the iron sequestrine solution may be prepared.

(Source: H.W. Nichols, In: J.R. Stein, *Phycological Methods: Culture Methods & Growth Measurements*; copyright © 1973 - with permission of Cambridge Universiity Press)

TABLE 3-17: BOZNIAK COMMUNITY pH 8.O FRESHWATER MEDIUM (BOZNIAK 1969).

Useful medium for culturing Chlorophyceae, Chrysophyceae, Cyanophyceae and Rhodophyceae.

ADDITIVE	CONCENTRATION
MACRONUTRIENTS (STOCK)	**ADD 1 ML EACH/L**
$Ca(NO_3)_2 \cdot 4H_2O$	24.4 g/l
$MgSO_4 \cdot 7H_2O$	1.0 g/l
$NaHCO_3$	16.5 g/l
$K_2HPO_4 \cdot 3H_2O$	0.8 g/l
KH_2PO_4	0.8 g/l
Na_2SiO_3	5.8 g/l
H_3BO_3	0.5 g/l
Na_2EDTA	10.0 g/l
CITRATE-CITRIC ACID (STOCK)	**ADD 1 ML EACH/L**
Ferric citrate	1.0 g/l
Citric acid	1.0 g/l
VITAMINS (STOCK)	**ADD 1 ML EACH/L**
Thiamine HCl	1×10^{-6} g/l
Cyanocobalamin	10×10^{-6} g/l
MICRONUTRIENTS	
K_2CrO_4	0.0037 g/l
$CoCl_2 \cdot 6H_2O$	0.020 g/l
$MnCl_2 \cdot 4H_2O$	0.890 g/l
$ZnCl_2$	0.0104 g/l
$VOSO_4 \cdot 2H_2O$	0.0039 g/l
MoO_3	0.0075 g/l
$CuSO_4$	0.00125 g/l

Prior to preparation, pH of water to 8.0.

Separate stock solutions of each macronutrient, the micronutrient solution, and the vitamin solution may be prepared and stored frozen.

The $Ca(NO_3)_2$ is autoclaved separately and added aseptically when cool to avoid precipitation of calcium phosphate.

The ferric citrate-citric acid and vitamin stocks are membrane filtered (0.22 μ). These are added after sterilization of the other constituents to the cooled medium.

(Source: H.W. Nichols, In: J.R. Stein, *Phycological Methods: Culture Methods & Growth Measurements*; copyright © 1973, with permission of Cambridtge University Press)

TABLE 3-18: CG 10 pH 8.0 FRESHWATER MEDIUM (Van Baalen, 1967)

Useful for culturing blue-green algae (Cyanophyceae)

ADDITIVE	CONCENTRATION
MACRONUTRIENTS	
$MgSO_4 \cdot 7H_2O$	0.25 g/l
K_2HPO_4	1.0 g/l
$Ca(NO_3)_2 \cdot 4H_2O$	0.025 g/l
KNO_3	1.0 g/l
Na_2 EDTA	0.010 g/l
$Fe_2(SO_4)_3 6H_2O$	0.004 g/l
Glycylglycine	1.0 g/l
MICRONUTRIENT STOCK 'A5'	**USE 1 ML/L**
H_3BO_3	2.86 g/l
$MnCl_2 \cdot 4H_2O$	1.81 g/l
$ZnSO_4 \cdot 7H_2O$	0.222 g/l
MoO_3 (85%)	0.0018 g/l
$CuSO_4 \cdot 5H_2O$	0.079 g/l
$CoCl2 \cdot 6H_2O$	0.010 g/l

Major nutrients are added as salts, or can be dissolved in stock solutions in appropriate concentrations. Aliquots of these stock solutions are then added to the final medium before bringing the medium up to volume. The micronutrient solution 'A5' is made as a single stock solution, and adjusted to pH 8.5.

(Source: H.W. Nichols, In: J.R. Stein, *Phycological Methods: Culture Methods & Growth Measurements;* 1973)

TABLE 3-19: BEIJERINCK pH 6.8 FRESHWATER MEDIUM (Stein 1966)

This medium is useful for culturing blue-green algae (Cyanophyceae). It is made from three stock solutions and a micronutrient solution.

ADDITIVE	CONCENTRATION
STOCK I	**USE 100 ML/L**
NH_4NO_3	1.5 g/l
K_2HPO_4	0.2 g/l
$MgSO_4 \cdot 7H_2O$	0.2 g/l
$CaCl_2 \cdot 2H_2O$	0.1 g/l
STOCK II	**USE 40 ML/L**
KH_2PO_4	9.07 g/l
STOCK III	**USE 60 ML/L**
K_2HPO_4	11.61 g/l
MICRONUTRIENTS[1]	**USE 1 ML/L**
H_3BO_3	1.0 g/100 ml
$CuSO_4 \cdot 5H_2O$	0.15 g/100 ml
EDTA	5.0 g/100 ml
$ZnSO_4 \cdot 7H_2O$	2.2 g/100 ml
$MnCl_2 \cdot 4H_2O$	0.5 g/100 ml
$FeSO_4 \cdot 7H_2O$	0.5 g/100 ml
$CoCl_2 \cdot 6H_2O$	0.15 g/100 ml
$(NH_4)_6Mo_7O_{24} \cdot 4H_2O$	0.10 g/100 ml

[1] The micronutrients are dissolved one at a time in 100 ml warm water. Following the addition of each, the pH is adjusted to 5 with KOH pellets. The final solution should have a pH @ 6.5 (iron precipitates at 7.0).

(Source: H.W. Nichols, In: J.R. Stein, *Phycological Methods: Culture Methods & Growth Measurements;* 1973)

TABLE 3-20: WOODS HOLE MBL pH 7.2 FRESHWATER MEDIUM (Guillard, In: J.R. Stein, 1973)

Useful for culturing Bacillariophyceae, Chlorophyceae, Chrysophyceae and Cyanophyceae.

ADDITIVE	CONCENTRATION
MACRONUTRIENTS	**USE 1 ML EACH/L**
$CaCl_2 \cdot 2H_2O$	36.76 g/l
$MgSO_4 \cdot 7H_2O$	36.97 g/l
$NaHCO_3$	12.60 g/l g/l
K_2HPO_4	8.71 g/l
$NaNO_3$	85.01 g/l
$Na_2SiO_3 \cdot 9H_2O$	28.42 g/l
MICRONUTRIENTS	**USE 1 ML EACH/L**
Na_2 EDTA	4.36 g/l
$FeCl_3 \cdot 6H_2O$	3.15 g/l
$CuSO_4 \cdot 5H_2O$	0.01 g/l
$ZnSO_4 \cdot 7H_2O$	0.022 g/l
$COCl_2 \cdot 6H_2O$	0.01 g/l
$MnCl_2 \cdot 4H_2O$	0.18 g/l
$Na_2MoO_4 \cdot 2H_2O$	0.006 g/l
VITAMINS	
Thiamine HCl	0.1 mg/l
Biotin	0.5 ug/l
Cyanocobalamin	0.5 ug/l
Tris	**USE 2 ML/L**
Tris(hydroxymethylaminomethane)[1]	50 g/200 ml (Adjust pH to 7.2 with HCl)

[1] Glycylglycine may be substituted for "TRIS" in aseptic cultures

(Source: H.W. Nichols, In: J.R. Stein, *Phycological Methods: Culture Methods & Growth Measurements*; 1973)

TABLE 3-21: VOLVOX pH 7.0 FRESHWATER MEDIUM (Darden 1966; Starr 1969).

Useful for culturing Chlorophyceae

ADDITIVE	CONCENTRATION
MACRONUTRIENTS (STOCK A)	**ADD 10 ML/L**
$Ca(NO_3)_2 \cdot 4H_2O$	0.118 g/l
$MgSO_4 \cdot 7H_2O$	0.04 g/l
Na_2 glycerophosphate $\cdot 5H_2O$	0.05 g/l
KCl	0.05 g/l
Glycylglycine	0.5 g/l
MICRONUTRIENTS (STOCK B)	**ADD 3 ML/L**
$FeCl_3 \cdot 6H_2O$	0.097 g/l
$McCl_2 \cdot 4H_2O$	0.041 g/l
$ZnCl_2$	0.005 g/l
$CoCl_2 \cdot 6H_2O$	0.002 g/l
Na_2MoO_4	0.004 g/l
Na_2EDTA	0.750 g/l
VITAMINS (STOCK C)	**ADD 1 ML/L**
Biotin	1.0 µg/l
Cyanocobalamin	1.0 µg/l

Adjust pH to 7.0 with 1N NaOH

(Source: H.W. Nichols, In: J.R. Stein, *Phycological Methods: Culture Methods & Growth Measurements*; 1973)

TABLE 3-22: CONCENTRATIONS OF ADDITIVES IN DEFINED FRESHWATER MEDIA

ADDITIVE	CONCENTRATION / LITER								
	BEIJERINCK	BOLD	BOZNIAK	CG 10	CHU NO.10	RODHE	VOLVOX	WARIS	WOODS HOLE
MACRONUTRIENTS (mM)									
$(NH_4)_2HPO_4$	-	-	-	-	-	-	-	-	0.15
NH_4NO_3	1.87	-	-	-	-	-	-	-	-
K_2HPO_4	4.11	0.43	0.03	0.57	0.06	0.029	-	-	0.05
KH_2PO_4	2.67	1.29	0.03	-	-	-	-	-	-
KNO_3	-	-	-	9.89	-	-	-	0.99	-
KCl	-	-	-	-	-	-	0.67	-	-
$NaNO_3$	-	2.94	-	-	-	-	-	-	-
$NaCl$	-	0.43	-	-	-	-	-	-	-
$NaHCO_3$	-	-	0.72	-	-	-	-	-	0.15
Na_2CO_3	-	-	-	-	0.19	-	-	-	-
Na_2SiO_3	-	-	0.21	-	0.20	0.164	-	-	0.1
$NaNO_3$	-	-	-	-	-	-	-	-	1.0
$MgSO_4$	0.08	0.30	0.04	1.01	0.10	0.042	0.16	0.08	0.15
$MnSO_4$	-	-	-	-	-	0.0002	-	-	-
$CaCl_2$	0.07	0.17	-	-	-	-	-	-	0.25
$Ca(NO_3)_2$	-	-	0.61	0.11	0.24	0.366	0.5	-	-
$CaSO_4$	-	-	-	-	-	-	-	0.37	-
$Fe_2(SO_4)_3$	-	-	-	0.008	-	-	-	-	-
Fe citrate	-	-	0.018	-	-	0.0041	-	-	-
Citric acid	-	-	0.005	-	-	0.0052	-	-	-
Na_2glycerophosphate	-	-	-	-	-	-	0.16	-	-
Glycylglycine	-	-	-	7.57	-	-	3.78	-	3.3[a]
Tris	-	-	-	-	-	-	-	-	4.13[a]
VITAMINS (nM)									
Biotin	-	-	-	-	-	-	0.41[b]	-	2.047
Cyanocobalamin	-	-	7.4	-	-	-	0.74[b]	-	0.368
Thiamine . HCl	-	-	3.0	-	-	-	-	-	297
MICRONUTRIENTS (µM)									
H_3BO_3	16.17	184.7	46.0	46.3	-	-	-	-	-
EDTA	7.65	171.09	-	-	-	-	-	17.86	-
$Na_2.EDTA$	-	-	26.86	26.86	-	-	12.09	-	11.71
$CoCl_2$	0.63	-	0.084	0.042	-	-	0.05	-	0.042
$Co(NO_3)_2$	-	16.8	-	-	-	-	-	-	-
$CuSO_4$	0.601	62.9	0.008	0.316	-	-	-	-	0.040
$FeCl_3$	-	-	-	-	4.9	-	2.15	-	11.65
$FeSO_4$	1.798	17.9	-	-	-	-	-	17.91	-
K_2CrO_4	-	-	0.002	0	-	-	-	-	-
$(NH_4)_6Mo_7O_{24}$	0.081	-	-	-	-	-	-	-	-
MoO_3	-	4.9	0.052	0.147	-	-	-	-	-
Na_2MoO_4	-	-	-	-	-	-	0.12	-	0.025
$MnCl_2$	2.56	7.3	4.498	9.147	-	-	1.24	-	0.910
$VOSO_4$	-	-	0.019	-	-	-	-	-	-
$ZnCl_2$	-	-	0.076	-	-	-	0.22	-	-
$ZnSO_4$	7.65	30.7	-	0.772	-	-	-	-	0.077
KOH	-	553	-	-	-	-	-	54.0	-

[a] Use glycylglycine or tris for buffer;

[b] Starr (1969) suggests: biotin 1.023 nM; cyanocobalamin, 0.11 nM

(Source: W. Nicholes, In: J. R. Stein (Ed.), *Phycological Methods: Culture Methods & Growth Measurements*; copyright © 1979 - with permission of Cambridge University Press)

TABLE 3-23: PROVASOLI'S ENRICHED SEAWATER (ES) (Provasoli, 1963, 1968; McLachlan, 1973)

Seawater is sterilized by filtration or autoclaving, enrichments assembled into a single solution, and added aseptically to the medium.

ADDITIVE	CONCENTRATION
NaNO$_3$ (STOCK A)	35 g/100 ml
Na$_2$glycerophosphate (STOCK B)	5 g/100 ml
Vitamin B$_{12}$ (STOCK C)	1 mg/100 ml
Thiamine (STOCK D)	50 mg/100 ml
Biotin (STOCK E)	0.5 mg/100 ml
Fe (as EDTA 1:1 molar) (STOCK F)	
Fe(NH$_4$)$_2$(SO$_4$) • 6H$_2$O	351 mg/100 ml
Na$_2$EDTA	300 mg/500 ml
P 11 TRACE METALS (STOCK G)	
H$_3$BO$_3$	1.14 g/l
FeCl$_3$ • 6H$_2$O	49 mg/l
MnSO$_4$ • 4H$_2$O	164 mg/l
ZnSO$_4$ • 7H$_2$O	22 mg/l
CoSO$_4$ • 7H$_2$O	4.8 mg/l
Na$_2$EDTA	1 g/l

Mix 10 ml of each stock solution A - E and 250 ml of each stock solution F and G and bring total volume to 1250 ml with distilled or deionized water. Add 20 ml of the above stock solution mixture to 1000 ml of filtered seawater to prepare full-strength medium.

(Source: H.C. Bold and M. J. Wynne, *Introduction to the Algae*; 1978)

TABLE 3-24: "f" ENRICHED SEAWATER MEDIA (Guillard and Ryther, 1962)
COMPOSITION PER LITER OF SEAWATER FOR "f/2"

This medium is widely used to culture a variety of marine phytoplankton. Pre-mixed modification of this formulation are available commercially.

ADDITIVE	CONCENTRATION
MAJOR NUTRIENTS	
NaNO$_3$	75 mg/l
NaH$_2$PO$_4$ • H$_2$O	5 mg/l
Na$_2$SiO$_3$ • 9H$_2$O	30 mg/l
TRACE METALS	
Na$_2$ EDTA	4.36 mg/l
FeCl$_3$ • 6H$_2$O	3.15 mg/l
CuSO$_4$ • 5H$_2$O	0.01 mg/l
ZnSO$_4$ • 7H$_2$O	0.022 mg/l
CoCl$_2$ • 6H$_2$O	0.01 mg/l
MnCl$_2$ • 4H$_2$O	0.18 mg/l
Na$_2$MoO$_4$ • 2H$_2$O	0.0006 mg/l
VITAMINS	
Thiamin HCl	0.1 mg/l
Biotin	0.5 mg/l
B-12	0.5 mg/l

TABLE 3-24 (cont.):

PREPARATION OF STOCK SOLUTIONS

ADDITIVE	CONCENTRATION
MAJOR NUTRIENTS STOCKS	
$NaNO_3$	7.5 g/100 ml
$NaH_2PO_4 \cdot H_2O$	0.5 g/100 ml
NH_4Cl	2.65 g/100 ml
$Na_2SiO_3 \cdot 9H_2O$	3.0 g/100 ml (heat to dissolve if necessary)
TRACE METAL PRIMARY STOCKS	
$CuSO_4 \cdot 5H_2O$	0.98 g/100 ml
$ZnSO_4 \cdot 7H_2O$	2.2 g/100 ml
OR $ZnCl_2$	1.05 g/100 ml
$CoCl_2 \cdot 6H_2O$	1.0 g/100 ml
$MnCl_2 \cdot 4H_2O$	1.8 g/100 ml
$Na_2MoO4 \cdot 2H_2O$	0.63 g/100 ml
VITAMIN PRIMARY STOCKS	
Biotin	10 mg/96 ml distilled H_2O
Vitamin B_{12}	10 mg/ 10 ml distilled H_2O

MAJOR NUTRIENT STOCKS are 10^3 more concentrated than in the final medium. Use 1 ml/liter of seawater to obtain medium "f/2", "h/2", or "f/2-beta".

TRACE METAL WORKING STOCK SOLUTIONS, EDTA CHELATED

Use 1 ml/l of **TRACE METAL WORKING STOCK** to make final "f/2" or "h/2" media.

1. "Ferric Sequestrene" as iron and chelator source.
Dissolve 5 g ferric sequestrene in 900 ml of distilled water, add 1 ml of each TRACE METAL PRIMARY STOCK, and bring to 1 liter. pH is @ 4.5.

2. Trace metal stock solution, using ferric chloride and di-sodium EDTA. Dissolve 3.15 g $FeCl_2 \cdot 6H_2O$ and 4.36 g Na_2 EDTA in 900 ml of distilled water; add 1 ml of each **TRACE METAL PRIMARY STOCK** and bring to one liter. pH is @ 2.0. The solution remains clear if left at pH 2.0. If titrated to @ pH 4.5 (taking ca. 7 ml of N_2NaOH), a precipitate will form, resembling that in the solution made with ferric sequestrene.

VITAMIN WORKING STOCK SOLUTION

Use 0.5 ml/l of **VITAMIN WORKING STOCK SOLUTION** for final "f/2" or "h/2" media.

Add 1.0 ml of **BIOTIN PRIMARY STOCK** and 0.1 ml of B_{12} **PRIMARY STOCK** to 100 ml distilled H_2O and add 20 mg of thiamine HCl (no primary stock of thiamine is needed).

(Source: R. Guillard, In: W.L. Smith and M.H. Chanley, *Culture of Marine Invertebrate Animals*; copyright ©1975 - with permission Plenum Publishers)

TABLE 3-25: MODIFIED F MEDIUM [1](Guillard & Ryther, 1962)

ADDITIVE	CONCENTRATION
N-P (STOCK A -500X)	MAKE UP TO ONE LITER DISTILLED H_2O
$NaNO_3$	42.07 g
$NaH_2PO_4 \cdot H_2O$	5.0 g
SODIUM METASILICATE (STOCK B -500X)	MAKE UP TO ONE LITER DISTILLED H_2O
$Na_2SiO_3 \cdot 9H_2O$	15.0 g
FERRIC CHLORIDE (STOCK C-500X)	MAKE UP TO ONE LITER DISTILLED H_2O
$FeCl_3 \cdot 6H_2O$	1.45 g

TABLE 3-25 (cont.):

ADDITIVE	CONCENTRATION
EDTA STOCK (STOCK D -1000X)	**MAKE UP TO ONE LITER DISTILLED H$_2$O**
Na$_2$ • EDTA	10.0 g
VITAMIN (STOCK E)	**MAKE UP TO ONE LITER DISTILLED H$_2$O**
Biotin	0.2 g
B12 primary stock (0.1 g/liter)	10 ml
Biotin primary stock (0.1 g/liter)	10 ml
TRACE METAL (STOCK F-1000X)	**MAKE UP TO ONE LITER DISTILLED H$_2$O**
TM primary stock A*	1 ml
TM primary stock B**	1 ml
TM primary stock C***	1 ml
TM primary stock D****	1 ml
Distilled water (to make)	1000 ml
***TM PRIMARY STOCK A**	**MAKE UP TO 100 ML DISTILLED H$_2$O**
CuSO$_4$ • 5H$_2$O	1.96 g
ZnSO$_4$ • 7H$_2$O	4.4 g
****TM PRIMARY STOCK B**	**MAKE UP TO 100 ML DISTILLED H$_2$O**
Na$_2$MoO$_4$ • 2H$_2$O	1.26 g
(NH$_4$)$_6$Mo$_7$O$_{24}$ • 2H$_2$O	0.9 g
*****TM PRIMARY STOCK C**	**MAKE UP TO 100 ML DISTILLED H$_2$O**
MnCl$_2$ • 4H$_2$O	36.0 g
******TM PRIMARY STOCK D**	**MAKE UP TO 100 ML DISTILLED H$_2$O**
CoCl$_2$ • 6H$_2$O	2.0 g

[1] 2 ml each of solutions A, B, and C plus 1 ml each of solutions D, E, and F per liter of seawater

(Source: Kongkeo, In: W. Fulks and K. L. Main, *Rotifer and Microalgal Culture Systems*; 1991 - reprinted with permission of Argent Laboratories)

TABLE 3-26: FORMULA OF WALNE MEDIUM FOR ALGAE CULTURE (Walne, 1974)

ADDITIVE	CONCENTRATION
STOCK A[1]	**MAKE UP TO ONE LITER DISTILLED H$_2$O**
FeCl$_3$ • 6H$_2$O	1.30 g
MnCl$_2$ • 4H$_2$O	0.36 g
H$_3$BO$_3$	33.60 g
EDTA (Na salt)	45.00 g
NaH$_2$PO$_4$ • 2H$_2$O	20.00 g
NaNO$_3$	100.00 g
TRACE METAL SOLUTION	1.0 ml
STOCK B[2]	**MAKE UP TO 100 ML DISTILLED H$_2$O**
Vitamin B$_{12}$ (Cyanocobalamin)	10 mg
Vitamin B$_1$ (Thiamin)	200 mg
STOCK C[3]	**MAKE UP TO 100 ML DISTILLED H$_2$O**
Na$_2$SiO$_3$ • 5H$_2$O	4.9 g
TRACE METAL SOLUTION[4]	**MAKE UP TO 100 ML DISTILLED H$_2$O**
ZnCl$_2$	2.1 g
CoCl$_2$ • 6H$_2$O	2.0 g
(NH$_4$)$_6$Mo$_7$O$_2$ • 4H$_2$O	0.9 g
CuSO$_4$ • 5H$_2$O	2.0 g

(continued on page 48)

(continued from page 47)

[1] Add 2 ml **STOCK A** per liter of seawater for *Chaetoceros calcitrans*, 1 ml **STOCK A** per liter of seawater for *Tetraselmis suecica*.

[2] This solution should be acidified to pH 4.5 before autoclaving. Add 0.1 ml **STOCK B** per liter of seawater.

[3] Add 2 ml **STOCK C** per liter of seawater for diatom culture only.

[4] Acidify with sufficient concentrated HCl to obtain a clear solution.

(Source: J.P. McVey, *CRC Handbook of Mariculture: Volume I - Crustacean Aquaculture*; copyright © 1983 with permission of CRC Press, Boca Raton, FL)

TABLE 3-27: ERDSCHREIBER MEDIUM (Starr, 1964)

CONSTITUENTS

(1.) Seawater autoclave 2 liters in a 3 liter borosilicate glass flat-bottomed-boiling flask with cotton wool plug at 1.06 kg/cm^2 for 20 min, stand for 2 days.

(2.) Soil extract: prepared as follows:

   A. 1 kg soil from a woodland or pasture area untreated with artificial fertilizers, insecticides, etc. with 1 liter freshwater

   B. Autoclave at 1.06 kg/cm^2 for 60 minutes.

   C. Decant

   D. Filter through Whatman No. I paper and then through glass fiber (GF/c) paper.

   E. Autoclave in 1 liter aliquots in polypropylene bottles at 1.06 kg/cm^2 for 20 minutes.

   F. Store in deep freeze until required.

   G. Autoclave 100 ml in 500 ml borosilicate glass flat-bottomed boiling flask with cotton wool plug at 1.06 kg/cm^2 for 20 minutes.

(3.) Nitrate/phosphate stock solution: dissolve 40 g $NaNO_3$, and 4 gm Na_2HPO_4 in 200 ml distilled water, autoclave in 500 ml flask at 1.06 kg/cm^2 for 20 minutes.

(4.) Silicate stock solution: dissolve 8 g $Na_2SiO_3 \cdot 5H_2O$ in 200 ml distilled water. Autoclave in 500 ml flask at 1.06 kg/cm^2 for 20 minutes.

PROCEDURE

Add 100 ml of soil extract (2) to 2 liters of seawater (1).

With sterile pipette add 2 ml nitrate/phosphate stock (3) and 2 ml silicate stock (4).

Decant 250 ml into eight empty autoclaved 500 ml flasks with cotton wool plugs.

Use bunsen burner to flame necks of flasks immediately before and after decanting/pipetting.

(Source: J.P. McVey, *CRC Handbook of Mariculture: Volume I - Crustacean Aquaculture*; copyright © 1983 with permission CRC Press, Boca Raton, FL).

TABLE 3-28: ENRICHED SEAWATER MEDIA (Guillard, 1975)

ADDITIVE	CONCENTRATION (µm/l)				
	f/2	h/2	f/2 beta	ES	SWM
INORGANIC MACRONUTRIENTS					
NaNO₃	880	-	880	660	500 - 2,000
NH₄Cl	-	500	-	-	-
NaH₂PO₄	36.3	36.3	36.3	=	50 - 100
Na₂glycerophosphate	-	-	-	25.0	-
Na₂SiO₃ • 9H₂O	54 - 107	54 - 107	54 - 107	-	200
INORGANIC MICRONUTRIENTS					
Fe EDTA	-	-	-	7,200	2.0
FeCl • 4H₂O	11.7	11.7	11.7	1.8	-
Na₂EDTA	11.7	11.7	11.7	26.9	48.0
CuSO₄ • 5H₂O	0.04	0.04	0.04	-	0.3
ZnSO₄ • 5H₂O	0.08	0.08	0.08	0.80	35.0
CoCl₂ • 4H₂O	0.05	0.05	0.05	0.17	0.30
MnCl₂ • 4H₂O	0.90	0.90	0.90	7.30	10.0
Na₂MoO₄ • 2H₂O	0.03	0.03	0.03	-	5.0
Boron	-	-	-	185	400
ORGANIC MICRONUTRIENTS					
Thiamine HCl (B₁)	100 µg	100 µg	100 µg	20 µg	-
Nicotinic acid	-	-	-	-	0.1 mg/l
Ca. pantothenate	-	-	-	-	0.1 mg/l
p-Aminobenzoic acid	-	-	-	-	10 µg/l
Biotin	0.5 µg	0.05 µg	0.5 µg	0.8 µg	1.0 µg/l
i-Inositol	-	-	-	-	5.0 mg/l
Folic acid	-	-	-	-	2.0 µg/l
Cyanocobalamin	0.5 µg	0.5 µg	0.5 µg	1.6 µg	1.0 µg/l
Thymine	-	-	-	-	3.0 µg/l
Tris	-	-	"	0.66	0 - 5000
Glycylglycine	-	-	-	-	5000
Soil extract	-	-	-	-	50 ml/l
Liver extract	-	-	-	-	10 mg/l

"h/2" media adds 26.5 mg (0.5 mM) NH₄Cl to the "f/2" Major Nutrient Stock for culturing those species which cannot grow well on nitrate. If NH₄Cl is added to seawater and autoclaved, 25 - 30 % of the ammonium is lost. NH₄Cl Stock, because of its low pH, can be autoclaved, and should be added aseptically to medium autoclaved separately.

"f/2-beta" differs from "f/2" in that citrate is used as the chelator. Dissolve 16.8 g citric acid (C₆H₈O₇ • H₂O) in 900 ml distilled water. Add 3.0 g ferric citrate (FeC₆H₅O₇ • 5H₂O) and 1 ml of each "f/2" TRACE METAL PRIMARY STOCK. Bring to one liter and autoclave (pH @ 2.3)

ES media = Provasoli Enriched Seawater (Provasoli, 1968)

SWM media (McLachlan 1964; Chen, Edelstein and McLachlan, 1969) has several variations, depending on the species under cultivation. The original reference should be consulted.

(Source: J.P. Mcvey, CRC Handbook of Mariculture: Volume I - Crustacean Aquaculture, copyright © 1983 - with permission of CRC Press, Boca Raton, FL)

TABLE 3-29: EFFECTS OF MATERIALS ON ALGAL CULTURES

Each letter represents the result of one test of a specific formulation or product, as reported by the indicated author (see reference list). Where more than one species was tested by an author, the most adverse result is tabulated. All tests were marine except those of references C and E. This table should only be used as a general guide, as specific manufacturer and product formulation, prior treatment of materials, and conditions of use can dramatically alter the acceptability of materials.

MATERIAL	SAFE	INHIBITORY	TOXIC
ACRYLIC (LUCITE; PERSPEX; PLEXIGLAS)	abdde	-	-
ALUMINIUM ALLOY	eeeeee	-	-
CHARCOAL (ACTIVATED)	-	bg	-
COPPER ALLOY	ee	e	eee
COTTON	b	-	-
EPOXY RESIN	ee	-	-
IRON	-	e	-
MEMBRANE FILTER (MILLIPORE)	ab	-	-
NYLON	be	a	ab
PARAFFIN	d	-	-
PLYWOOD	-	d	-
POLYCARBONATE	f	-	-
POLYETHYLENE (BLACK)	e	a	-
POLYETHLENE (WHITE; CLEAR)	ab	-	-
POLYPROPYLENE	abe	aa	e
POLYSTYRENE	b	-	-
POLYTETRAFLUOROETHYLENE (TEFLON)	abe	-	-
POLYURETHANE FOAM	-	e	-
POLYVINYL CHLORIDE	aa	aab	aaaabee
POLYVINYL CHLORIDE (CLEAR TYGON)	abcee	d	-
POLYVINYL CHLONDE (BLACK TYGON)	-	-	-
RUBBER (WHITE; BLACK; NEOPRENE)	e	abd	aaaaaaabcee
SILICONE (STOPPERS; TUBING; ETC)	abeee	-	-
SILICONE (SEALANT)	e	b	e
SOLDER (SILVER)	ee	-	-
SOLDER (SOFT)	e	-	-
STAINLESS STEEL	eeeee	a	-
TITANIUM	e	-	-
ZINC	-	e	-

REFERENCES:

a. Bernhard, M., A. Zattera and P. Filesi (1966)

b. Blankley, W.F., Unpublished observations

c. Davis, E.A., J. Dedrick, C.S. French, H.W. Milner, J. Myers, J.H.C. Smith and H.A. Spoehr (1953)

d. Doty, M.S and M. Oguri (1959)

e. Dyer, D.L. and D.E. Richardson (1962)

f. Lewin, J. (1966)

g. Ryther, J.H. and R.R. Guillard (1962)

(Source: W.F. Blankley, In: J. R. Stein (Ed.), *Phycological Methods: Culture Methods & Growth Measurements*; copyright © 1979 - with permission of Cambridge University Press)

TABLE 3-30: SPECIFIC GROWTH RATES [1] OF ALGAE CULTURED AT LOW AND HIGH TEMPERATURES

Marine Strain: Salinity = 33 ‰, Light:Dark = 24:0, f/2 Medium
Freshwater Strain: Salinity = O ppt, Light:Dark = 24:0, Complesal medium

LIGHT INTENSITY (lux)	LOW TEMPERATURE (°C)				HIGH TEMPERATURE (°C)			
	5		10		25		30	
	5000	2500	5000	2500	5000	2500	5000	2500
Caloneis schroderi	-0.01	-0.50	-0.10	0.25	0.28	0.30	0.38	0.29
Chaetoceros gracilis	-0.01	-0.05	0.16	0.07	0.52	0.39	0.62	0.73
Chaetoceros simplex	-0.06	-0.01	0.27	0.55	0.47	0.52	0.54	0.63
Cyclotella sp. [1]	0.16	0.08	0.00	-0.10	0.00	0.00	-0.18	-0.19
Cyclotella sp. [2]	-0.06	-0.20	0.12	0.16	0.57	0.58	0.42	0.26
Hanzchia marina	-0.36	-0.07	-0.68	-0.21	0.21	0.15	0.11	0.06
Navicula incerta	0.16	0.20	0.37	0.22	0.39	0.37	0.48	0.35
Navicula sp.	0.08	0.02	0.31	-0.01	0.29	0.26	0.30	0.43
Nitzschia sp.	-0.10	0.12	0.20	0.23	0.23	0.29	0.46	0.40
Phaeodactylum tricornutum [3]	0.37	0.39	0.71	0.63	0.85	0.84	-0.53	-0.53
Phaeodactylum tricornutum [4]	0.22	0.26	0.66	0.51	-0.03	-0.43	-0.45	-0.46
Skeletonema costatum	0.29	0.34	0.66	0.61	0.60	0.67	0.48	0.43
Thalassiosira fluviatilis	0.03	0.06	0.10	0.11	-0.03	0.10	0.28	0.23
Thalassiosira sp.	-0.29	-0.17	-0.04	-0.12	0.13	0.15	0.40	0.41
Boekelovia sp.	0.04	0.01	0.18	0.27	0.67	0.62	0.01	0.27
Isochrysis galbana	-0.57	-0.07	0.42	0.25	0.55	0.55	0.52	0.51
Isochrysis aff. galbana	-0.62	-0.67	0.06	-0.14	0.78	0.81	0.76	0.74
Nannochloris oculata	0.01	-0.03	0.01	0.04	0.85	0.92	1.09	1.14
Nannochloropsis salina	0.06	0.07	0.04	0.54	-0.32	-0.34	0.46	0.06
Chlorella ellipsoidea	0.03	0.02	0.58	0.53	0.88	0.85	0.98	0.94
Chlorella stigmatophora	0.01	0.24	0.36	0.24	0.78	0.56	0.66	0.39
Chlorella vulgaris *	-0.25	0.00	0.21	0.17	0.68	0.67	-0.29	-0.20
Dunaliella tertiolecta	0.28	0.17	0.43	0.42	0.60	0.64	0.60	0.45
Eudorina elegans	-0.24	-0.23	0.15	0.13	0.28	0.35	0.05	0.05
Gloeocystis sp.	-0.20	0.01	0.14	0.25	-0.19	0.09	0.03	0.20
Heterosigma sp.	0.02	0.00	0.43	0.38	0.92	0.78	1.02	0.60
Oocystis pusilla	0.14	0.14	0.21	0.19	0.16	0.15	0.17	0.27
Palmella mucosa	-0.01	0.02	0.76	0.57	0.93	1.10	0.93	1.05
Scenedesmus sp. *	-0.18	-0.45	-0.26	-0.36	-0.25	-0.18	-0.16	-0.13
Tetraselmis suecica	0.04	-0.02	0.42	0.28	0.46	0.44	0.55	0.51
Platymonas subcordiformis	0.04	0.11	0.38	0.39	0.30	0.30	0.23	0.28
Microcystis aeruginosa	0.08	0.04	0.10	0.03	0.94	0.87	0.91	0.59
Protogonyaulax sp.	-0.03	0.07	0.37	0.39	0.74	0.57	0.70	0.78
Euglena sp.	-0.24	-0.61	0.51	0.38	0.65	0.51	0.52	0.55

[1] $k \text{ (divisions/day)} = 3.322 \dfrac{\log \frac{N_1}{N_0}}{t_2 - t_1}$ (Guillard 1973)

* Freshwater strains, [1] NFUP-9, [2] NFUP-13, [3] NFUP-2, [4] NFUP-10

(Source: S.B. Hur, In: W. Fulks and K.L. Main, *Rotifer and Microalgae Culture Systems*; copyright © 1991 - Argent Laboratories)

TABLE 3-31: REPRESENTATIVE LIGHT SATURATION VALUES FOR MACROALGAE

SPECIES	INTENSITY (lux)	SOURCE
GREEN MACROALGAE		
Acetabularia sp.	16,200	Colinvaux *et al.* (1965)
Batophora sp.	16,200	Colinvaux *et al.* (1965)
Cauleurpa sp.	16,200	Colinvaux *et al.* (1965)
Cymopolia sp.	16,200	Colinvaux *et al.* (1965)
Dictyosphaeria sp.	16,200	Colinvaux *et al.* (1965)
Halimeda sp.	16,200	Colinvaux *et al.* (1965)
Neomeris sp.	16,200	Colinvaux *et al.* (1965)
Penicillus sp.	16,200	Colinvaux *et al.* (1965)
Udotea sp.	16,200	Colinvaux *et al.* (1965)
Ulva fasciata	2,500	Mohsen *et al.* (1973)
Valonia sp.	16,200	Colinvaux *et al.* (1965)
BROWN MACROALGAE		
Laminaria hyperboria	3,400	Luning (1971)
RED MACROALGAE		
Chondrus crispus	10,800	Burns and Mathiesen (1972)
Eucheuma nudum	3,240	Dawes *et al.* (1976)
Gigartina exasperata	11,880	Waaland (1973)
Gigartina stellata	21,600	Burns and Mathiesen (1972)
Iridaea cordata	11,880	Waaland (1973)
Pilaylella littoralis	1,620	West (1967)
Pleonosporium squarrulosum	1,620	Murray and Dixon (1973
Rhodochorton purpureum	2,160	West (1974)

(Source: S. Spotte, *Seawater Aquariums: The Captive Environment*; copyright © 1979 - reprinted by permission of John Wiley & Sons, Inc.)

TABLE 3-32: GENERAL LIGHTING SPECIFICATIONS FOR MARINE MACROALGAE USING COOL WHITE FLUORESCENT LAMPS

ALGAL GROUP	INTENSITY (lux)
GREEN (tropical)	13,000-16,200
GREEN (temperate)	7,560-10,800
BROWN	7,560-10,800
RED	2,160-8640

(Source: S. Spotte, *Seawater Aquariums: The Captive Environment*; copyright ©1979 - reprinted with permission of John Wiley & Sons, Inc.)

TABLE 3-33: TECHNICAL DATA ON LIGHT BULBS (Weast,1987; Lundegaard, 1985; Osborne, 1983)

BULB TYPE	WATTS	INITIAL LUMENS	INITIAL EFFIENCY	COLOR TEMP (°K)	HOURS OF LIFE	CRI [1]	LENGTH INCHES
INCANDESCENT							
60 INCAN	60	870	14.5	3000	1000	99	N/A
100 INCAN	100	1750	17.5	3000	750	99	N/A
200 INCAN	200	4010	20.0	3000	750	99	N/A
1000 INCAN	1000	23740	23.7	3000	1000	99	N/A
FLUORESCENT							
COOL WHITE	40	3150	78.8	4150	20000	62	48
WARM WHITE	40	3200	80.0	3000	20000	52	48
DAYLIGHT	40	2600	65.0	6250	20000	75	48
COOL WHITE HO	60	4300	62.0	4160	20000	62	48
VITA-LITE~ FS	40	2400	60.0	5500	20000	91	48
CHRONA 50~) FS	40	2210	55.3	5000	20000	90	48
CHRONA 75(~) FS	40	2000	50.0	7500	20000	92	48
COLORTONE 503 FS	40	2200	55.0	5000	20000	92	48
VERILUX(~ FS	40	2168	54.2	6200	20000	93	48
SP30	40	3325	83.1	3000	15000	70	48
SP35	40	3325	83.1	3500	15000	73	48
SP41	40	3265	81.6	4100	15000	70	48
GREEN							
COOL GREEN	40	2850	71.3	6450	20000	68	48
SEALUX	40	1600	40.0	5326	20000	G	48
AQUARILUX	40	880	22.0	10500	20000	F	48
ACTINIC 03	40		-	-	-	-	48
PLANT LIGHT	40	850	21.3	6750	20000	2	48
PLANT WS	40	1950	48.8	3050	20000	90	48
COOLWHITE	75	6300	84.0	4150	12000	62	60
WARM WHITE	75	6500	86.7	3000	12000	52	60
DAYLIGHT	75	5450	72.7	6250	12000	75	60
COOL WHITE	110	9200	83.6	4150	12000	62	96
WARM WHITE	110	9200	83.6	3050	12000	52	96
DAYLIGHT	110	7800	71.0	6250	12000	75	96
DAYLIGHT	215	13300	61.8	6250	12000	75	
HIGH INTENSITY DISCHARGE - MERCURY VAPOR							
CLEAR	400	21000	52.5	N/A	24000	G	N/A
DELUXE WHITE	400	22500	56.3	N/A	24000	G	N/A
WARM WHITE	400	19500	48.8	N/A	24000	G	N/A
CLEAR	1000	57000	57.0	N/A	24000	G	N/A
DELUXE WHITE	1000	63000	63.0	N/A	24000	G	N/A
WARM WHITE	1000	58000	58.0	N/A	24000	G	N/A
HIGH INTENSITY DISHARGE - METAL HALIDE [2]							
CLEAR E-23 1/2	175	16600	94.9	N/A	10000	E	N/A
DIFFUSE E-23 1/2	175	15750	90.0	N/A	10000	E	N/A
PHOSPHOR E-28	175	14000	80.0	N/A	10000	E	N/A
CLEAR E-28	250	20500	82.0	N/A	10000	E	N/A
PHOSPHOR E-28	250	20500	82.0	N/A	10000	E	N/A
CLEAR E-37	400	36000	90.0	N/A	20000	E	N/A
PHOSPHOR E-37	400	36000	90.0	N/A	20000	E	N/A
CLEAR BT-56	1000	110000	111.0	N/A	12000	E	N/A
PHOSPHOR BT-56	1000	105000	105.0	N/A	12000	E	N/A

[1] CRI (Color Rendering Index) Signifies the spectral distribution of light sources. The CRI of sunlight, as a standard, is 100. The higher the CRI of a bulb, the more all colors appear natural to the human eye.

(continued on page 54)

(continued from page 53)

CRI: F - Fair, G - Good, E - Excellent, 90 - 100 = Excellent

[2] Color temperature of metal halide bulbs varies 4300 to 5500 K.

To determine wattage of lights required to produce a specific amount of Lux over a tank:

$$WATTS = \frac{(LUX)\ (AREA\ IN\ SQUARE\ METERS)}{(BULB\ EFFICIENCY)\ (UTILIZATION\ FACTOR)} \quad OR \quad W = \frac{LA}{EU}$$

WHERE:

L = Total lux desired over surface area

A = Total surface area of tank in m^2

E = Efficacy of bulb in lumens/watt

U = Utilization factor (a constant 0.5 represents fraction of light striking water surface)

EXAMPLE:

If 15,000 lux are desired over a surface area of 0.6 meters, using 40 watt cool white flourescent bulbs, the required wattage would be:

$$W = \frac{(15,000)\ (0.6)}{(78.8)\ (0.5)} = 228.4$$

Six 40 watt cool white bulbs would be required to provide the desired illumination.

(Source: M. Moe, *The Marine Aquarium Reference*; copyright © 1989 - with permission of Green Turtle Publications)

TABLE 3-34: COMPARATIVE CHARACTERISTICS OF THE PHYSIOLOGICAL EFFECTIVENESS OF VARIOUS INCANDESCENT LAMPS

Kind of lamp	Perform-ance output of light source in watts	% incident radiation physiol. effective	Absorption of light energy by leaf meso-phyll plastid pigments (% incident energy)		Theoretical coefficient of utilization of incident energy		Quantum amounts in energy units of physiological radiation		Theor. effective. of light source compared to 300 W. incand. lamp	Theor. specific perform. essential for plant growth (KW/m^2)
			For all pigments	For chlorophyll alone	% output absorbed per light source	In relative units compared to 300 w incand. lamp	In/ergs	In relative units compared to 300 w incand. lamp		
Incand. vacuum (ir-free	10	-	59.0	53.7	-	-	3.36×10^{11}	1.08	-	-
	25	-	59.7	54.3	-	-	3.32×10^{11}	1.07	-	-
	40	-	60.0	54.5	-	-	3.28×10^{11}	1.06	-	-
	60	-	61.5	55.6	-	-	3.22×10^{11}	1.04	-	-
Incand. gas-filled	75	8.7	60.8	54.0	4.7	0.77	3.13×10^{11}	1.02	0.79	0.63 - 1.52
	100	9.1	61.1	54.1	4.9	0.80	3.13×10^{11}	1.00	0.80	0.62 - 1.50
	200	10.6	62.0	54.6	5.8	0.95	3.11×10^{11}	1.00	0.95	0.52 - 1.26
	300	11.1	63.0	55.0	6.1	1.00	3.10×10^{11}	1.00	1.00	0.50 - 1.20
	500	11.6	63.2	55.1	6.5	1.06	3.09×10^{11}	1.00	1.06	0.47 - 1.13
	1000	12.5	63.6	55.5	7.1	1.16	3.08×10^{11}	1.00	1.16	0.43 - 1.03
	2000	13.5	64.5	56.2	7.6	1.24	3.07×10^{11}	0.99	1.23	0.40 - 0.98
Projection lamp	1000 up to 3000	17.5	68.3	57.4	10.1	1.69	$3.06 - 10^{11}$	0.99	1.62	0.31 - 0.74

(Kleschin, 1960.)

(Source; E. Bickford and S. Dunn, *Lighting for Plant Growth*; copyritight © 1972 - with permission of Kent State University Press)

TABLE 3-35: COMPARATIVE CHARACTERISTICS OF THE PHYSIOLOCICAL EFFECTIVENESS OF FLUORESCENT AND MERCURY LAMPS (Kleschnin, 1960)

Kind of lamp	Performance output of light source in watts	% incident radiation physiologically effective	Absorption of light energy by leaf mesophyll plastid pigments (% incident energy) For all pigments	For chlorophyll alone	Theoretical coefficient of utilization of incident energy % output absorbed per light source	In relative units compared to 300 w incand. lamp	Quantum amounts in energy units of physiological radiation In/ergs	In relative units compared to 300 w incand. lamp	Theoretical effectiveness of light source compared to 300 W. incand. lamp	Theoretical specific performance essential for plant growth (KW/m²)
HIGH VOLTAGE LAMPS										
DAYLIGHT	50	18.4	78.0	58.5	10.7	1.76	2.76×10^{11}	0.89	1.51	0.33-0.80
Warm White	50	17.2	76.0	61.5	10.6	1.74	2.86×10^{11}	0.92	1.60	0.31-0.75
White	100	12.0	75.5	68.0	8.2	1.34	2.90×10^{11}	0.93	1.25	0.40-0.96
LOW VOLTAGE LAMPS										
DAYLIGHT										
DS-15	15	13.3	80.0	59.4	7.9	1.30	2.75×10^{11}	0.88	1.15	0.43-1.04
DS-20	20	15.3	80.0	59.4	9.1	1.49	2.75×10^{11}	0.88	1.32	0.38-0.91
DS-30	30	16.5	80.0	59.4	9.8	1.61	2.75×10^{11}	0.88	1.43	0.35-0.84
DS-40	40	18.1	80.0	59.4	10.7	1.75	2.75×10^{11}	0.88	1.55	0.32-0.78
DS-65	65	11.1	80.0	59.4	6.6	1.08	2.75×10^{11}	0.88	0.90	0.52-1.25
DS-100	100	13.5	80.0	59.4	8.0	1.31	2.75×10^{11}	0.88	1.16	0.43-1.04
WARM WHITE										
MBS- 15	15	12.3	84.1	61.9	7.6	1.25	2.82×10^{11}	0.91	1.14	0.04-1.05
MBS-20	20	13.6	84.1	61.9	8.4	1.38	2.82×10^{11}	0.91	1.25	0.40-0.96
MBS-30	30	15.5	84.1	61.9	9.6	1.57	2.82×10^{11}	0.91	1.43	0.35-0.84
MBS-40	40	16.0	84.1	61.9	9.9	1.62	2.82×10^{11}	0.91	1.47	0.34-0.82
WHITE										
BS-15	15	13.8	73.5	67.5	9.3	1.52	2.98×10^{11}	0.99	1.51	0.33-0.80
BS-20	20	16.0	73.5	67.5	10.8	1.77	2.98×10^{11}	0.99	1.76	0.28-0.68
BS-30	30	17.0	73.5	67.5	11.4	1.87	2.98×10^{11}	0.99	1.86	0.27-0.65
BS-40	40	18.8	73.5	67.5	12.6	2.06	2.98×10^{11}	0.99	2.05	0.24-0.59
BS-65	65	11.4	73.5	67.5	7.7	1.26	2.98×10^{11}	0.99	1.25	0.400.96
BS-100	100	14.9	73.5	67.5	10.1	1.65	2.98×10^{11}	0.99	1.64	0.30-0.73
BLUE (CaWO₄)[1]	15	10.9	88.3	54.3	5.9	0.97	2.41×10^{11}	0.78	0.76	0.65-1.58
	20	11.9	88.3	54.3	6.5	1.06	2.41×10^{11}	0.78	0.84	0.60-1.48
	30	13.4	88.3	54.3	7.3	1.20	2.41×10^{11}	0.78	0.91	0.55-1.32
GREEN (ZnSiO₃)	15	13.3	66.2	43.0	5.7	0.93	2.57×10^{11}	0.83	0.77	0.65-1.56
	20	14.4	66.2	43.0	6.2	1.02	2.57×10^{11}	0.83	0.85	0.59-1.4
	30	16.7	66.2	43.0	7.2	1.18	2.57×10^{11}	0.83	0.99	0.51-1.21
RED (CdB₂O₃)	15	9.0	74.5	72.5	6.5	1.07	3.03×10^{11}	0.98	1.05	0.47-1.14
	20	9.9	74.5	72.5	7.2	1.18	3.03×10^{11}	0.98	1.15	0.43-1.04
	30	11.3	74.5	72.5	8.2	1.34	3.03×10^{11}	0.98	1.31	0.38-0.92
GOLD	15	8.3	70.5	69.4	5.7	0.93	3.04×10^{11}	0.98	0.92	0.54-1.30
	20	8.0	70.5	69.4	6.2	1.02	3.04×10^{11}	0.98	1.00	0.50-1.20
	30	10.3	70.5	69.4	7.1	1.16	3.04×10^{11}	0.98	1.14	0.44-1.05
HIGH-PRESSURE MERCURY										
HgL-300	75	9.9	74.0	65.8	6.5	1.06	2.84×10^{11}	0.91	0.97	0.51-1.22
HgL-500	120	10.5	74.0	65.8	6.9	1.13	2.84×10^{11}	0.91	1.03	0.49-1.17

[1] Phosphor in glass wall.

(Source; E. Bickford and S. Dunn, *Lighting for Plant Growth*; copyright © 1972 - with permission of Kent University Press)

TABLE 3-36: USING A HEMACYTOMETER TO DETERMINE PHYTOPLANKTON DENSITY

The hemocytometer was originally developed as a medical tool for counting blood cells, but it is widely used in aquaculture to determine phytoplankton cell counts. It consists of two parts; a) a base which is a thick slide of thermal and shock resisting glass with an H-shaped trough cut into it. Two precisely measured shoulders rise 0.1 millimeter above each side of the trough, b) the second component is a thick cover glass (0.4 millimeters) which rests on top of the shoulders forming the top of the counting chamber.

Upon the base glass is fused a thin metallic film which is precisely etched into a pattern of nine squares, each one millimeter on the side. These are divided into 16 smaller squares, and the center square is further subdivided into 4 other squares each measuring 0.05 millimeters on a side. As hemacytometers may vary in dimensions, consult the manufacturers instructions to determine the depth of the counting chamber and the precise area of the counting grids before calculating phytoplankton density.

PREPARING A SAMPLE FOR COUNTING

1) Thoroughly clean the base slide and cover slip with distilled water and and lens paper.

2) Collect a sample of the algae culture and make serial dilutions if required. Addition of Lugol's stain may be added to kill and immobilize the cells.

3) Introduce a single drop of solution to the V-groove on the side of the hemocytometer, allowing the solution to spread through the counting chamber by capillary action. Avoid overfilling the counting chamber and allowing the cover slip to float, as this will result in high cell counts.

4) Place the hemocytometer on the stage of a compound microscope and at the lowest power of magnification focus onto the center grid.

5) Beginning at the upper righthand corner count all cells within the grid. It is suggested to count only cells on the upper line of the grid; cells intersecting the lower lines of the grid should be counted on the next lower grid. This method avoids duplicating counts.

CALCULATING CELL COUNTS

$$\text{CELLS/mm}^3 = \text{CELLS/mm}^2 \text{ X 10 X DILUTION}$$

WHERE:

$$\text{CELLS/mm}^2 = \text{AVERAGE CELLS COUNTED/ AREA COUNTED (mm}^2)$$

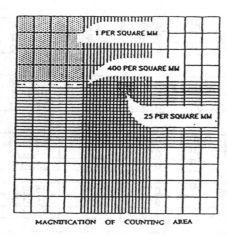

MAGNIFICATION OF COUNTING AREA

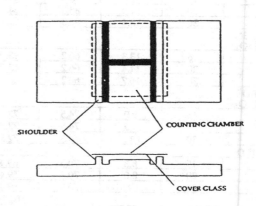

56

IV. Hatchery Systems and Methods

TABLE 4-1: PRELIMINARY WATER QUALITY SCREENING AND PRODUCTION LEVELS FOR MARINE APPLICATIONS

PARAMETER	SCREENING LEVEL	PRODUCTION LEVEL
AMMONIA (except for plants)	$< 1 \mu g/l \ NH_3$-N	$< 1 \mu g/l \ NH_3$-N research
		$< 10 \mu g/l \ NH^3$-N production
		$< 40 \ ug/l \ NH_3$-N holding (little or no feeding)
NITRITE	$< 0.05 \ mg/l \ NO_2$-N	$<0.10 \ mg/l \ NO_2$-N
DISSOLVED OXYGEN (except for plants)	90% of saturation	$> 6 \ mg/l$
TOTAL GAS PRESSURE	$<76 \ mm \ Hg$	$<20 \ mm \ Hg$
CARBON DIOXIDE (except for plants)	$5 \ mg/l \ CO_2$	$<10 \ mg/l \ CO_2$
HYDROGEN SULFIDE	$2 \mu g/l$ as H_2S	$<1 \mu g/l$ as H_2S
CHLORINE RESIDUAL	$10 \mu g/l$	$<1 \mu g/l$
pH	7.9-8.2	<7.9-8.2
TEMPERATURE	Depends on life stage and species	1 to 40°C Temperature
SALINITY	Depends on life stage and species	1 to 40 g/kg
METALS (total)		
CADMIUM	$1 \mu g/l$	$3 \mu g/l$
CHROMIUM	$10 \mu g/l$	$25 \mu g/l$
COPPER	$1 \mu g/l$	$3 \mu g/l$
IRON	$300 \mu g/l$	$100 \mu g/l$
MERCURY	$0.05 \mu g/l$	$0.1 \mu g/l$
MANGANESE	$50 \mu g/l$	$25 \mu g/l$
NICKEL	$2 \mu g/l$	$5 \mu g/l$
LEAD	$2 \mu g/l$	$4 \mu g/l$
ZINC	$10 \mu g/l$	$25 \mu g/l$

(Source: J. Huegenin and J. Colt, *Design and Operating Guide for Aquaculture Seawater Systems*; 1989)

TABLE 4-2: SUGGESTED CHEMICAL VALUES FOR HATCHERY WATER SUPPLIES (freshwater)

PARAMETER	CONCENTRATION (mg/l, ppm)	
	TROUT	WARM WATER FISH
DISSOLVED OXYGEN	5 to saturation	5 to saturation
CARBON DIOXIDE	0-10	0-15
TOTAL ALKALINITY (as $CaCO_3$)	10-400	50-400
% as phenolphthalein	0-25	0.40
% as methyl orange	75-100	60-100
% as ppm hydroxide	0	0
% as ppm carbonate	0-25	0-40
% as ppm bicarbonate	75-100	75-100
pH	6.5-8.0	6.5-9.0
TOTAL HARDNESS (as $CaCO_3$)	10-400	50-400
CALCIUM	4-160	10-160
MAGNESIUM	Needed for buffer system	
MANGANESE	0-0.15	0-0.5
FERROUS IRON	0	0
FERRIC IRON	0.5	0-0.5
PHOSPHOROUS	0.01-3.0	0.01-3.0
NITRATE	0-3.0	0-3.0
ZINC	0-0.05	same
HYDROGEN SULFIDE	0	0

(Source: US Fish and Wildlife Service, *Fish Hatchery Management*; 1982)

TABLE 4-3: THRESHOLD OF TOXICITY AND MAXIMUM PERMISSIBLE CONCENTRATION OF TOXIC SUBSTANCES IN THE WATER SUPPLY OF INDOOR FISH HATCHERIES.

SUBSTANCE	THRESHOLD CONCENTRATION (mg/l)	MAXIMUM PERMISSIBLE CONCENTRATION mg/l
AMMONIA	0.2 - 2.0	0.05
DDT	0.02 - 0.1	ABSENT
CALCIUM BISULPHATE	30 - 60	-
CALCIUM CHLORIDE	7,000 - 12,000	-
POTASSIUM CHLORIDE	700 - 5,200	-
POTASSIUM SULPHATE	800 - 1,000	-
MAGNESIUM CHLORIDE	5000 - 15,000	20
MAGNESIUM NITRATE	10,000	15
MAGNESIUM SULPHATE	30,000	50
MANGANESE (NITRATE;CHLORIDE;SULPHATE)	75 - 200	5
COPPER (COMPOUNDS)	0.08 - 0.8	0.005
SODIUM BICARBONATE	5,000	-
SODIUM CARBONATE	200 - 500	-
SODIUM CHLORIDE	7,000 - 15,000	-
CADMIUM	3 - 20	0.003
OZONE	0.02	-
MERCURY	0.1 - 0.9	-
ROTENONE	0.01 - 0.012	ABSENT
SULPHIDES	0.4 - 10	0.1
HYDROGEN SULPHIDE	1.0	0.1
IRON (COMPOUNDS)	0.9 - 2.0	0.01
PHENOL	6 - 17	0.0005
FORMALDEHYDE	15 - 30	-
TANNIN	15	5
PARAQUINONE	0.1 - 10	-
CHLORINE	0.05 - 0.4	ABSENT
CARBOLINEUM	7	-
ZINC (COMPOUNDS)	0.1 - 2.0	0.005

(Source: T.V.R. Pillay; *Aquaculture Principles and Practices*; 1990)

TABLE 4-4: FILTRATION EQUIPMENT FOR USE WITH SEAWATER SYSTEMS

MAXIMUM FILTRATION	FLOW RATE			
	1 gpm or less	1-10 gpm	10-100 gpm	100-1000 gpm
1 μ OR LESS	Cartridge Filter Diatomaceous Earth	Cartridge Filter Diatomaceous Earth	Diatomaceous Earth	Diatomaceous Earth
1-10 μ	Cartridge Filter Centrifuges and Cyclones	Cartridge Filter Centrifuges and Cyclones Sand Filters	Centrifuges and Cyclones Sand Filters	Sand Filters
10-75 μ	Filter Bags Centrifuges and Cyclones	Filter Bags Centrifuges and Cyclones Sand Filters	Filter Bags Centrifuges and Cyclones Sand Filters	Filter Bags Sand Filters
75-150 μ	Filter Bags Microscreens	Filter Bags Microscreens Sand Filters	Filter Bags Sand Filters Sand Filters	Microscreens
150-1000 μ	Screen Bags Microscreens	Screen Bags Microscreens Sedimentation	Screen Bags Sedimentation Sedimentation	Microscreens

The generic identifications used in this table encompass a wide variety of equipment with many different specifications. The table is only intended as a general guide to the major areas of applicability and most probable use.

(Source: J. Huegenin and J. Colt, *Design and Operating Guide for Aquaculture Seawater Systems*; 1989)

TABLE 4-5: SIEVE SEPARATIONS Of 16 GRADED TEST SANDS (% BY WEIGHT OF PARTICLES RETAINED ON SIEVE) (Hulbert and Feben, 1933)

SAND NUMBER	SIZE OF SIEVE OPENING (mm)										
	1.356	1.132	0.930	0.746	0.655	0.548	0.461	0.3831	0.339	0.243	0.149
9	0	0	0.01	0.09	0.40	8.90	33.70	23.03	20.32	12.21	1.34
10	0.90	1.60	7.00	19.00	22.20	28.05	14.91	3.59	1.62	0.96	1.17
11	9.80	7.80	18.10	27.7	20.0	14.43	1.95	0.06	0.02	0.10	0.04
12	0.50	1.00	4.40	15.90	21.50	31.40	20.14	4.25	0.27	0.60	0.04
13	0	0.60	8.50	29.70	6.60	15.30	18.90	8.07	6.87	4.94	0.52
14	5.90	5.00	14.6	18.2	13.1	12.5	15.8	7.70	4.00	3.16	0.04
15	0	0.10	1.3	12.8	15.3	22.50	29.32	13.17	3.94	1.58	0
16	1.10	4.10	23.30	38.80	17.30	8.70	5.01	0.95	0.20	0.54	0
17	0	0.20	16.00	50.60	8.80	14.66	8.49	0.73	0.15	0.37	0
18	5.50	13.8	33.2	27.4	8.9	3.23	4.55	1.47	0.47	1.48	0
19	0	0.30	13.0	45.3	10.27	19.63	9.95	0.85	0.22	0.48	0
20	0.10	0.20	10.95	33.35	24.8	19.30	9.18	1.52	0.06	0.47	0.07
21	0	0.10	5.40	18.40	4.70	10.90	25.30	14.7	11.06	8.69	0.75
23	0	0.10	5.40	19.5	18.3	22.3	20.70	9.88	2.36	1.40	0.06
24	0	0.30	8.5	27.0	21.6	20.2	14.2	5.63	1.28	1.22	0.07
25	0	0.10	4.50	18.50	18.20	22.80	22.15	10.69	2.16	1.24	0.06
MEDIAN (50 % size)	1.59	1.23	1.02	0.83	0.70	0.60	0.50	0.42	0.36	0.28	0.15

(Source: F.W. Wheaton, *Aquacultural Engineering*; 1985)

TABLE 4-6: DISINFECTANTS AND THEIR APPLICATION (after Hnath, 1983).

DISINFECTANT	WORKING STRENGTH	APPLICATION	EFFECTIVE AGAINST
CHLORINE	1-2%	Concrete; fibreglass; butyl-lined ponds; nets; footbaths	Bacteria; fungi; viruses
SODIUM HYDROXIDE	1% with 0.1% teepol; 0.5 gallon/m^2 (2.28 1/m^2)	Earthen ponds; concrete; fibreglass; butyl-lined ponds; nets; footbaths	Bacteria; viruses; protozoa
IODOPHORS	250 ppm	Concrete; fiberglass; butyl-lined ponds; nets; angling equipment; clothing; hands	Bacteria; viruses; protozoa
	100 ppm	Ova	-
QUATERNARY AMMONIUM COMPOUNDS (Hyamine; Roccal; etc.)	As manufacturers' instructions	Nets; clothing; hands	Bacteria
CALCIUM OXIDE	As powder; 380 g/m^2	Earthen ponds; fibreglass; concrete; butyl-lined ponds	Protozoa (whirling disease)

(Source: T.V.R. Pillay, *Aquaculture Principles and Practices*; 1990)

TABLE 4-7: ULTRAVIOLET ENERGY OF 2537 A WAVE-LENGTH TO INHIBIT COLONY FORMATION IN 90 AND 100 PERCENT OF TEST ORGANISMS (μW/s/cm^2) (Phillips and Hanel, 1960)

ORGANISM	UV ENERGY (μW/s/cm^2)	
	90%	100%
BACTERIA		
Bacillus anthracis	4,250	8,700
S. enteritidis	4,000	7,600
B. megaterium sp. (veg)	1,300	2,500
B. megaterium sp. (spores)	2,730	5,200
B. paratyphosus	3	200
B. subtilis	5,800	11,000
B. subtilis spores	11,600	22,000
Corynebacterium diphtheria	3,370	6,500
Eberthella typhosa	2,140	4,100
Escherichia coli	3,000	6,600
Micrococcus candidus	6,050	12,300
Micrococcus sphaeroides	10,000	15,400
Neisseria catarrhalis	4,400	8,500
Phytomonas tumefaciens	4,400	8,500
Proteus vulgaris	3,000	6,600
Pseudomonas aeruginosa	5,500	10,500
Pseudomonas fluorescens	3,500	6,600
S. typhimurium	8,000	15,200
Sarcina lutea	19,700	26,400
Serratia marcescens	2,420	6,160
Dysentery bacilli	2,200	4,200
Shigella paradysenteriae	1,680	3,400
Spirillum rubrum	4,400	6,160
Staphylococcus albus	1,840	5,720
Staphylococcus aureus	2,600	6,600
Streptococcus hemolyticus	2,160	5,500

TABLE 4-7 (cont.)

ORGANISM	UV ENERGY (μW/s/cm^2)	
	90%	100%
Streptococcus lactis	6,150	8,800
Streptococcus viridans	2,000	3,800
YEAST		
Saccharomyces ellipsoideus	6,000	13,200
Saccharomyces sp.	8,000	17,600
Saccharomyces cerevisiae	6,000	13,200
BREWER'S YEAST	3,300	6,600
BAKER'S YEAST	3,900	8,800
COMMON YEAST CAKE	6,000	13,200
MOLD SPORES		
Penicillium roqueforti (green)	13,000	26,400
Penicillium expansum (olive)	13,000	22,000
Penicillium digitatum (olive)	44,000	88,000
Aspergillus glaucus (bluish green)	44,000	88,000
Aspergillus flavus (yellowish-green)	60,000	99,000
Aspergillus niger (black)	132,000	330,000
Rhizopus nigricans (black)	111,000	220,000
Mucor racemosus A (white-gray)	17,000	35,200
Mucor racemosus B (white)	5,000	11,000
Oospora lactis (white)	5,000	11,000

Consult manufacturers of UV sterilizing lights for energy output at specified flow rates.

(Source: F.W. Wheaton, *Aquacultural Engineering*; copyright © 1985 - with permission of John Wiley & Sons, Inc.)

TABLE 4-8: ULTRAVIOLET ENERGY FOR 100 PERCENT KILL (μW s/cm^2) (Kelly, 1974)

ORGANISM	UV ENERGY (μW s/cm^2)
MOLD SPORES	
Penicillium roqueforti	26,400
Aspergillus niger	330,000
YEASTS	
BREWER'S YEAST	6,600
BAKER'S YEAST	8,800
COMMON YEAST CAKE	13,200
BACTERIA	
Streptococcus hemolyticus	5,500
Staphylococcus aureus	6,600
Escherichia coli	7,000
Proteus vulgaris	7,500
Bacillus subtillis	11,000
Bacillus subtillus spores	22,000
VIRUS	
BACTERIOPHAGE (*E. coli*)	6,600
INFLUENZA VIRUS	3,400
NEMATODE EGGS	40,000

Consult manufacturers of UV sterilizing lights for energy output at specified flow rates.

(Source: F. W. Wheaton, *Aquacultural Engineering*; copyright © 1985 - with permission of John Wiley & Sons, Inc.)

TABLE 4-9: SIZES AND MLD OF UV RADIATION FOR SOME MICROORGANISMS FREE-LIVING OR PARASITIC IN AQUARIUM OR HATCHERY WATER. (from Hoffman, 1974)

MICROORGANISM	LIFE STAGE	SIZE (µm)	MLD (µW sec/cm²)
Trichodina sp.	-	16 x 20	35,000
Trichodina nigra	-	22 x 70	159,000
Saprolegnia sp.	ZOOSPORE	4 x 12	35,000
Saprolegnia sp.	HYPHA	8 x 24	10,000
Oodinium ocellatum [1]	DINOSPORE	8 x 12	
Sarcina lutea	-	1.5	26,400
Ichthyophthirius sp.	TOMITE	20 x 35	336,000
Ichthyophthirius sp.	TOMITE	20 x 35	100,000
Cryptocaryon irritans [2]	TOMITE	35 x 56.5	
Chilodonella cyprini	-	35 x 70	1,008,400
Paramecium sp.	-	70 x 80	200,000

[1] From data in Nigrelli (1936).

[2] From data in Nigrelli and Ruggieri (1966).

(Source: S. Spotte, *Fish and Invertebrate Culture*; copyright © 1979 - with permission of John Wiley & Sons, Inc.)

TABLE 4-10: BIOLOGICAL ACTIVITY OF ANTIBIOTICS COMMONLY USED IN AQUACULTURE

ANTIBIOTIC	SPECTRA [1] G+	RANGE [2] µg/ml G-	BACTERIA	FUNGI	PROTOZOA
CHLORTETRACYCLINE [3]	1	1	0.002-50		25-1000
BACITRACIN	1	2	0.002-125	-	-
CARBOMYCIN	1	2	0.01-12		32-250
CHLORAMPHENICOL [3]	1	1	0.06-50		125-2000
COLISTIN	3	1	0.5-50	20	125
CANDICIDIN	-	-	-	0.5-50	-
ERYTHROMYCIN	1	2	0.003-200	-	-
KANAMYCIN	1	1	0.3-500	-	-
NYSTATIN	-	-	-	1-13	250
NEOMYCIN	1	1	0.2-100	-	43-3000
NOVOBIOCIN	1	2	0.02-200	10-1000	125
OLEANDOMYCIN	1	1	0.08-50	-	-
PENICILLIN	1	2	1-5000	-	-
POLYMIXIN B	3	1	0.02-50	125-250	125
STREPTOMYCIN [3]	1	1	0.5-300	-	-
TETRACYCLINE [3]	1	1	0.05-7.5	-	62-250
OXYTETRACYCLINE [3]	1	1	0.002-50	-	31-250
TRICHOMYCIN	-	-	-	0.6-10	250

[1] Effectiveness of antibiotic against gram-positive (G+) and gram-negative (G-) bacteria; (3) no activity; (2) little activity, effective against a few representative species; (1) activity, effective against most representative species.

[2] Range in ug/ml of antibiotic which inhibits microbial growth.

[3] Known as broad-spectrum antibiotic because active against gram-negative, gram-positive and other microorganisms.

(Source: W.L. Smith and M.H. Chanley, *Culture of Marine Invertebrate Animals*; copyright © 1975 - with permission of Plenum Publishing)

TABLE 4-11: STABILITY AND ACTIVITY OF ANTIBIOTICS COMMONLY USED IN AQUACULTURE

ANTIBIOTICS	pH[1]	STABILITY IN AQUEOUS SOLUTIONS			DEVELOP RESIST.[2]	CROSS RESIST.[3]
		DAYS	pH	°C		
1. CHLORTETRACYCLIN	6.0-6.6	14	2.5-3	25	slow	4;16;17
2. BACITRACIN	6.0-6.6	14	5-7	35-37	slow	none
3. CARBOMYCIN		11	5-7	25	yes	7;12
4. CHLORAMPHENICOL	7.4-8.0	30	6-8	30	slow	1;14;16;17
5. COLISTIN	7.0-8.0	16	7-7.8	20	slow	-
6. CANTICIDIN		7	7	4	-	-
7. ERYTHROMYCIN	7.4-8.0	1	7-8	25	rapid	3;12
8. KANAMYCIN	7.4-8.0	30	7.8	4	rapid	10
9. NYSTATIN	-	-	-	-	-	-
10. NEOMYCIN	7.4-8.0	30	2-9	25	slow	8;15
11. NOVOBIOCIN	6.0-6.6	60	7-10	24	rapid	none
12. OLEANDOMYCIN	6.0-6.6	1	5-7	25	rapid	3;7
13. PENICILLIN	6.0-6.6	3	6-7	25	yes	-
14. POLYMIXIN		365	6-7	37	slow	5
15. STREPTOMYCIN	7.4-8.0	90	3-7	25	rapid	8;10
16. TETRACYCLINE	6.0-6.6	3	7	37	slow	1;14;16;17
17. OXYTETRACYCLINE	6.0-6.6	7	7	25	slow	1;16
18. TRICHOMYCIN	-	-	-	-		none

[1] pH range for maximal antimicrobial activity.

[2] Bacterial strains develop resistance to the antibiotic.

[3] Microorganisms which have become resistant to an antibiotic numbered in column one also acquire resistance to antibiotics whose numbers are listed in this column.

Antibiotics with long half-lives include streptomycin, chloramphenicol, kanamycin and neomycin.

Effective antibiotics with short half-lives include chlortetracycline, oxytetracycline, bacitracin and carbomycin are most active at pH 6.0-6.6, and are much less effective in seawater. Colistin remains effective in seawater.

Antibiotics with very short half-lives include oleandomycin, wide-spectrum tetracycline, erythromycin and penicillin. Although effective against sensitive organisms, they are less effective for long-term control in cultures.

(Source: W.L. Smith and M.H. Chanley, *Culture of Marine Invertebrate Animals*; copyright © 1975 - with permission of Plenum Publishing)

TABLE 4-12: TEMPERATURE OPTIMA FOR *NITROSOMONAS* (NS) AND *NITROBACTER* (NB).

SPECIES	RANGE	TEMPERATURE (°C) OPTIMUM	INHIBITORY	SOURCE
NS	30-36	30-36	-	Buswell *et al.* (1953)
NS	10-40	30-35	-	Kawai *et al.* (1965)
NS	-	-	< 5	Buswell *et al.* (1953)
NB	8-24	28	-	Nelson (1931)
NB	4-45	34-35	-	Deppe & Engle (1960)
NB	-	42	-	Laudelout *et al.* (1960)
NB	10-40	30-35	-	Kawaietal. (1965)
NB	-	-	< 4	Depp & Engle (1960)
NB	-	-	>45	Depp & Engle (1960)

(Source: Wheaton, Hochheimer and Kaiser, In: D.E. Brune and J.R. Tomasso (Eds.), *Aquaculture and Water Quality*; 1991)

TABLE 4-13: THE pH RANGES GIVING THE BEST NITRIFICATION RATES FOR *NITROSOMONAS* (NS) AND *NITROBACTER* (NB).

SPECIES	RANGE TESTED	pH OPTIMA	SOURCE
NS		8.0 - 9.0	Hofman & Lees (1952)
NS	7.0 - 9.0	8.0	Engle & Alexander (1958)
NS		6.0 - 9.0	Winogradsky & Winogradsky (1933)
NS	5.0-10.0	9.0	Kawai *et al.* (1965)
NS		7.2 - 7.8	Loveless & Painter (1969)
NB		8.3 - 9.3	Mererhof (1917)
NB	7.0 - 8.6	7.8	Boon & Laudelout (1962)
NB		6.3 - 9.4	Winogradsky & Winogradsky (1933)
NB	5.0-10.0	9.0	Kawa *et al.* (1965)
BOTH	7.0 - 9.5	8.5	Kholdenbarin & Oertli (1977)

(Source: Wheaton, Hochheimer and Kaiser, In: D.E. Brune and J.R. Tomasso (Eds.), *Aquaculture and Water Quality*; 1991)

TABLE 4-14: GROWTH REQUIREMENTS AND TOXICITY OF VARIOUS COMPOUNDS TO *NITROBACTER* (NB) AND *NITROSOMONAS* (NS).

SPECIES	COMPOUNDS	CONCENTRATIONS	INHIBITION	SOURCES
NS	Ni	0.25 mg/l	mod	1
NS	Cr	0.25 mg/l	mod	1
NS	Cu	0.1 mg/l	slight	1

TABLE 4-14 (cont.):

SPECIES	COMPOUNDS	CONCENTRATIONS	INHIBITION	SOURCES
NS	Cu	0.5 mg/1	complete	1
NS	Stainless EN58B	presence	none	1
NS	Fe	< 0.5 mg/l	limits growth	1
NS	L-Arginine	5×10^{-3}M	slight	2
NS	L-Histidine	5×10^{-3}M	complete	2
NS	L-Phenylalanine	5×10^{-3}M	slight	2
NS	L-Tyrosine	5×10^{-3}M	slight	2
NS	DL-Methinoninesulphoxide	5×10^{-3}M	slight	2
NS	Creatinine	5×10^{-3}M	slight	2
NS	Glycocyamine	5×10^{-3}M	slight	2
NS	2,4 Dinitophenol	5×10^{-3}M	none	2
NS	Ethylurethane	5×10^{-3}M	moderate	2
NS	Calcium	50 mg/l	stimulated	
NB	Iron	7 mg/l	stimulated	3
NB & NS	Phosphate	310 mg/1-P	stimulated	4
NB	Magnesium	5 mgll	stimulated	3
NB	Phosphorus	5 mg/l	stimulated	3
NB	Molybdenum	1 mg/l	stimulated	3
NB	Zinc	1 mg/l	stimulated	3
NB	Molybdenum	0.1 - 1 mg/l	stimulated	5
NS	Magnesium	5.0 mg/l	stimulated	1
NS	Phosphate	0.28 mg/l	stimulated	1
NS	Iron	2 mg/l	stimulated	1
NS	Ni	0.1 mg/l	stimulated	1
NS	Cr	0.1 mg/l	stimulated	1
NS	Co	1.0 mg/l	stimulated	1
NS	Mn	1.0 mg/l	stimulated	1
NS	Zn	1.0 mg/l	stimulated	1

SOURCES:

[1] Skinner and Walker (1961)

[2] Lees (1952)

[3] Aleem (1959)

[4] Laudelout *et al.* (1967) (not referenced in source)

[5] Finstein and Delwiche (1965)

(Source: Wheaton, Hochheimer and Kaiser, In: D.E. Brune and J.R. Tomasso (Eds.), *Aquaculture and Water Quality*; 1991)

TABLE 4-15: EFFECTS OF COMMONLY USED ANTIBACTERIAL AGENTS AND PARASITICIDES ON NITRIFICATION IN FRESHWATER AQUARIUMS AT THERAPEUTIC LEVELS. (from Collins et al. 1975; 1976 and Levine and Meade 1976)

COMPOUND	CONCENTRATION (mg/l)	% INHIBITION	SOURCE
CHLORAMPHENICOL	50	0	B
	50	84	C
OXYTETRACYCLINE	50	0	B
SULFAMERAZINE	50	0	B
SULFANILAMIDE	25	65	C

TABLE 4-15 (cont.):

COMPOUND	CONCENTRATION (mg/l)	% INHIBITION	SOURCE
ERYTHROMYCIN	50	100	B
NIFURPIRINOL	1	0	B
	0.1	20	C
	4	44	C
CHLOROTETRACYCLINE	10	76	C
FORMALIN	25 [1]	0	A
	15	27	C
MALACHITE GREEN	0.1	0	A
FORMALIN + MALACHITE GREEN	25 + 0.1	0	A
METHYLENE BLUE	5	100	A
COPPER SULFATE	1 [2]	0	A
	5	0	C
POTASSIUM PERMANGANATE	4	0	A
	1	86	C

[1] Equivalent to 10 mg/l formaldehyde

[2] Hardness = $CaCO_3$ at 30 mg/l

SOURCES:

A = Collins *et al.* (1975); B = Collins *et al.* (1976); C = Levine and Meade (1976)

(Source: S. Spotte, *Fish and Invertebrate Culture*; copyright © 1979 - with permission of John Wiley & Sons, Inc.)

TABLE 4-16: SEA SALT QUANTITIES (IN POUNDS) NEEDED TO ADJUST FRESH WATER TO THE DESIRED SALINITY LEVEL IN A RECIRCULATING SYSTEM.

VOLUME (gallons)	DESIRED SALINITY (ppt)						
	5	10	15	20	25	30	35
100	4	8	13	17	21	25	29
200	8	17	25	33	42	50	58
300	13	25	38	50	63	75	88
400	17	33	50	67	83	100	117
500	21	42	63	83	104	125	146
600	25	50	75	100	125	150	175
700	29	58	88	117	146	175	204
800	33	67	100	133	167	200	233
900	38	75	113	150	188	225	263
1000	42	83	125	167	208	250	292
1100	46	92	138	183	229	275	321
1200	50	100	150	200	250	300	350
1300	54	108	163	217	271	325	379
1400	58	117	175	233	292	350	409
1500	63	125	188	250	313	375	438
1600	67	133	200	267	333	400	467
1700	71	142	213	283	354	425	496
1800	75	150	225	300	375	450	525
1900	79	158	238	317	396	475	554
2000	83	167	250	333	417	500	584

(Source: R. Malone, *Design of Recirculating Blue Crab Shedding System*; 1991)

TABLE 4-17: SEA WATER AND PUMP REQUIREMENTS FOR DIFFERENT PENAEID HATCHERY CAPACITIES [1]

HATCHERY CAPACITY	MAXIMUM TOTAL CONSUMPTION VOLUME (l/day)	RECOMMENDED RESERVOIR VOLUME (L)	TOTAL WATER CONSUMPTION PER RUN [2]	THEORETICAL PUMP HP REQUIREMENT WITH 6 M HEAD (kW)	RECOMMENDED PUMP HP (kW)
1 TANK	3,000	1,500	36,900	0.025 (0.018)	0.1(0.0746)
5 TANKS	12,000	6,000	149,000	0.075 (0.056)	0.3 (0.224)
10 TANKS	25,000	12,500	296,000	0.150(0.112)	0.6(0.448)
15 TANKS	37,000	18,500	433,000	0.200 (0.149)	0.8 (0.597)

[1] 2,000 liter tank capacity

[2] including 15% of actual for washing

(Source: A.N. Bose et al., Coastal Aquaculture Engineering; copyright © 1991 - with permission of Cambridge University Press)

TABLE 4-18: COMPRESSED AIR REQUIREMENTS FOR DIFFERENT PENAEID HATCHERY CAPACITIES [1]

HATCHERY CAPACITY	THEORETICAL HP REQUIREMENT (kW)	RECOMMENDED ACTUAL HP (kW) AT 0.2 kg/cm^2	TOTAL AIR VOLUMETRIC REQUIREMENT (l/min)
1 TANK	0.16 (0.12)	0.64 (0.48)	430
5 TANKS	0.305 (0.228)	1.22 (0.91)	820
10 TANKS	0.533 (0.398)	2.13 (1.59)	1430
15 TANKS	0.71	(0.53)	2.84 (2.12)

[1] 2,000 liter tank capacity

(Source: A.N. Bose et al., Coastal Aquaculture Engineering; copyright © 1991 - with permission of Cambridge University Press)

TABLE 4-19: AREA REQUIREMENTS OF A PENAEID HATCHERY INSTALLATION FOR DIFFERENT CAPACITIES (m^2) [1]

HATCHERY CAPACITY	A	B	C	D	E	F	TOTAL
1 TANK	6	8	4	10	45	7	80
5 TANKS	6	12	6	30	60	16	180
10 TANKS	9	12	9	50	90	23	183
15 TANKS	9	12	12	70	120	25	248

[1] 2,000 liter larval rearing tanks; 1,000 liter algal culture tanks

A = Algae culture room, inside building

B = Monitoring area, inside building

C = Storage compartment, inside building

D = Larval rearing area, inside building

E = Outdoor algal and Brachionus production area

F = Combined area for compressor, reservoir, and sand filter

(Source: A.N. Bose et al., Coastal Aquaculture Engineering; copyright © 1991 - with permission of Cambridge University Press)

TABLE 4-20: METHODS FOR DECAPSULATING *ARTEMIA* CYSTS

Decapsulation of *Artemia* cysts strips the corions (shell) down to the hatching membrane without damaging the embryo. This allows for effective disinfection of the cysts, provides a more edible and digestible food product, and requires less energy for the *Artemia* embryo to break out of the hatching membrane, yielding a higher caloric content per nauplii.

METHOD 1 - SODIUM HYPOCHLORITE INGREDIENTS:

1) 500 grams *Artemia* cysts

2) 4.77 liters of sodium hypochlorite solution (household bleach: 6.5% chlorine)

3) 75 grams sodium hydroxide (NaOH)

4) 2.2 liters of seawater or freshwater

Dissolve sodium hydroxide in 2.2 liters of water and allow to cool. Rehydrate artemia cysts in fresh water under vigorous aeration for 30 to 60 minutes and drain. Add artemia cysts and hypochlorite to sodium hydroxide solution and stir continuously. The color of the cysts in solution will begin to change from brown to gray, to orange. When the orange color is detected, usually in 4 to 5 minutes, quickly pour the decapsulated cysts through a sieve and rinse with fresh water.

To assure the decapsulation reaction has ceased, stir the cysts in 0.1 normal hydrochloric acid (HCL) for 30 to 60 seconds (0.1 N HCL = 65 mls of 36° baume HCl in 6.5 liters of water). After neutralizing the chlorine, pour the cysts through a seive and rinse thoroughly again in freshwater.

Incubate the cysts in a cone-bottomed tank with vigorous aeration.

During the decapsulation process the chemical reaction generates heat so the temperature of the solution should be monitored and not allowed to exceed 37 ° C. Addition of ice cubes to the solution can help maintain the temperature.

METHOD 2 - CALCIUM HYPOCHLORITE INGREDIENTS (70% active calcium hypochlorite)

1) 500 grams cysts

2) 325 grams calcium hypochlorite

3) 340 grams soda ash

4) 6.75 liters seawater or freshwater

Dissolve 325 grams calcium hypochlorite into 3.38 liters of seawater or freshwater and, in a separate container, dissolve 340 grams soda ash and 3.38 liters of water. Allow for the insoluble precipitates to settle, decant the supernatant, and discard the precipitants. After rehydrating the *Artemia* cysts, pour the two solutions together and add the cysts. Then proceed as in Method I. Decapsulation should occur in four to seven minutes, then chlorine should be neutralized with 0.1 N HCl as in Method I. Proceed to incubate the cysts.

Calculate hatching efficiency by taking replicate 1 ml samples from the aerated hatching chamber. Place the nauplii on filter paper for counting. The average count per ml should then be multiplied by the volume of water in the hatching chamber to determine the total number of nauplii. The number of nauplii in the hatching chamber, divided by the cyst count/g, yields the percent hatchout. Suppliers of *Artemia* cysts usually provide the number of cysts/g on the package label.

(Source: San Francisco Bay Brand, Inc.; 1988)

68

TABLE 4-21: SIZE AND INDIVIDUAL DRY WEIGHT OF *ARTEMIA* NAUPLII HATCHED IN STANDARD CONDITIONS (35 ‰, 25 °C) FROM DIFFERENT CYST SOURCES (Vanhaecke and Sorgeloos, 1980; 1983)

CYST SOURCE	NAUPLIUS LENGTH (µm)	NAUPLIUS DRY WEIGHT (µg)
San Francisco Bay; Calif. (SFB)	428	1.63
Macau; Brazil (BRAZIL)	447	1.74
Great Salt Lake; Utah (GSL)	486	2.42
Shark Bay; Australia (AUSTR)	458	2.47
Chaplin Lake; Canada (CAN)	475	2.04
Buenos Aires; Argentina (ARG)	431	1.72
Lavalduc; France (FRANCE)	509	3.08
Tientsin; PR China (CHINA)	515	3.09
Margherita di Savoia; Italy (ITALY)	517	3.33
Reference *Artemia* Cysts (RAC)	448	1.78

(Source: Soorgeloos, P. *et al.*, In: J.P. McVey, CRC *Handbook of Mariculture: Volume 1 - Crustacean Aquaculture*; copyright © 1983 - with permission of CRC Press, Boca Raton, FL.)

TABLE 4-22: IMPROVED HATCHING QUALITY OF *ARTEMIA* CYSTS AS A RESULT OF DECAPSULATION (Vanhaecke and Sorgeloos, 1983; Bruggeman et al., 1980)

CYST SOURCE	HATCHABILITY[1]	NAUPLIAR DRY WEIGHT[1]	HATCHING OUTPUT[1]
San Francisco Bay; CA	+15	+7	+23
Macau; Brazil	+12	+2	+14
Great Salt Lake; Utah	+24	-2	+21
Shark Bay; Australia	+4	+6	+10
Chaplin Lake; Canada	+132	+5	+144
Buenos Aires; Argentina	+35	+10	+49
Lavalduc; France	+2	+0	+2
Tientsin; PR China	+4	-1	+2
Margherita di Savoia; Italy	+10	+8	+19
Galera Zamba; Colombia	+14	+0	+13
Barotac Nuevo; Philippines	+11	+6	+19
Reference *Artemia* Cysts	+59	+1	+29

[1] Percent difference with untreated cysts.

HATCHING EFFICIENCY (HE) = number of hatched nauplii obtained from 1 g *Artemia* cysts

HATCHING OUTPUT = total naupliar biomass and energy produced from 1 g of *Artemia* cysts

(Source: J.P. McVey, CRC *Handbook of Mariculture: Volume 1 - Crustacean Aquaculture*; copyright © 1983 - with permission of CRC Press, Boca Raton, FL)

TABLE 4-23: DATA ON HATCHING EFFICIENCY (HE), HATCHING PERCENTAGE (H%),HATCHING RATE (T$_0$, T$_{90}$) HATCHING OUTPUT, INDIVIDUAL DRY WEIGHT, AND ENERGY CONTENT OF *ARTEMIA* NAUPLII HATCHED IN STANDARD CONDITIONS (35 ‰, 25°C) FROM DIFFERENT *ARTEMIA* CYST SOURCES

Abbrev. used	Batch no. or year of harvest	Hatching efficiency (nauplii/g)	Hatching percentage (H%)	HATCHING RATE T° (hr)	T^{90} (hr)	Individual dry weight (in µg)	Individual energy content (10^{-3} joule)	Hatching naupliar biomass (mg/g cysts)	Output naupliar energy (joule/g cysts)
SFB	288-2596	267,200	71.4	15.0	20.5	1.63	366	435.5	9,780
	288-2606	259,200	-	16.4	23.2	-	-	-	-
	236-2016	249,600	-	25.8	37.6	-	-	-	-
SPB	1628	259,200	84.3	13.9	20.1	1.92	429	497.7	11,120
BRAZIL	871172	304,000	82.0	15.7	23.7	1.74	392	529.0	11,917
	87500	182,400	-	16.0	29.1	-	-	-	-
	May 1978	297,600	-	16.4	21.9	-	-	-	-
PHIL	1978	214,000	78.0	14.7	22.0	1.68	382	359.5	8,175
GSL	1977	106,000	43.9	14.1	21.7	2.42	541	256.6	5,735
AUSTR	114	217,600	87.5	20.3	28.1	2.47	576	537.5	12,534
CAN	1978	65,600	19.5	14.3	33.0	2.04	448	133.8	2,937
ARG	1977	193,600	62.8	16.1	22.6	1.72	379	333.0	7,337
FRANCE	1979	182,400	75.8	19.5	30.5	3.08	670	561.8	12,221
CHINA	1978	129,600	73.5	16.0	37.2	3.09	681	400.5	8,826
ITALY	1977	137,600	77.2	18.7	25.3	3.33	725	458.2	9,976
R.A.C.		211,000	45.7	18.0	32.2	1.78	403	375.6	8,503

SFB = San Francisco Bay, California, USA

SPB = San Pablo Bay, California, USA

BRAZIL = Macau, Brazil

PHIL = Barotac Nuevo, Panay, Philippines

GSL = Great Salt Lake, Utah, USA

AUSTR = Shark Bay, Australia

CAN = Chaplin lake, Canada

FRANCE = Lavalduc, France

CHINA = Tien-Tsin, People's Republic of China

ITALY = Margherita Di Savoia, Italy

R.A.C. = Reference *Artemia* Cysts

HATCHING EFFICIENCY (HE) = number of hatched nauplii obtained from 1 gram of Artemia cysts

HATCHING PERCENTAGE (H%) = number of hatched nauplii against the total number of cysts used.

T$_0$ = incubation time until appearance of first nauplii

T$_{90}$ = incubation time until 90% of nauplii have hatched

HATCHING OUTPUT = total naupliar biomass and energy produced from 1 gram of *Artemia* cysts

(Source: J.P. McVey, CRC *Handbook of Mariculture: Volume 1 - Crustacean Aquaculture*; copyright © 1983 - with permission of CRC Press, Boca Raton, FL)

TABLE 4-24: REPRESENTATIVE BIOCHEMICAL AND NUTRITIONAL INFORMATION FOR NEWLY HATCHED *ARTEMIA SALINA* NAUPLII

COMPONENT	MEASUREMENT
SELECTED BIOCHEMICAL COMPOSITION [1]	
INDIVIDUAL DRY WT (mg)	1.48
ASH WT (% dry wt)	11.28
TOTAL LIPID (% dry wt)	13.7
FALTY ACIDS (% dry wt)	10.9
CALORIC CONTENT PER GRAM ASH-FREE DRY WT (cal)	5.503
INDIVIDUAL CALORIC CONTENT (μcal)	7.30
SELECTED BIOCHEMICAL COMPOSITION [2]	
WATEr (% wet wt)	90.85
DRY MATTER (% wet wt)	9.15
CARBON (% dry wt)	27.5
NITROGEN (% dry wt)	8.09
PHOSPHORUS (% dry wt)	1.24
RANGE OF CERTAIN ESSENTIAL AMINO ACIDS BETWEEN SOURCES [3] (g/100 gram PROTEIN)	
THREONINE	4.8 - 6.0
VALINE	3.1 - 5.5
METHIONINE	2.2 -.3.7
ISOLEUCINE	4.9. - .6.8
LEUCINE	7.9 - 10.1
PHENYLALANINE	5.1 - 10.4
HISTIDINE	2.7 - 4.9
LYSINE	8.7 - 11.7
ARGININE	9.7 - 11.5
RANGE OF SELECTED FATTY ACIDS BETWEEN SOURCES [4] (mg/gram DRY WEIGHT LIPID)	
18:0	2.79 - 6.83
18:1ω9	26.97 - 31.2
18:2ω6	3.69 - 9.59
18:3ω3	4.87 - 33.59
18:4ω3	0.96 - 4.88
20:1ω9	0.35 - 0.52
20:2ω6/ω9	0.06 - 0.24 [5]
20:3ω6	0.05 - 2.76
20:3ω3/20:4ω6	1.48 - 2.69 [5]
20:5ω3	1.68 -13.63
22:6ω3	0.06 - 0.26 [5]

[1] Benijts *et al.* (1976)

[2] Oppenheimer and Moreira (1980)

[3] Seidel *et al.* (1980)

[4] Schauer *et al.* (1980)

[5] May be totally absent from certain sources.

TABLE 4-25: DIETS FOR ENRICHMENT AND ARTEMIA PRODUCTION

ADDITIVE	ENRICHMENT MIXTURE g/100 gram diet	ARTEMIA DIET g/100 gram diet
FISH AUTOLYSATE	69	-
BREWERS' YEAST	-	79.4
VITAMIN PREMIX	11	3.6
COD LIVER OIL[1]	10	10
CHOLIN (50%)	4	2
D.L. METHIONIN	2	1
MINERAL PREMIX	4	2

VITAMIN PREMIX	CONCENTRATION
VITAMIN A	200 IU/g
VITAMIN D$_3$	700 IU/g
VITAMIN E	20 IU/g
VITAMIN K	3 IU/g
THIAMIN	3 mg/g
RIBOFLAVIN	7 mg/g
PYRIDOXIN	4 mg/g
ASCORBIC ACID	300 mg/g
FOLIC ACID	3 mg/g
VITAMIN B$_{12}$	.04 mg/g
INOSITOL	400 mg/g
BIOTIN	1-2 mg/g
CA PANTOTHENATE	15 mg/g
NIACIN	50 mg/g
B.H.T.	6 mg/g
CaHPO$_4$	22.9 mg/g

MINERAL PREMIX	G/100 GRAM OF PREMIX
CaHPO$_4$	50
CaCl$_2$	21.5
MgCO$_3$	12.4
KCl	9
NaCl	4
FeSO$_4$ · 7H$_2$O	2
ZnSO$_4$ · H$_2$O	1.6
CuSO$_4$ · 5H$_2$O	0.3
MnSO$_4$ · H$_2$O	0.3
NaF	0.1
Na$_2$SeO$_3$ · H$_2$O	0.01
KI	0.004
CoSO$_4$ · 7H$_2$O	0.002

[1] 2 ppt Vitamin E is added to oil.

(Source: J.M. McVey, CRC *Handbook of Mariculture: Volume II - Finfish Aquaculture*; copyright © 1991 - with permission of CRC Press, Boca Raton, FL)

TABLE 4-26: PERCENT COMPOSITION OF THE 6 MAJOR FATTY ACIDS IN CULTURED *ARTEMIA*.

Data are expressed as percentage of total fatty acid methyl esters for each sample (Leger *et al.*, 1986)

ARTEMIA STRAIN	FOOD SOURCE	FATTY ACID					
		16:0	16:1ω7[5]	18:1ω9[5]	18:2ω6	18:3ω3	20:5ω3
SFB[1]	*CHAETOCEROS*[4]	15.5	19.4	30.6	2.8	3.9	12.7
	MICROENCAPSULATED DIETS						
	LIPID FREE	13.6	6.8	43.2	8.2	7.0	1.6
	COD LIVER OIL	9.4	7.2	43.7	7.8	6.9	9.2
	TAPES OIL	9.4	5.6	40.1	5.5	6.3	8.0
	SOYBEAN OIL	12.4	2.9	35.1	20.7	7.5	3.4
SFB	WHEAT FLOUR EXTRACT	9.6	6.9	28.9	22.8	7.9	2.3
	RICE BRAN EXTRACT	24.4	4.9	34.3	26.1	4.5	2.2
	MILLED RICE EXTRACT	12.8	4.0	23.4	10.1	11.2	7.7
SFB	RICE BRAN	15.2	10.9	33.6	21.6	1.7	0.8
(N 236-2016)	RICE BRAN + COD LIVER OIL	12.2	14.4	36.4	9.1	1.2	9.2
SFB (N 1628)	RICEBRAN	14.4	9.0	30.2	16.5	4.8	1.6
	RICEBRAN + COD LIVER OIL	11.0	10.7	32.8	6.2	4.1	8.8
GSL[2]	CORN	10.6	5.8	39.5	32.0	1.6	2.2
	COPRA	14.1	11.3	32.9	8.0	0.9	1.3
	RICE BRAN	11.9	6.7	39.1	29.1	1.9	1.2
	SOYBEAN	8.9	4.2	37.3	33.1	3.5	1.0
	CHAETOCEROS	11.7	22.5	17.2	5.0	0.9	18.6
	DUNALIELLA	14.7	2.4	27.3	13.4	20.2	4.7
	CORN BYPRODUCT A	12.0	6.1	33.1	35.8	1.5	0.5
	CORN BYPRODUCT B	12.0	9.6	31.2	27.4	2.1	1.1
	DEFATTED RICE BRAN	13.3	9.1	36.1	23.5	1.8	0.9
N.S[3]	*CHAETOCEROS*	11.6	44.9	18.4	0.7	0.5	12.0
N.S.	*CHLAMYDOMONAS*	12.0	4.4	14.0	7.7	11.9	4.6
N.S.	*MONOCHRYSIS*	12:9	13.4	17.8	6.5	4.4	17.3
N.S.	*PHAEODACTYLUM*	9.8	9.2	21.6	10.0	9.0	11.0
N.S.	*PLATYMONAS*	12.0	5.0	14.7	6.5	13.9	9.2

[1] San Francisco Bay, CA-USA

[2] Great Salt Lake, UT-USA

[3] Source not specified

[4] Only polar lipid fraction given for SFB series fed with *Chaetoceros* and with microencapsulated diets

[5] May include other monoenes

(Source: P. Sorgeloos, *et al.*, *Manual for the Culture and Use of Brine Shrimp Artemia in Aquaculture*, 1986))

TABLE 4-27: MILLILITERS OF WATER DISPLACED BY 50 EGGS CONVERTED TO NUMBER OF EGGS PER FLUID OUNCE.

ML DISPLACED	NUMBER/OZ.	ML DISPLACED	NUMBER/OZ	ML DISPACED	NUMBER/OZ
3.0	492.88	7.1	208.25	11.2	132.00
3.1	477.00	7.2	205.35	11.3	130.89
3.2	462.10	7.3	202.55	11.4	129.70
3.3	448.10	7.4	199.80	11.5	128.60
3.4	434.90	7.5	197.15	11.6	127.45
3.5	422.45	7.6	194.55	11.7	126.40
3.6	410.75	7.7	192.05	11.8	125.30
3.7	399.65	7.8	189.55	11.9	124.25
3.8	389.10	7.9	187.15	12.0	123.20
3.9	379.15	8.0	184.83	12.1	122.20
4.0	369.65	8.1	182.55	12.2	121.20
4.1	360.65	8.2	180.30	12.3	120.20
4.2	352.05	8.3	178.15	12.4	119.25
4.3	343.85	8.4	176.05	12.5	118.30
4.4	336.05	8.5	173.95	12.6	117.35
4.5	328.60	8.6	171.95	12.7	116.45
4.6	321.45	8.7	169.95	12.8	115.50
4.7	314.60	8.8	168.05	12.9	114.60
4.8	308.05	8.9	166.15	13.0	113.75
4.9	301.75	9.0	164.30	13.1	112.85
5.0	295.75	9.1	162.50	13.2	112.00
5.1	289.95	9.2	160.70	13.3	111.20
5.2	284.35	9.3	159.00	13.4	110.35
5.3	279.00	9.4	157.30	13.5	109.55
5.4	273.80	9.5	155.65	13.6	108.70
5.5	268.85	9.6	154.05	13.7	107.95
5.6	264.05	9.7	152.45	13.8	107.15
5.7	259.40	9.8	150.90	13.9	106.40
5.8	254.95	9.9	149.35	14.0	105.60
5.9	250.60	10.0	147.85	14.1	104.85
6.0	246.45	10.1	146.40	14.2	104.15
6.1	242.40	10.2	144.95	14.3	103.40
6.2	238.50	10.3	143.55	14.4	102.70
6.3	234.70	10.4	142.15	14.5	102.00
6.4	231.05	10.5	140.80	14.6	101.30
6.5	227.50	10.6	139.50	14.7	100.60
6.6	224.05	10.7	138.20	14.8	99.90
6.7	220.70	10.8	136.90	14.9	99.25
6.8	217.45	10.9	135.65	15.0	98.60
6.9	214.30	11.0	134.40	-	-
7.0	211.25	11.1	133.20	-	-

(Source: US Fish and Wildlife Service, *Fish Hatchery Management*; 1982

TABLE 4-28: MODIFIED VON BAYER TABLE FOR THE ESTIMATION OF THE NUMBERS OF FISH EGGS IN A LIQUID QUART.

NO. OF EGGS	EGG DIAMETER	NO. OF EGGS PER 12 INCH TROUGH	LIQUID QUART
35	0.343	1677	52
36	0.333	1833	57
37	0.324	1990	62
38	0.316	2145	67
39	0.308	2316	72
40	0.300	2606	78
41	0.292	2690	84
42	0.286	2893	90
43	0.279	3116	97
44	0.273	3326	104
45	0.267	3556	111
46	0.261	3806	119
47	0.255	4081	128
48	0.250	4331	135
49	0.245	4603	144
50	0.240	4895	153
51	0.235	5214	163
52	0.231	5490	172
53	0.225	5862	185
54	0.222	6185	193
55	0.218	6531	204
56	0.214	6905	216
57	0.211	7204	225
58	0.207	7630	238
59	0.203	8089	253
60	0.200	8459	264
61	0.197	8851	277
62	0.194	9268	290
63	0.191	9712	304
64	0.188	10184	318
65	0.185	10638	334
66	0.182	11225	351
67	0.179	11799	359
68	0.177	12203	381
69	0.174	12348	401
70	0.171	13533	423
7	0.169	14020	438
72	0.167	14529	454
73	0.164	15341	479
74	0.162	15916	497
75	0.160	16621	516
76	0.158	17157	536
77	0.156	17825	557
78	0.154	18528	579
79	0.152	19270	602

(Source: US Fish and Wildlife Service, *Fish Hatchery Management*; 1982)

TABLE 4-29: POLLUTION LOAD MEASURED AS BIOLOGICAL OXYGEN DEMAND

Expressed as biological oxygen demand (mg O₂/min) as a function of fish mass and feeding rate.

BODY MASS (g)	FEEDING RATE (PERCENT % OF BODY MASS/DAY)				
	0.0%	2.5%	5.0%	7.5%	10.0%
30	0.06	0.10	0.14	0.18	0.22
40	0.07	0.13	0.18	0.23	0.28
50	0.08	0.15	0.21	0.28	0.34
60	0.09	0.17	0.25	0.32	0.40
80	0.11	0.21	0.31	0.41	0.52
100	0.12	0.25	0.38	0.50	0.63
150	0.15	0.34	0.54	0.73	0.92
200	0.18	0.43	0.69	0.94	1.20
250	0.20	0.52	0.84	1.16	1.48
300	0.22	0.61	0.99	1.37	1.75
400	0.26	0.77	1.28	1.79	2.30
500	0.29	0.93	1.57	2.21	2.84
600	0.32	1.09	1.85	2.62	3.38
800	0.38	1.40	2.42	3.44	4.46
1000	0.43	1.70	2.97	4.25	5.53
1500	0.53	2.45	4.36	6.27	8.18
2000	0.62	3.17	5.72	8.27	10.80
3000	0.78	4.60	8.43	12.30	16.10
4000	0.91	6.01	11.10	16.20	21.30
5000	1.03	7.40	13.80	20.20	26.50
6000	1.14	8.79	16.40	24.10	31.70
8000	1.33	11.50	21.70	31.90	42.10
10,000	1.50	14.20	27.00	39.70	52.50
20,000	2.19	27.7	53.20	78.70	104.10
30,000	2.72	40.90	79.20	117.50	155.70
40,000	3.19	54.20	105.20	156.19	207.20

Pollution Load, expressed as biological oxygen demand (mg O₂/min), calculated as:

$$X = \sum_{j=1}^{q} B_j^{\,0.544} \times 10^{-2}) + 0.051F$$

WHERE:

X = Demand in mg O₂/min⁻¹⁾
B = Biomass of Individual Animals (g)
F = Amount of Food Entering System Per Day
q = The Number of Animals Maintained

(Source: S. Spotte, *Fish and Invertebrate Culture*; copyright © 1979 - reprinted with permission of John Wiley & Sons, Inc.)

TABLE 4-30: SOME OXYGEN CONSUMPTION VALUES FOR FARMED FISH.

SPECIES	SIZE (g)	TEMP. (°C)	FEED RATE	OXYGEN CONSUMED (g O_2 / kg fish / hr)	SOURCE
Salmo gairdneri	100	15	?	0.3	Liao (1971)
	100 l	15	production levels	0.3	Muller-Fuega *et al* (1978)
O. nerka	28.J	15	unfed	0.23	Brett and Zala (1975)
	28.6	15	3% body wt day^{-1}	0.28	
Ictalurus punctatus	100	30	unfed	0.56	Andrews and Matsuda (1975)
	100	30	satiation	0.81	Andrews and Matsuda (1975)
Cyprinus carpio	100	10	fed	0.17	Beamish (1964)
		20	fed	0.48	Beamish (1964)
		25	fed	0.70	Beamish (1964)
Hypohthalmichthys molitrix	15	20	-	0.20	Muhamedova (1977)[1]
	240	23	fed	0.25	Vetskanov (1975)[1]
Oreochromis niloticus	50	25	unfed	0.16	Ross and Ross (1984)
	50	30	unfed	0.24	Ross and Ross (1984)
	50	35	unfed	0.40	Ross and Ross (1984)
Macrobrachium rosenbergii	0.5g	24	unfed	36.0	Nelson *et al* (1977)
	0.5	24	satiation [2]	43.0	Nelson *et al.* (1977)

[1] From ADCP, 1984.   [2] Fed Purina marine ration #2

(Source: M.C.M. Beveridge, *Cage Aquaculture*; 1987)

TABLE 4-31: TYPICAL OXYGEN TRANSFER RATES OF VARIOUS DEVICES USED IN FISH CULTURE SYSTEMS (modified from Colt and Tchobanoglous, 1981).

AERATION SYSTEM	TRANSFER RATE kg O_2 kW / h STANDARD[a]	6 mg l^{-1} O_2[b]
DIFFUSED AIR SYSTEM		
A: FINE BUBBLE	1.2-2.0	0.25-0.42
B: MEDIUM BUBBLE	1.0-1.6	0.21-0.34
C: COARSE BUBBLE	0.6-1.2	0.13-0.25
SURFACE AERATOR (low speed)	1.2-2.4	0.25-0.80
SURFACE AERATOR (high speed)	1.2-2.5	0.25-0.50
VENTURI AERATOR	1.2-2.4	0.25-0.50
FLOATING ROTOR AERATOR	1.2-2.4	0.25-0.5
U-TUBE AERATOR		
A: ZERO HEAD	4.5	0.95
B: 1-FOOT HEAD	45.6	9.58
GRAVITY AERATOR	1.2-1.8	0.25-0.38
PURE OXYGEN SYSTEM		
A: FINE BUBBLE	-	1.2-1.80
B: MECHAN. SURF. AERATION		1.0-1.2
C: TURBIN-SPARGER		1.2-1.5

[a] 20°C, tapwater, D.O. = 0 mg/l   [b] 20°C, D.O. = 6 mg/l

Note: kg/kW•h x 1.6440 = lbO_2/hp•h

(Source: J. Colt and W. Tchobanoglous, In: *Proceedings of the Bio-Engineering Symposium for Fish Culture*; copyright © 1981 - with permission of American Fisheries Society)

TABLE 4-32: AERATION EFFECTIVENESS (AF) OF TYPICAL GRAVITY AERATORS.

TYPE HEIGHT (cm)	AF(%)	REFERENCE
LOW HEAD TYPES		
CASCADE	**45° ANGLE**	
25	22 - 26	Tebbun (1972)
50	36 - 38	
CORRUGATED INCLINED PLANE	**20° ANGLE**	
30	18 - 29	Chesness and Stephens (1971)
60	30 - 50	
HORIZONTAL PERFORATED TRAYS		
110	95 -100	Strasburg (1964)
LATTICE AERATOR		
30	29 - 37	Cheness and Stephens (1971)
60	48-61	
SIMPLE WEIR		
30	7 -10	Haskell *et al.* (1960)
SPLASH BOARD		
30	23 - 25	Chesness and Stephens (1971)
60	36 - 41	
PACKED COLUMN		
30	94 - 96	Hackney and Colt (1982)
60	96 - 98	
HIGH HEAD TYPES [1]		
ALFALFA GATE	61	Moore and Boyd (1984)
ELL ASPIRATOR	83	
GATE VALVE (half-open)	76	
SCREEN	59	
SCREEN COVER WITH ROCKS	63	
SCREEN COVER	52	
SCREEN EXTENSION	51	
SLOTTED CAP	65	
SPLASHBOARD	51	
SPLASHBOARD WITH HOLES	53	
STRAIGHT PIPE	25	
TEE ASPIRATOR	72	

[1] Pressure head ranged from 9.9 to 17 cm.

AERATION EFFECTIVENESS (AF) is used as a rating parameter for gravity aerators, and is defined as:

$$AF = \frac{C_{out} - C_{in}}{C^* - C_{in}}$$

WHERE:

C = solubility

C^* = solubility at saturation

(Source: Colt and Orwicz, In: D.E. Brune and J. R. Tomasso, *Aquaculture and Water Quality*; 1991)

TABLE 4-33: TYPICAL STANDARDIZED AERATOR EFFICIENCY (SAE) AND EQUIVALENT AERATION DEPTHS (D$_E$) FOR AERATORS USED IN AQUACULTURE.

TYPE	SAE (kg O$_2$kW · hr)	D$_e$ (%)	REFERENCE
SURFACE AERATORS			
LOW-SPEED SURFACE	1.2 - 2.4	0	Metcalf and Eddy (1979)
LOW-SPEED SURFACE WITH DRAFT TUBE	1.2 - 2.4	0	
HIGH-SPEED SURFACE	1.2 - 24	0	
PADDLEWHEEL			
TRIANGULAR BLADES	2.7 - 2.9	0	Ahmad and Boyd (1988)
PVC PIPE BLADES	1.2 - 1.9	0	Boyd *et al.* (1988)
TRACTOR POWERED	13-2.0	0	Buschetal. (1984)
GRAVITY AERATORS			
CASCADE WEIR(45°)	1.5 - 1.8	0	Tebbutt (1972)
CORRUGATED INCLINED PLANE (20°)	1.0 - 1.9	0	Chesness *et al.* (1971; 1972)
HORIZONTAL SCREENS	1.2 - 2.6	0	Hartman (1983)
LATTICE AERATOR	1.8 - 2.6	0	Chesness *et al.* (1971; 1972)
PACKED COLUMN			
ZERO HEAD	1.2 - 2A	0	Hackney and Colt (1982)
0.5 - 1.0 M HEAD	10 - 80[1]	0	
SUBSURFACE AERATORS			
AERATION CONE	2.5	-	Speece *et al.* (1971)
AIR-LIFT PUMP	2.0 - 2.1	30[1]	Nagy (1979); Cornacchia and Colt (1984); Reinemann and Timmons (1989)
DIFFUSED-AIR			
FINE-BUBBLE	1.2 - 2.0	36[2]	Metcalf amd Eddy Inc. (1979); Colt and Weslers (1982)
MEDIUM-BUBBLE	1.0 - 1.6	32[2]	Metcalf amd Eddy (1979)
COARSE-BUBBLE	0.6-12	28[2]	
NOZZLE AERATOR	13 - 26	-	Chesness *et al.* (1973); Dijkstra *et al.* (1979)
PROPELLER-ASPIRATOR-PUMP	1.7 - 1.9	-	Boyd and Martinson (1984)
STATIC TUBE	1.8 - 2.4	29[2]	Ban and Carnpbell (1974); Colt and Wester (1982)
U-TUBE			
ZERO HEAD	0.72 - 2.3	-	Speece (1970)
0.5 - 1.0 M HEAD	10 - 40	-	
VENTURI AERATOR	2.0 - 3.3	51[2]	Toerber and Mandt 1979; Colt and Westers 1982

[1] Does not include pumping power
[2] Only an estimate

FIELD AERATION EFFICIENCY (FAE) for aquaculture systems will be signficantly less than SAE values due to the necessity of maintaining a dissolved oxygen concentration of 5 to 7 mg/l in the system.

(Source: Colt and Orwicz, In: D.E. Brune and J. R. Tomasso, *Aquaculture and Water Quality*; 1991)

TABLE 4-34: TRANSPORT CONDITIONS FOR VARIOUS FISH SPECIES.

Data from Piper *et al* (1982), Horvath *et al* (1984), in Beveridge (1987).

SPECIES	SIZE	STOCKING DENSITY (g/l)	DURATION (hr)	TEMPERATURE (°C)
CHINOOK SALMON	40 mm	60-120	8-10 max	5-10
	60mm	120-240	8-10 max	5-10
COHO SALMON	100-130 mm	240-360	8-10 max	5-10
TROUT (rainbow-brook-brown)	200-280mm	300-420	8-10 max	5-10
CHANNEL CATFISH	100g	350-600	8-16 max	18
	10g	250-400	8-16 max	18
	4g	200-350	8-16 max	18
	2g	150-200	8-16 max	18
LARGEMOUTH BASS	20g (100 mm)	120	12 max	18-30
	5g (75 mm)	80	12 max	18-30
	1g (50 mm)	60	12 max	18-30
COMMON AND BIGHEAD CARP	<100g	280	-	5
		50	-	30
SILVER CARP	<100g	90	-	5
		25	-	25
TILAPIAS	0.5-200g	100-200	24 max	8-28

(Source: M.C.M. Beveridge, *Cage Aquaculture*; 1987)

TABLE 4-35: RECOMMENDED LOADING RATES FOR TRANSPORTING FISH WITHOUT ARTIFICIAL AERATION.

SPECIES	SIZE	RATIO OF FISH WEIGHT: WATER WEIGHT
BROWN TROUT	Fingerlings	1:70 - 200
	2.0-2.7 in (5-7 cm)	1:50 - 200
	3.9-5.1 in (10-13 cm)	1:45 - 90
COMMON CARP	less than 5 cm)	1:38 - 75
	3.5-4.7 in (9-12 cm)	1:13 - 23
	5.9-7.0 in (15-18 cm)	1:9 - 13
	0.55 lb (250 gm)	1:7 - 12
	1.65 lb (750 gm)	1:5 - 10
	3.3 lb (1.5 kg)	1:5 - 7

(Source: Courtesy of Dr. William McLarney, *The Freshwater Aquaculture Book*; copyright © 1984 - with permission of Hartley & Marks, Inc.)

TABLE 4-36: TYPICAL PACKING DENSITIES AND SHIPPING DURATIONS FOR VARIOUS SHRIMP STAGES [1]

SHRIMP STAGE	SHIPPING DURATION	WATER TEMPERATURE (°C)	PACKING DENSITY
NAUPLII	3 hours	24	@ 114 shrimp/ml
	8 hours	24	@ 21 shrimp/ml
	24 hours	24	@ 10 shrimp/ml
POSTLARVAE	24 hours	22	15 shrimp/ml
JUVENILES (1-5g)	8 hours	18	55 g/l
	24 hours	18	7 g/l
BROODSTOCK (> 40 g)	8 hours	18	31 g/l
	>24 hours	15	25 g/l

[1] Based on packing in filtered seawater (33 ‰ salinity) with 100 mg/l Tris HCL and 250 mg/l activated charcoal added.

(Source: J. Wyban and J. Sweeney, *Intensive Shrimp Production Technology*; copyright © 1991 - with permission of Oceanic Institute, Inc.)

TABLE 4-37: CONCENTRATIONS OF SOME DRUGS USED TO TRANQUILIZE FISH FOR TRANSPORT

DRUG	CONCENTRATION
SODIUM BICARBONATE	642 mg/l
TERTIARY AMYL ALCOHOL	2 ml/gallon
METHYLPARAFYNOL (Dormison)	1-2 ml gallon
CHLORAL HYDRATE	3.0-3.5 g/gallon
SODIUM BARBITAL	6.7-7.7 micrograms/l
URETHANE	1-4 g/l
METHANE TRICAINESULFONATE (MS-222)	40-150 mg/l
QUINALDINE	2-20 mg/l
MIXTURE OF MS-222 AND QUINALDINE	20-30 mg/l MS-222 and 5 mg/l of quinaldine

(Source: Courtesy of Dr. William McLarney, *The Freshwater Aquaculture Book*; copyright © 1984 - with permission of Hartley & Marks, Inc.)

V. Plumbing and Materials

TABLE 5-1: PVC PIPE SPECIFICATIONS - SCHEDULE 40 (20 FT. LENGTHS)

NOMINAL PIPE SIZE (in)	APPROXIMATE WEIGHT PER 100 FT		OUTSIDE DIAMETER (in)	INSIDE DIAMETER (in)	WALL THICKNESS (in)
	PVC, TYPE 1	CPVC			
1/2	16.4	19.0	0.840	0.622	0.109
3/4	21.8	25.2	1.050	0.824	0.113
1	32.1	37.5	1.315	1.049	0.133
1 1/4	43.4	50.7	1.660	1.380	0.140
1 1/2	51.8	60.7	1.900	1.610	0.145
2	69.5	81.5	2.375	2.067	0.154
2 1/2	109.6	129.3	2.875	2.469	0.203
3	143.5	169.1	3.500	3.068	0.216
4	204.3	232.9	4.500	4.026	0.237
6	360.9	409.6	6.625	6.065	0.280
8	545.3	-	8.625	7.981	0.322
10	791.3	-	10.750	10.020	0.365
12	1,035.2	-	12.750	11.938	0.406

TABLE 5-2: PVC PIPE SPECIFICATIONS - SCHEDULE 80 (20 FT. LENGTHS)

NOMINAL PIPE SIZE	APPROXIMATE WEIGHT PER 100 FT				OUTSIDE DIAMETER (in)	INSIDE DIAMETER (in)	WALL THICKNESS (in)
	PVC, TYPE 1	CPVC	PROPYLENE	PVDF			
1/4	10.1	11.9	-	-	.540	.302	.119
1/2	20.5	24.3	14.0	24.4	.840	.546	.147
3/4	27.8	32.9	18.9	33.0	1.050	.742	.154
1	40.4	48.5	27 1	48.7	1.315	.957	.179
1 1/4	56.7	66.9	37.9	-	1.660	1.278	.191
1 1/2	68.9	81.1	44.8	81.4	1.990	1.500	.200
2	94.9	108.5	62.3	112.6	2.375	1.939	218
2 1/2	144.9	165.4	-	-	2.875	2.323	.276
3	193.8	221.3	126.6	-	3.500	2.900	.300
4	283.3	323.4	185.2	-	4.500	3.826	.337
6	541.1	616.8	359.9	-	6.625	5.761	.432
8	821 9	905.8	-	-	8.625	7.625	.500
10	1,227.7	-	-	-	10.750	9.564	.593
12	1,710.4	-	-	-	12.750	11.376	.687

PVC = Poly (vinyl chloride) - a resin prepared by the polymerization of vinyl chloride with or without the addition of small amounts of other monomers.

CPVC = Chlorinated Poly (Vinyl Chloride) plastics - made by combining chlorinated poly (vinyl chloride) with colorants, fillers, plasticizers, stabilizers, lubricants and other compounding ingredients.

POLYPROPYLENE - a polymer prepared by the polymerization of propylene as the sole monomer.

PVDF - a crystalline, high molecular weight polymer of vinylidene floride, containing 59 % fluorine by weight.

(Source: Chemtrol, *Plastic Piping Handbook*; copyright © 1979 - with permission of Nibco, Inc.)

TABLE 5-3: ESTIMATED SOLVENT CEMENT REQUIREMENTS (Number of Joints Per ...[1])

PIPE SIZE (in.)	PINT	QUART	GALLON
1/2	130	260	1040
3/4	80	160	640
1	70	140	560
1 1/4	50	100	400
1 1/2	35	70	280
2	20	40	160
2 1/2	17	34	136
3	15	30	120
4	10	20	80
5	8	16	64
6	N/R	8	24
8	N/R	3	12
10	N/R	N/R	10
12	N/R	N/R	6

Each joint represents one socket in a fitting.   N/R = Not recommended

[1] The estimated PVC and CPVC solvent cement requirements should only be considered as a guideline for usage and could vary according to a wide variety of installation conditions. Further, these estimates should in no way be used to restrict the liberal cement application instructions recommended for the pipe.

TABLE 5-4: DRILL SIZES FOR PIPE TAPS

TAPS (in.)	THREADS/IN	DRILL DIA. (in.)	TAPS (in.).	THREADS/IN	DRILL DIA.(in.)
1/8	27	11/32	2	11 1/2	2 3/16
1/4	18	7/16	2 1/2	8	2 9/16
3/8	18	37/64	3	8	3 3/16
1/2	14	23/32	3 1/2	8	3 11/16
3/4	14	59/64	4	8	4 3/16
1	11 1/2	1 5/32	4 1/2	8	4 3/4
1 1/4	11 1/2	1 1/2	5	8	5 5/16
1 1/2	11 1/2	1 49/64	6	8	6 5/16

TABLE 5-5: TAP AND DRILL SIZES (American Standard Coarse)

DRILL SIZE (in.)	TAP SIZE (in.)	THREAD/IN	DRILL SIZE (in.)	TAP SIZE (in.)	THREAD/IN
7	1/4	20	49/64	7/8	9
F	5/16	18	53/64	15/16	9
5/16	3/8	16	7/8	1	8
U	7/16	14	63/64	1 1/8	7
27/64	1/2	13	1 7/64	1 1/4	7
31/64	9/16	12	1 13/64	1 3/8	6
17/32	5/8	11	1 11/32	1 1/2	6
19/32	11/16	11	1 29/64	1 5/8	5 1/2
21/32	3/4	10	1 9/16	1 3/4	5
23/32	13/16	10	1 11/16	1 7/8	5
-	-	-	1 25/32	2	4 1/2

(Source: Chemtrol, *The Plastic Piping Handbook*, copyright © 1979 - with permission of Nibco, Inc.)

TABLE 5-6: RECOMMENDED SUPPORT SPACING IN FEET FOR PVC PIPE

PIPE SIZE (in.)	SCHEDULE 40 TEMPERATURE (°F)								PIPE SIZE (in.)	SCHEDULE 80 TEMPERATURE (°F)							
	20	40	60	80	100	120	140	160		20	40	60	80	100	120	140	160
1/4	5 1/2	5	5	4 1/2	4 1/2	3 1/2	3	1 1/2	1/4	6	6	5 1/2	5 1/2	4	3	2	
3/8	5 1/2	5	5	4 1/2	4 1/2	3 1/2	3	1 1/2	3/8	6 1/2	6 1/2	6	5 1/2	5 1/2	4 1/2	3 1/2	2
1/2	6	5 1/2	5 1/2	5	4 1/2	4	3	1 1/2	1/2	7	6 1/2	6	6	5 1/2	4 1/2	3 1/2	2
3/4	6	5 1/2	5 1/2	5	4 1/2	4	3	1 1/2	3/4	7	7	6 1/2	6	5 1/2	5	3 1/2	2
1	6 1/2	6 1/2	6	5 1/2	5 1/2	4 1/2	3 1/2	2	1	7 1/2	7	7	6 1/2	6	5	4	2
1 1/4	6 1/2	6 1/2	6	5 1/2	5 1/2	4 1/2	3 1/2	2	1 1/4	8	7 1/2	7	6 1/2	6 1/2	5 1/2	4	2 1/2
1 1/2	7	6 1/2	6	6	5 1/2	4 1/2	3 1/2	2	1 1/2	8	7 1/2	7 1/2	7	6 1/2	5 1/2	4	2 1/2
2	7	7	6 1/2	6	5 1/2	5	3 1/2	2	2	8	8	7 1/2	7	6 1/2	5 1/2	4	2 1/2
2 1/2	8	7 1/2	7 1/2	7	6 1/2	5 1/2	4	2 1/2	2 1/2	9 1/2	9	8 1/2	8	7 1/2	6 1/2	5	3
3	8	8	7 1/2	7	6 1/2	5 1/2	4 1/2	2 1/2	3	10	9 1/2	9	8 1/2	8	6 1/2	5	3
3 1/2	8 1/2	8	7 1/2	7 1/2	7	6	4 1/2	2 1/2	3 1/2	10 1/2	10	9 1/2	9	8	7	5 1/2	3
4	9	8 1/2	8	7 1/2	7	6	4 1/2	2 1/2	4	10 1/2	10	9 1/2	9	8 1/2	7	5 1/2	3
5	9	8 1/2	8 1/2	8	7 1/2	6	4 1/2	2 1/2	5	11	10 1/2	10	9	8 1/2	7	5 1/2	3
6	9 1/2	9	8 1/2	8	7 1/2	6 1/2	5	3	6	12	11 1/2	10 1/2	10	9 1/2	8	6	3 1/2
8	10 1/2	10	9 1/2	9	8 1/2	7	5 1/2	3	8	12 1/2	12	11 1/2	11	10	8 1/2	6 1/2	4
10	11	10 1/2	10	9	8 1/2	7	5 1/2	3	10	14	13 1/2	12 1/2	12	11	9 1/2	7	4
12	11 1/2	11	10 1/2	10	9	8	6	3 1/2	12	15	14 1/2	13 1/2	13	12	10	8	4 1/2

THE NECESSARY SUPPORT SPACING INDICATED IN THE ABOVE TABLE ASSUMES:

a. Uniform placement of supports

b. No concentrated loads

c. Maximum fluid weight of 85 lb/cu. ft.

Support spacing may be doubled for vertical runs. Valves and heavy fittings should be individually supported.

(Source: Houdaille Industries, Inc. Cedar Falls, Iowa 50613)

TABLE 5-7: SYMBOLS USED TO DESIGNATE STANDARD PIPE FITTINGS

TYPES OF CONNECTIONS

SCREWED ENDS

FLANGED ENDS

BELL - AND - SPIGOT ENDS (B&S)

WELDED AND BRAZED ENDS

SOLDERED ENDS

PIPELINE INTERSECTIONS

PIPELINE CROSSOVERS

ELBOWS

ELBOW, 90 DEGREES

ELBOW, 45 DEGREES

ELBOW, SPECIFIED ANGLE

30°

ELBOW LONG RADIUS

LR

ELBOW, REDUCING

ELBOW, SIDE OUTLET, OUTLET DOWN

(continued page 86)

(continued from page 85)

ELBOW, SIDE OUTLET, OUTLET UP

ELBOW, TURNED DOWN

ELBOW, TURNED UP

ELBOW, UNION

TEES

TEE

TEE, DOUBLE SWEEP

TEE, OUTLET DOWN

TEE, OUTLET UP

TEE, SINGLE WEEP, OR PLAIN TPY

VALVES

	SCREWED	FLANGED	B & S	WELDED	SOLDERED
GATE VALVE					
GLOBE VALVE					
SAFETY VALVE					

TABLE 5-8: LOSS COEFFICIENTS FOR STANDARD PIPE FITTINGS

These are approximate loss coefficients for abrupt (standard) fittings and transitions. There is no allowance for biofouling. Values are not completely independent of pipe diameter. The smaller the diameter the higher the "K" value. Flanged fittings will have lower values than threaded fittings. The loss coefficients are dimensionless.

FITTINGS	LOSS COEFFICIENT (K)
COUPLINGS AND UNIONS	0.04 - 0.08
45° ELBOWS	0.35 - 0.40
90° ELBOWS	0.75 - 0.90
180° RETURN BEND	0.4 - 1.5
"T" FLOW ALONG RUN	0.04 - 0.9
"T" USED AS ELBOW (entering run)	1.3 - 1.6
"T" USED AS ELBOW (entering branch)	1.5 - 1.8
GATE VALVE (open)	0.17 - 0.2
BUTTERFLY VALVE (open)	0.24
ROOT VALVE (high values at low velocities)	0.8 - 15.0
SWING TYPE CHECK VALVE [1]	2.0 - 3.0
DISK TYPE CHECK VALVE [1]	10.0
BALL TYPE CHECK VALVE [1]	70.0

[1] May have even higher values at low velocities

FRICTIONAL HEAD LOSS FOR FITTINGS IS CALCULATED AS:

$$h = K \frac{V^2}{2g}$$

WHERE:

h = Head Loss (in feet or meters)

K = Loss Coefficient (nondimensional)

V = Average Velocity in Adjoining Pipe (ft/s or m/s)

g = Gravitational constant (32.2 ft/s^2 or 9.81 m/s^2)

EXAMPLE:

Calculate head loss for water traveling at 2.0 m/s through a Tee used as elbow (entering branch).

$$h = 1.5 \frac{2.0^2}{2(9.81)} = 0.30 \ m$$

(Source: J. Huegenin and J. Colt, *Design and Operating Guide for Aquaculture Seawater Systems*, 1989)

TABLE 5-9: VALUES OF *L/D* RATIOS FOR SELECTED FITTINGS FOR DETERMINING FLOW RESISTANCE (FRICTIONAL LOSS) COVERTED TO "EQUIVALENT LENGTH OF PIPE"

FITTING	L/D
GLOBE VALVE - PERPENDICULAR STEM (OPEN)	340
GLOBE VALVE - Y PATTERN (OPEN)	160
ANGLE VALVE (OPEN)	145
GATE VALVE (OPEN)	13
GATE VALVE (3/4 OPEN)	35
GATE VALVE (1/2 OPEN)	160
GATE VALVE (1/4 OPEN)	900
WING CHECK VALVE (OPEN)	135
CHECK VALVE (OPEN)	150
PLUG COCK - FULL PORT;TWO-WAY (OPEN)	18
PLUG COCK - REDUCED PORT;THREE-WAY (STRAIGHT THROUGH)	44
ELBOW (90° STANDARD)	30
ELBOW (90° LONG RADIUS)	20
ELBOW (45° STANDARD)	16
ELBOW (90° SQUARE CORNER)	57
TEE - STANDARD (THROUGH RUN)	20
TEE - STANDARD (THROUGH BRANCH)	60
RETURN BEND (CLOSE PATTERN)	50

EXAMPLE:

One inch PVC pipe has a total length of 10 feet with two 90° elbows, one check valve (open) and one gate valve (open). Deterimine the "equivalent length" of 1" PVC pipe which corresponds to this plumbing configuration:

The inside diameter of 1" PVC pipe (SCH 40) is 1.049 inches . Since the equivalent length is desired in feet, d = 1.049/12 = 0.0874.

THEREFORE:

$$TOTAL\ EQUIVALENT\ LENGTH\ (ft) = \frac{L}{d} \times \frac{1.049}{12} \times NUMBER\ OF\ FITTINGS$$

TWO 90° ELBOWS = 30 x 0.0874 x 2 = 5.2

ONE OPEN CHECK VALVE = 150 x 0.0874 x 1 = 13.11

ONE OPEN GATE VALVE = 13 x 0.0874 x 1 = 1.14

TOTAL EQUIVALENT LENGTH (ft) = 5.2 + 13.11 + 1.14 + 10 = 29.45 ft

(Source: T. B. Hardison, *Fluid Mechanics for Technicians*; copyright © 1977 - with permission of Reston Publishers)

TABLE 5-10: FRICTIONAL LOSSES IN PLASTIC SCHEDULE 80 PIPES

These values should be typical of clean used plastic pipe and seawater. Note assumptions inherent in calculations. They do not include allowance for any significant biofouling, which can create much higher losses. Since schedule 80 pipe has relatively small inside diameters, these values should be conservative for schedule 40 or other thinner walled pipes. Loss are given in velocity (ft/s) and pressure (ft of head) per 100 ft of pipe.

| FLOW GPM | FLOW LPM | NOMINAL PIPE SIZES (IN.) | | | | | | | | | | | | | | |
| | | 1 | | 1.5 | | 2 | | 3 | | 4 | | 6 | | 8 | | 12 | |
		VEL	LOSS	VEL	LOSS	VEL	LOSS	VEL	LOSS	VEL	LOSS	VEL	LOSS	VEL	LOSS	VEL	LOSS
10	0.63	4.4	8.8	1.8	1.1	1.1	0.31	0.5	0.04								
20	1.26	8.9	32.0	3.6	3.7	2.2	1.1	1.0	0.16	0.6	0.04						
30	1.89	13.3	72.1	5.5	7.9	3.2	2.3	1.5	0.32	0.8	0.09						
50	3.15			9.1	20.0	5.4	5.6	2.4	0.81	1.4	0.21	0.6	0.03				
70	4.42			12.7	37.1	7.6	10.7	3.4	1.5	2.0	0.39	0.9	0.05				
100	6.31			18.2	71.7	10.8	20.2	4.8	2.9	2.8	0.74	1.2	0.10	0.7	0.03		
150	9.46					16.2	42.8	7.3	5.9	4.2	1.5	1.9	0.21	1.1	0.06	0.5	0.008
200	12.62							9.7	10.2	5.6	2.6	2.5	0.36	1.4	0.09	0.6	0.014
250	15.77							12.1	15.5	7.0	4.0	3.1	0.54	1.8	0.14	0.8	0.020
300	18.93							14.5	21.7	8.4	5.5	3.7	0.74	2.1	0.19	1.0	0.028
350	22.08							17.0	29.5	9.8	7.4	4.3	0.98	2.5	0.26	1.1	0.036
400	25.24							19.4	37.4	11.2	9.7	4.9	1.25	2.8	0.32	1.3	0.047
500	31.55									13.9	14.5	6.2	1.9	3.5	0.49	1.6	0.070
600	37.85									16.7	21.5	7.4	2.7	4.2	0.68	1.9	0.097
700	44.16									19.5	26.9	8.6	3.6	4.9	0.90	2.2	0.127
800	50.47											9.9	4.6	5.6	1.16	2.5	0.166
900	56.78											11.1	5.7	6.3	1.4	2.8	0.202
1000	63.09											12.3	6.9	7.0	1.8	3.2	0.242
1200	75.71											14.8	9.7	8.4	2.4	3.8	0.336

GPM = Gallons per minute

LPM = Liters per minute

VEL = Velocity in ft/s

ASSUMPTIONS:

Seawater 70°F (21°C)

Kinematic viscosity = 1.1×10^{-5} ft^2/sec

Equivalent Sand Roughness = 4.2×10^{-5} ft, PVC schedule 80 pipe inside diameters, Darcy-Weisbach Equation.

(Source: J. Huguenin and J. Colt, *Design and Operating Guide for Aquaculture Seawater Systems*; 1989)

TABLE 5-11: FRICTIONAL LOSSES FOR AIR AS A FUNCTION OF FLOW AND PIPE SIZE

Calculations based on Schedule 40 PVC pipe internal diameters, pipe equivalent sand roughness of 4.2×10^5 ft, average pressure of 3 psi, average air temperature of 90° F and an air kinematic viscosity of 1.7×10^{-4} ft²/s Velocities are in ft/s and frictional losses in psi per 100 ft of pipe. Under this temperature and pressure, the SCFM volume is 13% greater than actual and the conversion for frictional head losses in feet of air to psi is loss(psi)= head(ft)/1661.

| FLOW RATE SCFM[1] | NOMINAL PIPE SIZE (in) | | | | | | | | | |
| | 0.5 | | 0.75 | | 1.0 | | 1.5 | | 2.0 | |
	VEL	LOSS	VEL	LOSS	VEL	LOSS	VEL	LOSS	VEL	LOSS
5	35.0	0.68	19.9	0.18	12.3	0.06	5.2	0.01	3.2	0.00
10	70.0	2.43	39.8	0.61	24.6	0.19	10.4	0.02	6.3	0.00
15	105	5.17	60.0	1.18	37.0	0.40	15.7	0.05	9.5	0.02
20	140	8.66	79.6	2.07	49.2	0.67	20.9	0.08	12.7	0.03
30			120	4.51	73.8	1.40	31.3	0.17	19.0	0.05
50			199	11.6	123	3.56	52.2	0.43	31.7	0.13
75					184	7.42	78.3	0.90	47.5	0.27
100					246	12.9	104	1.54	63.3	0.46
125							121	1.99	79.1	0.68
150							157	3.26	95.0	0.93
200							209	5.63	127	1.62
250									158	2.44

[1] SCFM - Standard Cubic Feet per Minute of gas, in this case air, at standard conditions of one atmosphere of pressure (atm) and temperature of 20 °C. (68°F)

(Source: J. Huegenin and J. Colt, *Design and Operating Guide for Aquaculture Seawater Systems*; 1989)

TABLE 5-12: DISCHARGE OF AIR THROUGH AN ORIFICE UNDER PRESSURE (Standard)

| PRESSURE BEFORE ORIFICE | | | ORIFICE DIAMETER (in) | | | | | | | | |
| PSI | IN. | FEET | 1/54 | 1/32 | 1/16 | 1/8 | 3/16 | 1/4 | 3/8 | 1/2 | 5/8 |
			AIR-FLOW RATE THROUGH ORIFICE IN SCFM (100% FLOW COEFFICIENT)								
1	27.7	2.3	0.028	0.112	0.450	1.80	4.03	7.18	16.2	28.7	45.0
2	55.4	4.6	0.040	0.158	0.633	2.53	5.80	10.1	22.8	40.5	63.3
3	83.0	6.9	0.048	0.194	0.775	3.10	7.0	12.4	27.8	49.5	77.5
4	110.7	9.2	0.056	0.223	0.892	3.56	8.10	14.3	32.1	57.0	89.2
5	138.4	11.5	0.062	0.248	0.993	3.97	9.0	15.9	35.7	63.5	99.3
6	166.0	13.8	0.068	0.272	1.09	4.34	9.80	17.4	39.1	69.5	109
7	193.8	16.1	0.073	0.293	1.17	4.68	10.5	18.7	42.2	75.0	117

FLOW THROUGH AN ORIFICE

$$Q = \frac{A \times C \times K \sqrt{P}}{144}$$

WHERE:

Q = Flow in CFM (feet³/minute)
C = Orifice Coefficient (0.65 for sharp edge; 0.85 for nozzles)
P = Pressure in inches of water across the orifice

A = Area of Orifice in Square Inches
K = Constant 4005
144 in.² = 1 ft²

EXAMPLE:

Using the table above, calculate the number of holes (1/16" diameter; sharp drilled) in a 50 foot long sparger at 27.7" depth (1 psi) to deliver air flow at 85 feet³/minute (CFM).

$$\frac{85}{.450 \times 0.65} = 290.6 \, holes$$

Or one hole approximately every 0.5 inches along the entire length of the sparge.

TABLE 5-13: FLOW RATE, Q (liters/min), FOR AIRLIFT PUMPS AS A FUNCTION OF LENGTH AND DIAMETER AT SUBMERGENCE VALUES OF 0.8, 0.9, AND 1.0

PIPE LENGTH (cm)	SUBMERGENCE	LIFT PIPE DIAMETER (cm)					
		1.0	2.0	3.0	4.0	6.0	8.0
30	0.8	1.7	7.7	18.7	35.3	86.1	162.1,
	0.9	2.0	9.2	22.4	42.2	102.9	193.7
	1.0	2.4	10.9	26.5	50.0	122.0	230.0
50	0.8	2.0	9.1	22.2	41.9	102.2	192.4
	0.9	2.4	10.9	26.6	50.0	122.1	229.8
	1.0	2.8	12.9	31.4	59.2	144.0	272.0
75	0.8	2.3	10.4	25.5	48.0	117.0	220.4
	0.9	2.7	12.5	30.4	57.3	139.8	263.2
	1.0	3.2	14.7	36.0	67.7	165.0	311.0
100	0.8	2.5	11.5	28.1	52.8	128.9	242.7
	0.9	3.0	13.7	33.5	63.1	153.9	289.8
	1.0	3.5	16.2	39.6	74.5	182.0	342.0
150	0.8	2.9	13.2	32.1	60.5	147.6	277.9
	0.9	3.4	15.7	38.4	72.2	176.2	331.9
	1.0	4.0	18.6	45.3	85.3	208.0	392.0
200	0.8	3.2	14.5	35.4	66.6	162.5	306.0
	0.9	3.8	17.3	42.2	79.5	194.0	365.0
	1.0	4.4	20.4	49.8	93.8	229.0	431.0
300	0.8	3.6	16.6	40.5	76.3	186.1	350.4
	0.9	4.3	19.8	48.4	91.0	222.2	418.3
	1.0	5.1	23.4	57.0	107.4	262.0	493.0

FLOW RATE DETERMINATION:,

$$Q = [0.758S^{3/2}L^{1/3} - 0.0752]D^{5/2}$$

WHERE:

Q = Maximum Flow Rate of Water When Air Flow is Optimum (liters / minute)

S = Submergence (fraction of total length below water surface)

L = Pipe Length (cm),

D = Pipe Diameter (cm)

(Source: S. Spotte, *Fish and Invertebrate Culture*; copyright © 1979 - reprinted with permission of John Wiley & sons, Inc.)

91

TABLE 5-14: PUMP SELECTION: SHALLOW WELLS, LOW PRESSURE

For pumps lifting water from depth of 25 feet or less and delivery at not more than 20 feet above level of pump. Capacities based on delivery against 20 psi.

WELL SIZE (inches diameter)	TOTAL LIFT[1] (feet)	PISTON PUMP	SHALLOW-WELL JET	SHALLOW-WELL TURBINE	STRAIGHT CENTRIFUGAL[2]
			RANGE IN PUMP CAPACITIES (gallons/hour)		
11/4	10	250-500	400-1,830	400-565	450-684
	15	250-500	375-1,650	390-555	350-672
	20	250-500	285-1,440	380-545	275-646
	25	250-500	240-780	370-535	-
11/2	10	250-500	440-2,800	400-1,330	450-2,300
	15	250-500	360-2,640	390-1,310	350-2,200
	20	250-500	285-2,500	380-1,290	275-2,050
	25	250-500	210-1,200	370-1,270	200-1,900
2 or larger [3]	10	250-500	440-3,660	400-1,330	450-3,500
	15	250-500	360-3,540	390-1,310	350-3,300
	20	250-500	285-3,420	380-1,290	275-3,100
	25	250-500	210-3,180	370-1,270	200-1,900

[1] Distance from lowest water level to pump)

[2] Many manufacturers of straight centrifugal pumps recommend limiting their use to 15 feet or less.

[3] Also includes dug wells, cisterns, springs, ponds, lakes, or marine intakes.

The range of pumping capacities in this table are not for any one pump, but rather for a number of pumps, with different capacities, operating under the conditions given. All figures are taken from manufacturers' published ratings on pumps.

(Source: AAVIM, *Planning for an Individual Water System*; 1973)

TABLE 5-15: PUMP SELECTION: SHALLOW WELLS, HIGH PRESSURE

For pumps lifting from depths of 25 feet or less and delivering higher than 20 feet above pump level.
Capacities based on 20 psi of tank pressure at delivery.

WELL SIZE (INCHES DIAMETER)	TOTAL LIFT (FEET)	PISTON PUMPS			CENTRIFUGAL (SINGLE AND MULTISTAGE STRAIGHT)		SHALLOW WELL TURBINE			SHALLOW WELL JET		
		HEIGHT FROM PUMP TO DELIVERY POINT (in feet)										
		25-90	TO 140	TO 200	25	40	25	100	200	25	40	90
		RANGE IN PUMP CAPACITIES (GALLONS/HOUR)										
1 1/4	10	260-540	260-580	260-580	900-1400	900-1200	0-227			530-1450	345-1080	345-450
	15	260-540	260-580	260-580	750-1200	750-1100	0-220			460-1270	300-930	300-400
	20	260-540	260-580	260-580	600-1000	600-1000	0-213			310-1080	240-750	240-350
	25	260-540	260-580	260-580	450-800	450-300	0-206			240-780	200-540	200-300
1 1/2	10	260-1020	260-1020	260-1020	2200-2500	1800-2100	227-1240	270-1460	285-1120	530-2610	345-1080	345-450
	15	260-1020	260-1020	260-1020	1950-2300	1450-1850	220-1220	265-1440	282-1115	460-2280	300-930	300-400
	20	260-1020	260-1020	260-1020	1800-2100	1200-1600	213-1200	260-1425	279-1110	370-1920	240-750	240-350
	25	260-1020	260-1020	260-1020	1600-2000	1000-1500	206-1180	255-1410	276-1105	240-1260	200-540	200-540
2.0	10	260-1680	260-1680	260-1020	2200-5600	1800-2100	227-1240	270-1460	285-1120	530-2610	345-1080	810-3350
	15	260-1680	260-1680	260-1020	1950-5250	1450-1850	220-1220	265-1440	282-1115	460-2280	300-930	600-2800
	20	260-1680	260-1680	260-1020	1800-2100	1200-1600	213-1200	260-1425	279-1110	370-1920	240-750	420-2340
	25	260-1680	260-1680	260-1020	1600-2000	1000-1500	206-1180	255-1410	276-1105	240-1260	200-540	0-1700
2 1/2	10	260-2640	260-2640	260-1020	2200-5600	1800-2100	227-1240	270-1460	285-1120	530-2610	345-1080	810-3350
	15	260-2640	260-2640	260-1020	1950-5250	1450-1850	220-1220	265-1440	282-1115	460-2280	300-930	600-2800
	20	260-2640	260-2640	260-1020	1800-2100	1200-1600	213-1200	260-1425	279-1110	370-1920	240-750	420-2340
	25	260-2640	260-2640	260-1020	1600-2000	1000-1500	205-1180	255-1410	276-1105	240-1260	200-540	0-1700
3.0 or larger	10	260-3960	260-2640	260-2640	2200-5600	1800-2100	227-1240	270-1460	285-1120	530-2610	345-1080	810-3350
	15	260-3960	260-2640	260-2640	1950-5250	1450-1850	220-1220	265-1440	252-1115	460-2280	300-930	600-2800
	20	260-3960	260-2640	260-2640	1800-2100	1200-1600	213-1200	260-1425	249-1110	370-1920	240-750	420-2340
	25	260-3960	260-2640	260-2640	1600-2000	1000-1500	206-1180	255-1410	246-1105	240-1260	200-540	0-1700

Range of pumping capacities in this table are not those of any one pump, but rather for a wide range of pumps with different capacities made by different manufacturers but operating under the conditions given. All figures are taken from manufacturers' published ratings on pumps offered for individual water supplies.

(Source: AAVIM, *Planning for an Individual Water System*; 1973)

TABLE 5-16: PUMPS FOR LIFTING WATER FROM MORE THAN 25 FEET DEPTH

Water sources include driven, drilled, bored and dug wells. Capacities based on 20 psi tank pressure at point of deliver.

| WELL SIZE (INCHES DIAMETER) | TOTAL LIFT (FEET) | PISTON PUMPS | | | CENTRIFUGAL JET | | | | | | CENTR. SUBMERS. GAL/HR | DEEP-WELL TURBINE GAL/HR |
| | | SINGLE ACTING | DOUBLE-ACTING | EUREKA CYLINDER | PUMP AND BUILDING SAME LEVEL | | 50 FEET | | 75 FEET | | | |
					1-PIPE	2-PIPE	1-PIPE	2-PIPE	1-PIPE	2-PIPE		
					RANGE IN PUMP CAPACITIES (GALLONS/HOUR)							
2.0	30	170-225	300-385	190-270	310-900	Not Adaptable	270-1032	Not Adaptable	250-1000	Not Adaptable	Not Adaptable	Not Adaptable
	50	170-225	300-385	190-270	245-620		220-690		190-790			
	100	170-225	300-385	190-270	165-350		0-315		120-490			
	125	170-225	300-385	190-270	140-250		0-220		110-405			
	150	170-225	300-385	194-270	120-180		0-192		0-360			
	200	170-225	300-385	194-270	100-125		0-100		0-100			
	250	170-225	300-385	190-270								
	300	170-225	300-385	194-270								
	350	170-225	300-385	194-270								
2.5	30	180-250	300-445	310-465	400-1480	Not Adaptable	360-1380	Not Adaptable	320-1020	Not Adaptable	Not Adaptable	Not Adaptable
	50	180-250	300-445	310-465	360-1270		310-1070		290-850			
	70	180-250	300-445	310-465	210-970		200-870		170-700			
	100	180-250	300-445	310-465	200-600		180-540		160-430			
	125	180-250	300-445	310-465	180-450		100-400		140-320			
	200	180-250	300-445	310-465	160-240		140-220		120-170			
	250	180-250	300-445	310-465	150-200		130-180		100-140			
	300	180-250	300-445	310-465	140-150		120-130		95-100			
3.0	30	180-285	300-480	465-625	400-2250	350-470	360-2030	0-582	280-1800	0-576	Not Adaptable	Not Adaptable
	50	180-285	300-480	465-625	360-1900	280-330	330-1700	0-384	250-1640	0-444		
	70	180-265	300-480	465-625	250-1600		220-1400	0-240	170-1120	0-384		
	100	180-285	300-480	465-625	180-1200		160-1100		130-880	0-252		
	125	180-285	300-480	465-625	160-900		140-810		110-650	0-175		
	150	180-285	300-480	465-625	150-170		130-670		100-520			
	200	180-285	300-480	465-625	140-500		0-450		0-360			
	250	180-285	300-480	465-625	130-330		0-290		0-230			
	300	180-285	300-480									
3.5	30	180 - 360	300 - 720	Requires Special Cylinder	Same As 3" Well	Not Adaptable	Same As 3" Well	Not Adaptable	Same As 3" Well	Same As 3" Well	Not Adapatable	Not Adaptable
	50	180-360	300-720									
	70	180-360	300-720									
	100	180-360	300-720									
	125	180-360	300-720									
	150	180-360	300-720									
	200	180-360	300-720									
	250	180-360	300-720									
	250	180-360	300-720									
	300	180-360	300-720									
	350	180-360	300-720									

TABLE 5-16 (cont.):

WELL SIZE (INCHES DIA METER)	TOTAL LIFT (FEET)	PISTON PUMPS			CENTRIFUGAL JET						CENTR. SUBMERS. GAL/HR	DEEP-WELL TURBINE GAL/HR
					PUMP AND BUILDING SAME LEVEL		50 FEET		75 FEET			
		SINGLE ACTING	DOUBLE-ACTING	EUREKA CYLINDER	1-PIPE	2-PIPE	1-PIPE	2-PIPE	1-PIPE	2-PIPE		
					**RANGE IN PUMP CAPACITIES (GALLONS/HOUR)**							
4.0	30	180-585	300-720	675-1100	480-3000	400-1400	450-1700	350-1260	450-4000	320-1120	640-4000	2160-7860
	50	180-585	300-720	675-1100	450-1900	330-1000	400-1700	290-900	410-3350	260-800	570-3600	1560-7560
	70	180-585	300-720	675-1100	400-1500	200-650	360-1500	230-580	350-2100	210-520	480-3350	1200-7200
	100	180-585	300-720	675-1100	300-1100	0-570	310-1200	210-510	290-1530	180-460	320-3120	120-6720
	125	180-585	300-720	675-1100	220-900	0-500	270-810	180-450	270-1100	160-400	160-2920	0-6300
	150	180-585	300-720	675-1100	0-750	0-400	220-670	170-360	220-670	150-320	0-2640	0-5880
	200	180-585	300-720	675-1100	0-500	0-300	180-450	160-270	180-450	140-240	0-2250	0-4320
	250	180-585	300-720	675-1100	0-330	0-270	170-290	150-210	170-290	130-180	0-2000	0-3840
	300	180-585	300-720	675-1100	0-230	0-180	160-210	140-160	160-200	120-140	0-1560	
	400	180-585	300-720								0-1130	
	500	180-535	300-720								0-820	
	600	180-535	300-720								0-650	
	700	180-585	300-720								0-590	
	800	180-585	300-720								0-510	
	900										0-400	
	1000										0-270	
5.0	30	180-825	300-1620	Not Adaptable	Not Adaptable	650-2950	1500-4075	650-2350	Not Adaptable	1000-2000	640-4000	2160-7860
	50	180-825	300-1620			540-2300	1085-3375	690-1800		790-1600	570-3600	1560-7560
	70	180-825	300-1620			450-1800	870-2875	420-1600		595-1440	480-3350	1200-7200
	100	180-825	300-1620			300-1300	610-2050	282-1100		468-1040	320-3120	120-6720
	125	180-825	300-1620			275-1000	390-1580	250-900		400-800	160-2920	0-6300
	150	180-825	300-1620			190 800	0-1340	192-720		220-640	0-2640	0-5880
	200	180-825	300-1620			0-550	0-750	0-440		190-440	0-2250	0-4320
	250	180-825	300-1620			0-350	0-340	0-310		0-280	0-2000	0-3840
	300	180-825	300-1620			0-260		0-230		0-210	0-1560	
	400	180-825	300-1620								0-1130	
	500	180-825	300-1620								0-820	
	600	180 825	300-1620								0-660	
	700	180-825	300-1620								0-590	
	800	180-825	300-1620								0-510	
6.0 and larger	30	180-1290	300-2160	Not Adaptable	Not Adaptable	930-3400	Not Adaptable	2000-4000	Not Adaptable	0-2700	640-4000	2160-7860
	50	180-1290	300-2160			930-3000		1800-3350		0-2400	570-3600	1560-7560
	70	180-1290	300-2160			930-2000		1620-2600		0-1600	480-3350	1200-7200
	100	180-1290	300-2160			680-1300		1080-1750		0-1040	320-3120	1120-6720
	125	180-1290	300-2160			590-1000		830-1200		0-800	160-2920	0-6300
	150	180-1290	300-2160			230-800		0-720		0-640	0-2640	0-5880
	200	180-1290	300-2160			0-550		0-500		0-440	0-2250	0-4320
	250	180-1290	300-2160			0-350		0-310		0-280	0-2000	0-3840
	300	180-1290	300-2160			0-260		0-230		0-210	0-1560	
	400	180-1290	300-2160								0-1130	
	500	180-1290	300-2160								0-820	
	600	180-1290	300-2160								0-660,	
	700	180-1290	300-2160								0-590	
	800	180-1290	300-2160								0-510	
	900										0-400	
	1000										0-270	

[1] Minimum capacity based on 8- stroke at 50 strokes per minute; maximum based on 9- stroke at 45 strokes per minute.

(Source: AAVIM, *Planning for an Individual Water System*; 1973)

TABLE 5-17: THE PHYSICAL PROPERTIES OF VARIOUS RIGID MESH MATERIALS USED IN CAGE FABRICATION

MATERIAL	MESH SIZE (mm)	GAUGE (mm)	DENSITY (kg m^2)	STRENGTH	SOURCE
NETLON	50	3.3	0.34	-	Milne (1970)
90:10 CU-NI EXPANDED METAL	10	1.3	1.86	3.3-4.3 x 10^8 N/m^2 [a]	Woods Hole Eng. Associates (1981)
GALVANISED STEEL WELD MESH	25	2.5	3.40	205kg [b]	Milne (1970)
GALVANISED STEEL CHAINLINK	25	2.0	2.03	127kg [b]	Milne (1970)
	32	2.6			
	45	3.2			
	55	4.0			
PLASTABOND (PVC COATED CHAINLINK)	76	2.5	3.25	127kg [b]	Milne (1970)

[a] Refers to yield strength of a panel
[b] Refers to tensile wet break strength

TABLE 5-18: SPECIFICATIONS OF 7-STRAND GALVANISED STEEL CABLE (Thomas *et al*, 1967)

DIAMETER (mm)	WEIGHT PER UNIT LENGTH (kg /100 m)	SAFE LOAD [1] (kg)
3.2	4.8	227
4.8	11.2	635
6.4	18.6	1,043
7.9	31.3	1,724
9.5	43.9	2,268
12.7	75.9	3,856.
25.4 [2]	296.82	14,300

[1] Safe load is approximately 0.25 times breaking load.
[2] Estimated value

TABLE 5-19: SPECIFICATIONS OF OPEN LINK PROOF STEEL CHAIN (from Thomas *et al*, 1967).

DIAMETER OF LINK (mm)	WEIGHT PER UNIT LENGTH (kg/m)	SAFE LOAD [1] (kg)
4.8	0.6	263
6.4	1.1	408
7.9	1.6	617
9.5	2.4	844
12.7	4.1	1,497
15.9	6.1	2,268
19.1	8.6	3,209
22.2	11.6	4,355
25.4	14.9	5,625
28.6	18.9	7,076
30.2	23.2	8,709

[1] Safe load is approximately 0.25 times working load.

(Source: Tables 5-17 through 5-19; M.C.M. Beveridge, *Cage Aquaculture*; 1987)

TABLE 5-20: CHARACTERISTICS OF SYNTHETIC FIBERS. (Modified from Milne, 1972 and Klust, 1982.)

CHARACTERISTIC	P.A.6.6[1]	P.A.6	P.E.S.[2]	P.E.[3]	PP[4]
FIBRE DENSITY (g/cm^3)	1.14	1.14	1.38	0.96	0.91
BREAKING STRENGTH	very high	very high	high	high	high [5]
BREAKING STRENGTH - WET (expressed as % dry breaking strength)	85-95[5]	85-95[5]	100	110	100
WEIGHT IN WATER (as % of air-dry weight)	12	12	28	0 (buoyant)	0 (buoyant)
EXTENSIBILITY (wet)	high	high	low	intermediate (PA and PES)	low
STIFFNESS	flexible	flexible	moderately stiff	stiff	stiff
SOFTNESS	soft	soft	moderate	moderate	hard
RESISTANCE AGAINST WEATHERING [6]	medium	medium	high	medium	low-medium
RESISTANCE TO FOULING	moderate	moderate	"	low	moderate

[1] PA = Polyamide - nylon
[2] PES = Polyester type Terylene, Dacron, Diolen, Tergal, Tetoron, and Trevira
[3] PE = Polyethylene type high density, polymerised at low pressure
[4] PP = Polypropylene
[5] Continuous filament form
[6] Without treatment or colouring

Note that the two types of nylon are referred to, although both have near identical properties.

TABLE 5-21: COMPARATIVE PROPERTIES OF THREE-STRAND FIBER ROPES. THE RESISTANCE TO SUNLIGHT REFERS TO SYNTHETIC FIBERS WHICH ARE STABILIZED BUT NOT COLOURED (after Klust, 1983).

ROPES MADE OF	PA (cont. filament)	PES (cont. filament.)	PP (monofil.)	PP (split fiber)	PE (monofil.)	MANILA
BREAKING LENGTH (in km) (dry)	30.6	19.2	28.8	28.8	21.2	10.8
WET STRENGTH TO PERCENTAGE OF DRY STRENGTH	80 to 90	100	100	100	100 to 115	105 to 120
STRENGTH RATIO TO MANILA	2.83	1.78	2.67	2.67	1.96	1
ELONGATION (DRY) AT 30% OF BREAKING STRENGTH (%)	23.3	5	12.7	8.5	9.9 to 15.1	4.7
ELASTICITY	high	high	medium	medium	low/creep	low
ELONGATION RATIO TO MANILA AT 30% OF BREAKING STRENGTH	4.9	1.1	2.7	1.8	2.1 to 3.2	1
TOUGHNESS	very high	high	high	high	medium	low
RESISTANCE TO SUSTAINED LOAD	very high	very high	high	high	medium	low
RESISTANCE TO REPEATED LOAD	very high	very high	high	high	low	low
RESISTANCE TO SHOCK LOADING	very high	high	high	high	medium	low
FLEXING ENDURANCE	very high	very high	medium	low	low	low
SHRINKAGE IN WATER	small/varies	none	none	none	none	great
RESISTANCE TO SUNLIGHT	medium	high	medium	medium	medium	medium
RESISTANCE TO ROTTING	very good	very good	very good	very good	very good	low
RESISTANCE TO AGEING	good	good	good	good	good	medium
SURFACE FEEL	smooth	smooth	smooth/stiff	harsh	smooth/waxy	harsh
FIBRE DENSITY (G/CM^3)	1.14	1.38	0.91 (floating)	0.91 (floating)	0.96 (floating)	1.35

(Source: M.C.M. Beveridge, *Cage Aquaculture*; 1987)

TABLE 5-22: MASS AND BREAKING STRENGTH (Kp) OF COMMON BRAIDED AND THREE STRAND LAID SYNTHETIC FIBER ROPES (after Klust, 1983).

NOM. DIA. (mm)	NYLON (PA) mass	Kp	PES mass	Kp	PP mass	Kp	PE mass	Kp	BRAIDED NYLON (PA) mass	$Kp^{(a)}$	$Kp^{(b)}$	BRAIDED PES mass	$Kp^{(a)}$	$Kp^{(b)}$
4	1.1	320	1.5	295	-	-	0.8	200	0.9	280	225	1.1	260	215
6	2.4	750	3.0	565	1.7	550	1.8	400	2.0	620	500	2.4	575	440
8	4.2	1,350	5.1	1,020	3.0	960	3.3	700	3.6	1,110	900	4.4	1,000	760
10	6.5	2,080	8.1	1,590	4.5	1,425	4.9	1,090	5.6	1,700	1,400	6.8	1,540	1,160
12	9.4	3,000	11.6	2,270	6.5	2,030	7.2	1,540	8.1	2,475	2,025	9.8	2,160	1,620
14	12.8	4,100	15.7	3,180	9.0	2,790	9.5	2,090	10.5	3,200	2,325	13.3	2,860	2,130
16	16.6	5,300	20.5	4,060	11.5	3,500	12.8	2,800	14.3	4,350	3,500	17.4	3,650	2,700
18	21.0	6,700	26.0	5,080	14.8	4,450	16.1	3,460	18.1	5,500	4,450	22.0	4,500	3,300
20	26.0	8,300	32.0	6,350	18.0	5,370	20.0	4,270	22.3	6,700	5,350	27.2	5,300	3,950
22	31.5	10,000	38.4	7,620	22.0	6,500	24.3	5,080	27.0	8,100	6,500	32.8	5,800	4,600
24	37.5	12,000	46.0	9,140	26.0	7,600	29.5	6,100	32.2	9,650	7,750	39.0	6,250	4,450
26	-	-	-	-	-	-	32.8	6,910	-	-	-	-	-	-
28	51.0	15,800	63.0	12,200	35.5	10,100	39.3	8,030	-	-	-	-	-	-
32	66.5	20,000	82.0	15,700	46.0	12,800	52.5	10,400	-	-	-	-	-	-
40	104.0	30,000	128.0	23,900	72.0	19,400	78.5	15,600	-	-	-	-	-	-

MASS = kg/100 m linear length

NB K_p = kilopound = 9.81 N.(newtons)

(a) round braided rope

(b) solid braided rope.

(Source: M.C.M. Beveridge, *Cage Aquaculture*; 1987)

TABLE 5-23: SUMMARY OF MECHANICAL PROPERTIES OF BAMBOO, COMPARED WITH TIMBER, CONCRETE, AND STEEL. (Chong, 1977 and Janssen 1981).

CHARACTERISTIC	BAMBOO	SOFT WOOD	HARD WOOD	CONCRETE	STEEL
DENSITY (kg m^{-3})	600-800	450-650	400-750	2,400	7,800
ULTIMATE TENSILE STRENGTH (N mm^{-2})	200-300	200-110	200-110	4	270-700
E-VALUE [1] TENSILE STRENGTH (N mm^{-2})	17,400-10,300	5,000-9,000	8,000-18,000	28,600	210,000
ULTIMATE SHORT-TERM BENDING STRESS (N mm^{-2})	84	-	-	-	-
E-VALUE BENDING STRESS (N mm^{-2})	20,500	-	-	-	-
ULTIMATE SHEAR STRESS (N mm^{-2})	2.25	11	14	-	-

[1] E-value = Modulus of elasticity = stress/strain (δ/E) ratio of load per unit area to deformation per unit length.

(Source: M.C.M. Beveridge, *Cage Aquaculture*; 1987)

TABLE 5-24: APPROXIMATE DENSITIES OF COMMON MATERIALS

Approximate maximum density given in lb/ft^3 for all materials. (ASHRAE, 1985; Kevgor Aquasystems, 1990))

MATERIAL	DENSITY (lb/ft^3)	MATERIAL	DENSITY (lb/ft^3)	MATERIAL	DENSITY (lb/ft^3)
CLAY (damp & plastic)	110	CYPRESS (southern)	32	REINFORCED CINDER	115
CLAY (dry)	63	DOUGLAS FIR	34	AEROCRETE	80
CLAY & GRAVEL (dry)	100	FIR (white)	27	LT.WT. CINDER FILL	60
EARTH (loose)	76	PINE	27	HAYDITE	100
EARTH (dry & packed)	95	HEMLOCK	29	NAILCODE	75
EARTH (moist & loose)	78	MAPLE (hard & black)	42	PERLITE	50
EARTH (moist & packed)	96	OAK (white)	47	PUMICE	90
EARTH (mud & packed)	115	SPRUCE (eastern & sitka)	28	VERMICULITE	60
SAND OR GRAVEL (dry)	120	POPLAR	28	EPS POLYSTYRENE	5
ALUMINUM (cast)	165	REDWOOD	28	POLYPROPYLENE	56
COPPER	556	WALNUT	38	MASONRY MORTAR	116
IRON (cast)	450	FIR PLYWOOD (0.125 - 0.75")	34	GYPSUM (sand)	12
IRON (wrought)	485	GYPSUM PLASTERBOARD	50	GYPSUM (perlite)	55
LEAD	710	PARTICLE BOARD (low density)	37	PORTLAND CEMENT (sand)	120
STAINLESS STEEL	510	PARTICLE BOARD (high density)	62.5	PORTLAND CEMENT (perlite)	55
STEEL (ROLLED)	490	LAMINATED PAPERBOARD	30	PORTLAND CEMENT (vermiculite)	55
BIRCH - RED OAK	44	REINFORCED STONE	150	WATER (@4°C-pure)	62 5
CEDAR (northern white)	22	PLAIN STONE	144	GRANITE	175
CEDAR (western red)	23	PLAIN SLAG	130	LIMESTONE	165
HARDBOARD (std. tempered)	63	MARBLE	165	SANDSTONE	147
MINERAL FIBER INSULATION	0.3-2.0	GLASS FIBER	4-9	SLATE	175
POLYSTYRENE BEADS	1.0	POLYURETHANE FOAM	1.5-2.5	COMMON BRICK	120
ASPHALT SHINGLES	70	ASPHALT ROLL ROOFING	70	FACING BRICK	130
FLAX FIBER INSULATION	4.9	KAPOK BETWEEN BURLAP	1.0	JUTE FIBER INSULATION	6.7
SAWDUST	12.0	WOOD PLANER SHAVINGS	8.8	SUGAR CANE FIBER (in asphalt)	13.8
ACOUSTICAL TILE	23.0	CORKBOARD	14.0	CORN FIBER INSULATION	15.00

TABLE 5-25: HEAT-TRANSFER COEFFICIENTS (U) FOR COMMON CONSTRUCTION MATERIALS

MATERIAL	COEFFICIENT (U) $(CAL\ SEC^{-1}\ CM^{-2\ O}C^{-1})$
SINGLE PLATE GLASS	1.60×10^{-4}
DOUBLE GLASS WITH AIR SPACE	0.70×10^{-4}
5 CM CONCRETE	0.91×10^{-4}
15 CM CONCRETE	0.52×10^{-4}
1 CM FIBERGLASS REINFORCED PLASTIC	0.52×10^{-4}
1 CM FIBERGLASS-REINFORCED PLASTIC WITH 2CM WET SOAKED PLYWOOD	0.80×10^{-4}
2 CM DRY PLYWOOD (EPOXY COATED)	0.80×10^{-4}
WATER SURFACE	2.30×10^{-4}

CONVERSIONS:

Q (cal sec^{-1}) X 14.3 = Btu hr^{-1}
1 Btu hr^{-1} = 3.41 watts

CALCULATING HEATING OR COOLING REQUIREMENTS
$$Q = [A1U1 + A2U2 + ...]dT$$

WHERE:

A = Heat Transfer Area in cm^2
U = Heat-transfer Coefficient
dT = Overall Temperature Gradient in Degrees Centigrade

EXAMPLE:

The dimensions of a 1/4" fiberglass tank are 200 cm long, 100 cm wide and 50 cm tall.
One open side of the tank is constructed of double-paned glass for viewing.
The top of the tank is uncovered. The water in the tank must be maintained at 11°C above the ambient room temperature.

1. What is the heat loss from the tank?

$Q = [A1U1 + A2U2 + ...]dT$
$Q = [2(0.52) + 0.5(0.52) + 0.5(0.52) + 1(0.52) + 1(0.7) + 2(1.6)]11$
$Q = (5.98)11 = 65.78$ cal sec^{-1}

2. Convert to Btu hr^{-1}

65.78 cal sec^{-1} X 14.3 = 940.65 Btu hr^{-1}

3. Convert heat flow to continuous wattage needed to maintain the desired temperature.

940.65 Btu hr^{-1} X 3.41 = 3207.63 watts

(Source: S. Spotte, *Seawater Aquariums: The Captive Environment,* copyright © 1979 - reprinted with permission of John Wiley & Sons, Inc.)

TABLE 5-26: FIXING COEFFICIENT, K, OF SANDBAG ANCHORS ON DIFFERENT SUBSTRATES AND VARYING MOORING CABLE LENGTH:WATER DEPTH RATIO OR SCOPE (L:D) [1] (from Nomura and Yamazaki, 1977).

NATURE OF SUBSTRATE	l:d				
	1	2	3	4	5
SAND (vertically pulled)	0.16	0.47	0.60	0.71	0.73
SAND (horizontally pulled)	0.21	0.59	0.65	0.68	0.74
MEAN VALUE	0.19	0.53	0.63	0.70	0.74
SANDY MUD (vertically pulled)	-	0.33	0.41	0.49	0.61
SANDY MUD (horizontally pulled)	-	0.31	0.31	0.43	0.62
MEAN VALUE	0.10	0.32	0.36	0.46	0.62
MUD (vertically pulled)	0.05	0.13	0.20	0.23	0.30
MUD (horizontally pulled)	0.05	0.33	0.34	0.48	0.52
MEAN VALUE	0.05	0.23	0.27	0.35	0.41

[1] Sandbag anchors are defined as any deadweight or block anchor (e.g. bags of sand, stones or concrete blocks, scrap metal, etc.)

THE FIXING COEFFICIENT, K, IS DEFINED AS THE THE RATIO R/W WHERE:

R = The Horizontal Force Exerted (kg/m^2)

W = The Mass of the Anchor (kg)

(Source: M.C.M. Beveridge, *Cage Aquaculture*; 1987)

TABLE 5-27: THE FIXING COEFFICIENT, K, OF KEDGE-TYPE ANCHORS ON DIFFERENT SUBSTRATES AND VARYING CABLE LENGTH:WATER DEPTH RATIOS (l:d) (Nomura and Yamazaki, 1977).

SUBSTRATE	l:d					
	1.0	1.5	2.0	3.0	4.0	5.0
SAND	0.26	1.10	1.90	4.37	"	5.83
SANDY MUD	0.23	1.90	3.27	4.40	5.50	5.15
MUD	0.11	0.60	1.99	3.29	5.11	6.46
MEAN	0.20	1.20	2.39	4.02	5.31	5.81

"Kedge-type" anchors include navy stockless, kedge, danforth, grapnel and mushroom type anchors.

The holding power of an embedding anchor is related to its frictional resistance in soil, and dependent upon fluke area, soil pentration, and the mechanical properties of the soil, rather than on the mass of the anchor. Anchor penetration is a function of fluke shape and the angle between the fluke and shank. Frictional resistance of the soil is dependent upon soil cohesiveness and shear strength.

(Source: M.C.M. Beveridge, *Cage Aquaculture*; 1987)

VI. Feeding and Nutrition

TABLE 6-1: RECOMMENDED AMOUNTS OF FOOD TO FEED RAINBOW TROUT (PERCENTAGE OF BODY WEIGHT PER DAY) FOR DIFFERENT SIZES AND WATER TEMPERATURES IN INTENSIVE CULTURE SYSTEMS (Post 1975)

TEMP. (°C)	0.18 or less	0.18 - 1.49	1.50 - 5.25	5.26 - 11.9	12.0 - 23.1	23.3 - 38.4	38.5 - 62.4	62.5 - 90.8	90.9 - 124	125 - 166	167 or more
					WEIGHT OF FISH IN GRAMS						
5.5	3.5	2.8	2.4	1.8	1.4	1.2	0.9	0.8	0.7	0.6	0.5
6.0	3.6	3.0	2.5	1.9	1.4	1.2	1.0	0.9	0.8	0.7	0.6
6.5	3.8	3.1	2.5	2.0	1.5	1.3	1.0	0.9	0.8	0.8	0.6
7.0	4.0	3.3	2.7	2.4	1.6	1.3	1.1	1.0	0.9	0.8	0.7
7.5	4.1	3.4	2.8	2.2	1.7	1.4	1.2	1.0	0.9	0.8	0.7
8.0	4.3	3.6	3.0	2.3	1.7	1.4	1.2	1.0	0.9	0.8	0.7
9.0	4.5	3.8	3.0	2.4	1.8	1.5	1.3	1.1	1.0	0.9	0.8
9.5	4.7	3.9	3.2	2.5	1.9	1.5	1.3	1.1	1.0	0.9	0.8
10.0	5.2	4.3	4.4	2.7	2.0	1.7	1.4	1.2	1.1	1.0	0.9
10.5	5.4	4.5	3.5	2.8	2.1	1.7	1.5	1.3	1.1	1.0	0.9
11.0	5.4	4.5	3.4	2.8	2.1	1.7	1.5	1.3	1.1	1.0	0.9
11.5	5.6	4.7	3.8	2.9	2.2	1.8	1.5	1.3	1.1	1.1	1.0
12.0	5.8	4.9	3.9	3.0	2.3	1.9	1.6	1.4	1.3	1.1	1.0
13.0	6.1	5.1	4.2	3.2	2.4	2.0	1.6	1.4	1.3	1.1	1.0
13.5	6.3	5.3	4.3	3.3	2.5	2.0	1.7	1.5	1.3	1.2	1.0
14.0	6.7	5.5	4.5	3.5	2.6	2.1	1.8	1.5	1.4	1.2	1.1
14.5	7.0	5.8	4.8	3.6	2.7	2.2	1.9	1.6	1.4	1.3	1.2
15.0	7.3	6.0	5.0	3.7	2.8	2.3	1.9	1.7	1.5	1.3	1.2
15.5	7.5	6.3	5.1	3.9	3.0	2.4	2.0	1.4	1.5	1.4	1.3

TABLE 6-2: FEEDING CHART FOR TILAPIA CULTURED IN SEMI-INTENSIVE, INTENSIVE, AND POLYCULTURE SYSTEMS IN GRAMS PER FISH PER DAY (Marek, 1975[1]; Zohar, 1986[2])

FISH WEIGHT (g)	POLYCULTURE[1]	SEMI - INTENSIVE[2] (less than 20,000/ha)	INTENSIVE[1] (more than 20,000/ha)
1 - 3	0.2	0.3	-
3 - 5	0.3	0.4	-
5 - 10	0.5	0.6	-
10 - 20	0.8	1.0	-
20 - 30	1.2	1.5	-
30 - 50	1.4	2.0	-
50 - 70	1.6	2.5	-
70 - 100	1.8	3.0	1.8
100 - 150	2.3	3.5	2.7
150 - 200	2.8	4.0	3.7
200 - 250	-	-	5.5
250 - 300	3.5	5.0	5.5
300 - 350	4.3	6.0	6.3
350 - 400	-	-	6.7
400 - 500	5.0	7.0	7.2
500 - 600	6.0	8.0	8.3
600 - 700	7.0	9.0	-

(Source: B. Hepher, *Nutrition of Pond Fishes*; copyright © 1988 - with permission of Cambridge University Press)

TABLE 6-3: FEEDING CHART FOR CHANNEL CATFISH AS PERCENTAGE OF BODY WEIGHT PER DAY (Foltz, 1982)

FISH WEIGHT (g)	\multicolumn{16}{c}{TEMPERATURE (°C)}															
	15	16	17	18	19	20	21	22	23	24	25	26	27	28	29	30
4	2.0	2.2	2.4	2.5	2.7	2.9	3.1	3.2	3.4	3.5	3.7	3.8	4.0	4.1	4.3	4.4
6	1.9	2.1	2.2	2.4	2.6	2.8	2.9	3.1	3.2	3.4	3.5	3.7	3.8	3.9	4.1	4.2
8	1.8	2.0	2.1	2.3	2.5	2.6	2.8	2.9	3.1	3.2	3.4	3.5	3.6	3.8	3.9	4.0
10	1.7	1.9	2.1	2.2	2.4	2.5	27	2.8	3.0	3.1	3.2	3.4	3.5	3.6	3.7	3.8
13	1.7	1.8	2.0	2.1	2.3	2.4	2.6	2.7	2.8	3.0	3.1	3.2	3.3	3.5	3.6	3.7
16	1.6	1.7	1.9	2.0	2.2	2.3	2.5	2.6	2.7	2.8	3.0	3.1	3.2	3.3	3.4	3.6
20	1.5	1.7	1.8	2.0	2.1	2.2	2.4	2.5	2.6	2.7	2.9	3.0	3.1	3.2	3.3	3.4
24	1.5	1.6	1.8	1.9	2.0	2.2	2.3	2.4	2.5	2.6	2.8	2.9	3.0	3.1	3.2	3.3
29	1.4	1.6	1.7	1.8	2.0	2.1	2.2	2.3	2.4	2.5	2.7	2.8	2.9	3.0	3.1	3.2
35	1.4	1.5	1.6	1.8	1.9	2.0	2.1	2.2	2.3	2.5	2.6	2.7	2.8	2.9	3.0	3.1
41	1.3	1.5	1.6	1.7	1.8	1.9	2.0	2.2	2.3	2.4	2.5	2.6	2.7	2.8	2.9	3.0
48	1.3	1.4	1.5	1.6	1.7	1.9	2.0	2.1	2.2	2.3	24	2.5	2.6	2.7	2.8	2.9
56	1.2	1.4	15	1.6	1.7	1.8	1.9	2.0	2.1	2.2	2.3	2.4	2.5	2.6	2.7	2.8
64	1.2	1.3	1.4	1.5	1.6	1.7	1.8	1.9	2.0	2.1	2.2	2.3	2.4	2.5	2.6	2.7
73	1.2	1.3	14	15	1.6	1.7	1.8	1.9	2.0	2.1	2.2	2.2	2.3	2.4	2.5	2.6
83	1.1	1.2	1.3	1.4	1.5	1.6	1.7	1.8	1.9	2.0	2.1	2.2	2.3	2.3	2.4	2.5
94	1.1	1.2	1.3	1.4	15	1.6	1.7	1.8	1.9	1.9	2.0	2.1	2.2	2.3	2.3	2.4
106	1.0	1.2	1.2	1.3	1.4	1.5	1.6	1.7	1.8	1.9	2.0	2.0	2.1	2.2	2.3	2.3
119	1.0	1.1	1.2	1.3	1.4	1.5	1.6	1.7	1.7	1.8	1.9	2.0	2.0	2.1	2.2	2.3
133	1.0	1.1	1.2	1.3	1.4	1.4	15	1.6	1.7	1.8	1.8	1.9	2.0	2.1	2.1	2.2
148	1.0	1.0	1.1	1.2	1.3	1.4	15	1.6	1.6	1.7	1.8	1.9	1.9	2.0	2.1	2.1
163	0.9	1.0	1.1	1.2	1.3	1.3	1.4	1.5	1.6	1.7	1.7	1.8	1.9	1.9	2.0	2.1
180	0.9	1.0	1.1	1.1	1.2	1.3	1.4	1.5	1.5	1.6	1.7	1.7	1.8	1.9	1.9	2.0
198	0.9	1.0	1.0	1.1	1.2	1.3	13	1.4	15	1.6	1.6	1.7	1.7	1.8	1.9	1.9
218	0.8	0.9	1.0	1.1	1.2	1.2	11	1.4	1.4	1.5	1.6	1.6	1.7	1.8	1.8	1.9
238	0.8	0.9	1.0	1.0	1.1	1.2	1.1	1.3	1.4	1.5	1.5	1.6	1.6	1.7	1.8	1.8
260	0.8	0.9	0.9	1.0	1.1	1.2	1.2	1.3	1.3	1.4	1.5	1.5	1.6	1.6	1.7	1.8
283	0.8	0.8	0.9	1.0	1.0	1.1	1.2	1.2	1.3	1.4	1.4	1.5	1.5	1.6	1.7	1.7
307	0.7	0.8	0.9	1.0	1.0	1.1	1.1	1.2	1.3	1.3	1.4	1.4	1.5	1.5	1.6	1.6
333	0.7	0.8	0.8	0.9	1.0	1.0	1.1	1.2	1.2	1.3	1.3	1.4	1.4	1.5	1.5	1.6
360	0.7	0.8	0.8	0.9	0.9	1.0	1.1	1.1	1.2	1.2	1.3	1.3	1.4	1.4	1.5	1.5
388	0.7	0.7	0.8	0.9	0.9	1.0	1.0	1.1	1.1	1.2	1.3	1.3	1.4	1.4	1.5	1.5
418	0.6	0.7	0.8	0.8	0.9	0.9	1.0	1.1	1.1	1.2	1.2	1.3	1.3	1.4	1.4	1.5
449	0.6	0.7	0.7	0.8	0.9	0.9	1.0	1.0	1.1	1.1	1.2	1.2	1.3	1.3	1.4	1.4
482	0.6	0.7	0.7	0.8	0.8	0.9	0.9	1.0	1.0	1.1	1.1	1.2	1.2	1.3	1.3	1.4
517	0.6	0.7	0.7	0.8	0.8	0.9	0.9	1.0	1.0	1.1	1.1	1.1	1.2	1.2	1.3	1.3
553	0.6	0.6	0.7	0.7	0.8	0.8	0.9	0.9	1.0	1.0	1.1	1.1	1.1	1.2	1.2	1.3

(Source: B. Hepher, *Nutrition of Pond Fishes*; 1988)

TABLE 6-4: FEEDING CHART FOR COMMON CARP CULTURED IN PONDS DURING SUMMER (water temperature over 20 °C). AMOUNTS ARE GIVEN IN GRAMS PER FISH (modified from Marek, 1975) [1]

FISH WEIGHT (g)	\multicolumn{14}{c}{DENSITY OF FISH PER HECTARE (X 1000)}													
	\multicolumn{2}{c}{2 - 4}		\multicolumn{2}{c}{4 - 6}		\multicolumn{2}{c}{6 - 8}		\multicolumn{2}{c}{8 - 10}		\multicolumn{2}{c}{12 - 15}		\multicolumn{2}{c}{15 - 20}		\multicolumn{2}{c}{20 - 50}		\multicolumn{2}{c}{+}
	g	%Prot.	g	%Prot.	g	%Prot.	g	%Prot.	g	%Prot.	g	%Prot.	g	%Prot.
20 - 50	1	12	2	12	2	12	3	12	3	12	3	18	3	18
50 - 100	2	12	3	12	4	12	4	18	4	18	4	18	4	18
100 - 200	6	12	6	18	9	18	9	18	9	18	9	25	9	25
200 - 300	10	12	11	18	11	18	12	25	12	25	12	25	12	25

TABLE 6-4 (cont.):

FISH WEIGHT (g)	DENSITY OF FISH PER HECTARE (X 1000)													
	2 - 4		4 - 6		6 - 8		8 - 10		12 - 15		15 - 20		20 - 50	
	g	%Prot.	g	%Prot.	g	%Prot.	g	%Prot.	g	%Prot.	g	%Prot.	g	%Prot.
300 - 400	11	18	13	25	14	25	15	25	15	25	15	25	15	25
400 - 500	14	18	15	25	16	25	17	25	17	30	17	30	17	30
500 - 600	15	25	17	25	18	25	19	30	19	30	19	30	19	30
600 - 700	15	25	18	45	19	30	19	30	19	30	19	30	19	30
700 - 800	16	25	18	25	20	30	19	30	19	30	19	30	19	30
800 - 900	17	25	18	25	19	30	20	30	20	30	20	30	20	30
900 - 1,000	17	25	19	25	20	30	21	30	21	30	21	30	21	30
1,000 - 1,100	18	25	20	30	21	30	22	30	22	30	22	30	22	30
1,100 - 1,200	18	25	20	30	21	30	22	30	22	.30	22	30	22	30

[1] Based on 4 pelleted diets of increasing protein content: 12, 18, 25 and 30%.

(Source: B. Hepher, *Nutrition of Pond Fishes*; 1988.

TABLE 6-5: PUBLISHED FEEDING TABLES FOR MARINE SHRIMP. FIGURES ARE % OF BODY WEIGHT FED DAILY.

SIZE (g)	SOURCE					
	1	2	3	4	5	6
1	16.0	6.0	-	10.0	15.0	14.0
2	11.7	4.8	5.5	8.0	13.0	8.2
3	8.6	4.2	4.7	6.0	10.0	6.2
4	7.2	3.8	4.2	4.0	9.0	5.2
5	6.2	3.4	3.9	3.8	8.0	4.5
6	4.8	3.2	3.6	3.6	7.0	3.9
7	4.4	2.9	3.3	3.4	6.0	3.6
8	4.0	2.8	3.0	3.2	5.5	3.3
9	3.9	2.7	2.9	3.0	5.0	3.0
10	3.6	2.6	2.8	2.8	4.5	2.8
11	3.5	2.4	2.6	2.6	4.0	2.6
12	3.3	2.3	2.6	2.5	3.8	2.5
13	3.1	2.2	2.5	2.4	3.5	2.3
14	3.0	2.1	2.4	2.3	3.3	2.2
15	2.9	2.0	2.3	2.2	3.0	2.1
16	2.7	1.9	2.3	2.2	2.8	2.0
17	2.5	1.9	2.2	2.1	2.6	2.0
18	2.4	1.8	2.1	2.1	2.5	1.9
19	-	1.7	2.0	2.0	2.4	1.8
20	-	1.7	2.0	2.0	2.3	1.8
21	-	1.7	1.9	-	-	1.8
22	-	1.6	1.8	-	-	1.8

KEY TO SOURCES:

1. Zendejas (1991); Juveniles, unspecified stocking density.

2. Villalon (1991); 6.5 - 9 juveniles/m^2.

3. Villalon(1991); 12.5-18.5PL's/m^2.

4. Akiyama and Chwang (1989); 5/m^2.

5. Akiyama and Chwang (1989); 10/m^2.

6. Nicovita/Nicolini Hermanos S.A., Peru; 30 PL's/

(Source: Clifford, H.C. III, In: J. Wyban (Ed.) *Proceedings of the Special Session on Shrimp Farming*. World Aquaculture Society; 1992)

TABLE 6-6: DIETARY PROTEIN REQUIREMENTS OF FISH AND SHRIMP (EXPRESSED AS % OF DRY DIET)

SPECIES	DIETARY PROTEIN REQUIREMENT	SIZE RANGE [1]	FEEDING REGIME [2]	CULTURE SYSTEM	REFERENCE
FISH					
Oreochromis mossambicus	40	Fingerling	6%bw/d	Indoor/tank	Jauncey (1982)
Oreochromis niloticus	35	Fry	15%bw/d	Indoor/tank	Santiago *et al.* (1982)
O. niloticus	28-30	Fry/fing.	6%bw/d	Indoor/tank	De Silva & Perera (1985)
O. niloticus	25	Fingerling	3.5%bw/d	Indoor/tank	Wang *et al.* (1985)
O. niloticus	35	Fingerling	4%bw/d	Indoor/tank	Teshima *et al.* (1985)
O. niloticus	19-29	Juvenile	3%bw/d	Outdoor/cage	Wannigama *et al.* (1985) [a]
O. niloticus/aureus hybrids	30	Grower	2-2.5%bw/d	Outdoor/pond	Viola & Zohar (1984) [b]
Oreochromis aureus	30	Fingerling	6%bw/d	Indoor/tank	Toledo *et al.* (1983)
O. aureus	36	Fingerling	8.8%bw/d	Indoor/tank	Davis & Stickney (1978)
O. aureus	56	Fry	20%bw/d	Indoor/tank	Winfree & Stickney (1981)
O. aureus	34	Fingerling	10%bw/d	Indoor/tank	Winfree & Stickney (1981)
Tilapia zilli	35	Fingerling	5%bw/d	Indoor/tank	Mazid *et al.* (1979)
T. zilli	35-40	Fingerling	4%bw/d	Indoor/tank	Teshima *et al.* (1978)
Cyprinus carpio	35	Grower	5%bw/d	Indoor/tank	Jauncey (1981)
C. carpio	34	Fingerling	*Ad.lib.*	Indoor/tank	Murai *et al.* (1985)
C. carpio	38	Fingerling	*Ad.lib.*	Indoor/tank	Ogino & Salto (1970)
Ctenopharyngodon idella	41-43	Fry	Fixed (?)	Indoor/tank	Dabrowski (1977)
Mugil capito	24	Fingerling	*Ad.lib.*	Indoor/tank	Papaparaskeva-Papoutsoglou & Alexis (1985)
Ictalurus punctatus	35	Grower	Fixed (1-4%bw/d)	Outdoor/cage	Lovell (1972) [c]
I. punctatus	29-42	Grower	Fixed (1-4%bw/d)	Outdoor/pond	Prather & Lovell (1973) [d]
I. punctatus	45	Grower	Fixed (34-45kg/ha/d)	Outdoor/pond	Lovell (1975) [e]
I. punctatus	25	Grower	*Ad.lib.*	Indoor/tank	Page & Andrews (1973)
I. punctatus	36	Fingerling	3%bw/d	Indoor/tank	Garling & Wilson (1976)
I. punctatus	25	Juvenile	Fixed (3-4%bw/d)	Outdoor/pond	Deyoe *et al.* (1968) [f]
I. punctatus	35	Juvenile/grow.	3%bw/d	Indoor/tank	Page & Andrews (1973)
Alosa sapidissima	42.5	Fingerling	*Ad.lib.*	Outdoor/tank	Murai et al. (1979)
Pangasius sutchi	25	Fry/fing.	10%bw/d	Indoor/tank	Chuapoehuk & Pthisoong (1985)
Chanos chanos	40	Fry	10%bw/d	Indoor/tank	Lim *et al.* (1979)
Channa micropeltes	52	Grower	2%bw/d	Indoor/tank	Wee & Tacon (1982)
Fugu rubripes	50	Fingerling	10%bw/d	Indoor/tank	Ranazawa *et al.* (1980)
Chrysophrys aurata	38.4	Fingerling/juv.	*Ad.lib.*	Indoor/tank	Sabaut & Luquet (1973)
Morone saxatilis	47	Fingerling	*Ad.lib*	Indoor/tank	Millikin (1983)
M. saxatilis	55	Fingerling	*Ad.lib.*	Indoor/tank	Millikin (1982) [g]
Anguilla japonica	44.5	Fingerling	*Ad.lib.*	Indoor/tank	Nose & Arai (1973)
Micropterus dolomieui	45.2	Fry/fing.	*Ad.lib.*	Indoor/tank	Anderson *et al.* (1981)
Micropterus salmoides	40-41	Fingerling	*Ad.lib.*	Indoor/tank	Anderson *et al.* (1981)
Pleuronectes platessa	50	Juvenile	*Ad.lib.*	Indoor/tank	Cowey *et al.* (1972)
Salvelinus alpinus	36-43.6	Juvenile/grow.	*Ad.lib.*	Indoor/tank	Jobling & Wandsvik (1983)
Salmo gairdneri	42	Grower	Fixed (?)	Indoor/tank	Austreng & Refstie (1979)
S. gairdneri	40	Fingerling/juv.	Fixed	Indoor/tank	Satia (1974) [h]
S. gairdneri	40-45	Fingerling/juv.	*Ad.lib.*	Indoor/tank	Zeltoun *et al.* (1973) [i]
PRAWN					
Macrobrachium rosenbergii	40	PL 0.15g	12-5%bw/d	Indoor/tank	Millikin *et al.* (1980)
M. rosenbergii	15	PL 0.12g	Fixed	Outdoor/tank	Boonyaratpalin & New (1982) [j]
M. rosenbergii	35	PL 0.10g	5%bw/t	Outdoor/tank	Balazg & Ross (1976) [k]
M. rosenbergii	27	PL 1.90g	5%bw/d	Outdoor/pond	Stanley & Moore (1983) [l]

TABLE 6-6 (cont.):

SPECIES	DIETARY PROTEIN REQUIREMENT	SIZE RANGE [1]	FEEDING REGIME [2]	CULTURE SYSTEM	REFERENCE
SHRIMP					
Penaeus indicus	30-40	PL 1-42 day	Fixed	Indoor/tank	Bhaskar & All (1984) [m]
P. indicus	43	PL 0.4-1.1g	10-15%bw/d	Indoor/tank	Colvin (1976)
Penaeus aztecus	40	PL 24-135mg	100-50%bw/d	Indoor/tank	Venkataramiah *et al.* (1975)
P. aztecus	43-51	PL 0.4-1.3g	Fixed (?)	Indoor/tank	Zein-Eldin & Corliss (1976) [n]
Penaeus setiferus	28-32	Juveniles 4g	5%bw/d	Indoor/tank	Andrews; Sick & Baptist (1972)
Penaeus merguiensis	50-55	Juv. 3-8g	Fixed (?)	Indoor/tank	AQUACOP (1978) [n]
P. merguiensis	34-42	PL 0.3g	Fixed (?)	Indoor/tank	Sedgwick (1979) [n]
Penaeus monodon	55	PL 2mg	Fixed (?)	Outdoor/tank	Bages & Sloane (1981) [o]
P. monodon	34	PL 5mg	100%bw/d	Indoor/tank	Khannapa (1977)
P. monodon	40	PL 25mg-0.7g	100-10%bw/d	Indoor/tank	Khannapa (1977)
P. monodon	40	Juv. 1-3g	Fixed (?)	Indoor/tank	AQUACOP (1977) [n]
P. monodon	45.8	PL 0.5-1g	Fixed (?)	Indoor/tank	Lee (1971) [n]
Penaeus vannamei	36	Juv. 4-20g	Fixed (?)	Indoor/tank	Smith *et al.* (1985) [n]
P. vannamei	30-35	PL 32mg-0.5g	(?)	Indoor/tank	Colvin & Brand (1977)
Penaeus stylirostris	30-35	PL 45mg	(?)	Indoor/tank	Colvin & Brand (1977)
P. stylirostris	44	PL 5mg	(?)	Indoor/tank	Colvin & Brand (1977)
Penaeus californiensis	44	PL 5mg	(?)	Indoor/tank	Colvin & Brand (1977)
P. californiensis	30	Juv. 1g+	(?)	Indoor/tank	Colvin & Brand (1977)
Penaeus japonicus	52-57	PL 0.8g	Ad.lib.	Indoor/tank	Deshimaru & Yone (1978)
P. japonicus	40	Juv. 1-2g	Fixed (?)	Indoor/tank	Balazs *et al.* (1973) [n]
P. japonicus	54	PL 0.6-1g	Ad.lib.	Indoor/tank	Deshimaru & Kuroki (1974)
Palaemon serratus	30-40	PL 0.1-0.2g	Fixed (?)	Indoor/tank	Forster & Beard (1973) [n]

[1] Fish size range: fry 0-0.5g; fingerling 0.5-10g; juvenile 10-50g; grower 50g and above.

[2] Feeding regime: %bw/d - fixed feed intake expressed as a percentage of body weight per day, or *Ad lib.* 2-4 times daily.

[a] No difference in protein requirement at three stocking densities of 400, 600 and 800 fish/m^2, using $5m^3$ cages.

[b] $200m^2$ earthen ponds, fish density of 2/m^2, ponds also fertilized with poultry litter at a rate of 5kg/pond/week.
[c] Fish stocking density of 300/m^3.

[d/e] Fish stocking density of 9880/hectare.

[f] Plastic lined ponds, with fish stocking density of 3000-3700/hectare.

[g] Increased dietary protein requirement reported for fingerling striped bass from 47 to 55% with an increase in water temperature from 20.5 to 24.5°C.

[h] Feed intake fixed within all groups to the lowest recorded *Ad libitum* feed intake observed.

[i] Protein requirement said to increase from 40 to 45% with increasing salinity.

[j] Outdoor concrete ponds, 5 animals/m^2, infrequent water exchange, all animals fed at same fixed rate based on highest recorded intake.

[k] Outdoor fiberglass tanks, 17 animals/m^2, high water exchange.

[l] Animals housed within pens in earthen pond, 10 animals/m^2.

(continued on page 107)

(continued from page 106)

(m) All animals fed at fixed rate of 5mg feed/larvae/day (PL 1-10), 15mg feed/larvae/day (PL 11-50), and 20mg feed/larvae/day (PL 24-42).

(n) All animals fed to excess once or twice daily.

(o) Diet formulated to 55% crude protein, but actual level after diet processing was 45%.

(Source: A.G.J. Tacon, *Standard Methods for the Nutrition of Farmed Fish and Shrimp*; 1990 - with permission of Argent Laboratories, Inc.)

TABLE 6-7: CALCULATED DIETARY ESSENTIAL AMINO ACID REQUIREMENTS OF FISH AND SHRIMP AT VARYING DIETARY PROTEIN LEVELS

Values are expressed as a percent (%) of the dry diet

EAA	DIETARY PROTEIN LEVEL (%)							CARCASS EAA PATTERN
	25	30	35	40	45	50	55	
FISH (1)								
ARGININE	1.07	1.29	1.51	1.72	1.94	2.15	2.37	12.3
HISTIDINE	0.45	0.55	0.64	0.73	0.82	0.91	1.00	5.2
ISOLEUCINE	1.28	1.53	1.79	2.04	2.30	2.55	2.81	8.0
LEUCINE	1.28	1.53	1.79	2.04	2.30	2.55	2.81	14.6
LYSINE	1.49	1.77	2.07	2.37	2.66	2.96	3.25	16.9
METHIONINE	0.48	0.58	0.67	0.77	0.87	0.96	1.06	5.5
CYSTINE *	0.17	0.21	0.24	0.28	0.31	0.35	0.38	2.0
PHENYLALANINE	0.73	0.87	1.02	1.16	1.31	1.45	1.60	8.3
TYROSINE *	0.58	0.69	0.81	.092	1.04	1.15	1.27	6.6
THREONINE	0.80	0.97	1.13	1.29	1.45	1.61	1.77	9.2
TRYPTOPHAN	0.15	0.18	0.21	0.24	0.27	0.30	0.33	1.7
VALINE	0.83	1.00	1.16	1.33	1.50	1.66	1.82	9.5
SHRIMP (2)								
ARGININE	1.36	1.63	1.90	2.17	2.44	2.71	2.98	15.5
HISTIDINE	0.38	0.46	0.54	0.62	0.69	0.77	0.85	4.4
ISOLEUCINE	0.59	0.71	0.83	0.95	1.07	1.19	1.31	6.8
LEUCINE	1.22	1.47	1.71	1.96	2.20	2.45	2.69	14.0
LYSINE	1.29	1.54	1.80	2.06	2.31	2.57	2.83	14.7
METHIONINE	0.47	0.57	0.66	0.76	0.85	0.95	1.04	5.4
CYSTINE *	0.24	0.28	0.33	0.38	0.42	0.47	0.52	2.7
PHENYLALANINE	0.67	0.81	0.94	1.08	1.21	1.35	1.48	7.7
TYROSINE *	0.68	0.82	0.96	1.09	1.23	1.37	1.50	7.8
THREONINE	0.84	1.01	1.18	1.34	1.51	1.68	1.85	9.6
TRYPTOPHAN	0.24	0.28	0.33	0.38	0.42	0.47	0.52	2.7
VALINE	0.74	0.89	1.04	1.19	1.34	1.49	1.64	8.5

(1) Carcass EAA pattern of whole fish tissue (Wilson & Conwey, 1985)

(2) Carcass EAA pattern of short-necked clam (Deshimaru *et al.*, 1985)

* Non-essential amino acids

(Source: A.G.J. Tacon, *Standard Methods for the Nutrition of Farmed Fish and Shrimp*; 1990 - with permission of Argent Laboratories, Inc.)

TABLE 6-8: RECOMMENDED DIETARY NUTRIENT LEVELS FOR OMNIVOROUS FISH SPECIES [1]

NUTRIENT LEVEL	FISH SIZE CLASS [2]				
	FRY	FINGERLING	JUVENILE	GROWER	BROOD
CRUDE LIPID (% min)	8	7	7	6	5
FISH:PLANT LIPID [3]	1:1	1:1	1:1	1:1	1:1
CRUDE PROTEIN (% min)	42	39	37	35	37
AMINO ACIDS (% min) [4]					
ARGININE	1.81	1.68	1.59	1.51	1.59
HISTIDINE	0.76	0.71	0.67	0.64	0.67
ISOLEUCINE	1.18	1.09	1.04	0.98	1.04
LEUCINE	2.15	1.99	1.89	1.79	1.89
LYSINE	2.48	2.31	2.19	2.07	2.19
METHIONINE	0.81	0.75	0.71	0.67	0.71
CYSTINE	0.29	0.27	0.26	0.24	0.26
PHENYLALANINE	1.22	1.13	1.07	1.02	1.07
TYROSINE	0.97	0.90	0.85	0.81	0.85
THREONINE	1.35	1.26	1.19	1.13	1.19
TRYPTOPHAN	0.25	0.23	0.22	0.21	0.22
VALINE	1.40	1.30	1.23	1.16	1.23
CARBOHYDRATE (% max)	30	35	40	40	40
CRUDE FIBRE (% max)	1.5	2	3	4	4
MAJOR MINERALS (%)					
CALCIUM (% max)	2.5	2.5	2	2	2
PHOSPHORUS (% min)	1.0	0.8	0.8	0.7	0.8
MAGNESIUM (% min)	0.08	0.07	0.07	0.06	0.07
ADDED DIETARY SUPPLEMENTS					
TRACE MINERALS (mg/kg min)					
IRON	60	50	40	30	60
ZINC	100	83	67	50	100
MANGANESE	50	42	33	25	50
COPPER	6	5	4	3	6
COBALT	I	0.84	0.67	0.5	
IODINE	6	5	4	3	6
CHROMIUM	0.S	0.42	0.33	0.25	0.5
SELENIUM	0.2	0.17	0-13	0.10	0.2
VITAMINS (IU/kg min) [5]					
VITAMIN A	3000 (6000)	2500 (5000)	2000 (4000)	1500 (3000)	3000 (6000)
VITAMIN D3	1500 (3000)	1250 (2500)	1000 (2000)	750 (1500)	1500 (3000)
VITAMINS (mg/kg min)					
VITAMIN E	120 (240)	100 (200)	80 (160)	60 (120)	120 (240)
VITAMIN K	10 (12)	8 (10)	6 (8)	5 (6)	10 (12)
THIAMINE	18 (36)	15 (30)	12 (24)	9 (18)	18 (36)
RIBOFLAVIN	24 (48)	20 (40)	16 (32)	12 (24)	24 (48)
PYRIDOXINE	18 (36)	15 (30)	12 (24)	9 (18)	18 (36)
PANTOTHENIC ACID	48 (144)	40 (120)	32 (96)	24 (72)	48 (144)
NICOTINIC ACID	108 (216)	90 (180)	72 (144)	54 (108)	108 (216)
BIOTIN	0.2 (0.4)	0.15 (0.3)	0.1 (0.2)	0.1 (0.2)	0.2 (0.4)

TABLE 6-8 (cont.):

NUTRIENT LEVEL	FISH SIZE CLASS [2]				
	FRY	FINGERLING	JUVENILE	GROWER	BROOD
FOLIC ACID	3 (6)	2.5 (5)	2 (4)	1.5 (3)	3 (6)
VITAMIN B$_{12}$	0.015 (0.03)	0.0125 (0.025)	0.01 (0.02)	0.0075 (0.015)	0.015 (0.03)
VITAMIN C [6]	300 (900)	250 (750)	200 (600)	150 (450)	300 (900)
CHOLINE [6]	1200 (2400)	1000 (2000)	800 (1600)	600 (1200)	1200 (2400)
INOSITOL	150 (300)	125 (250)	100 (200)	75 (150)	150 (300)

[1] Dietary nutrient levels recommended for clear-water 'intensive' aquaculture systems (ie. tanks, cages, raceways)

[2] Fish size class: fry 0.0.5 g; fingerling 0.5-10g; juvenile 10-50 g; grower 50+; brood fish 1kg+ (varies with species)

[3] Equal proportions of plant and fish oil should be used so as to satisfy essential fatty acid requirements

[4] Amino acid requirement based on carcass essential amino acid pattern of whole fish tissue

[5] Suggested minimum dietary vitamin levels required to prevent deficiency signs. (Values in parentheses indicate suggested dietary vitamin levels taking into account processing, storage and leaching losses

[6] These vitamins should be added separately to the diet and not included in the form of a multivitamin premix; vitamin C Should be added in fat coated form so as to minimize losses through diet processing and leaching.

(Source: A.G.J. Tacon, *Standard Methods for the Nutrition of Farmed Fish and Shrimp*; 1990 - with permission of Argent Laboratories, Inc.)

TABLE 6-9: RECOMMENDED DIETARY NUTRIENT LEVELS FOR CARNIVOROUS FISH SPECIES [1]

NUTRIENT LEVEL	FISH SIZE CLASS [2]				
	FRY	FINGERLING	JUVENILE	GROWER	BROOD FISH
CRUDE LIPID (% min)	16	14	14	12	10
FISH:PLANT LIPID [3]	7:1	7:1	7:1	7:1	7:1
CRUDE PROTEIN (% min)	52	49	47	45	47
AMINO ACIDS (% min)[4]					
ARGININE	2.24	2.11	2.02	1.94	2.02
HISTIDINE	0.95	0.89	0.85	0.82	0.85
ISOLEUCINE	1.46	1.37	1.32	1.26	1.32
LEUCINE	2.66	2.50	2.40	2.30	2.40
LYSINE	3.08	2.90	2.78	2.66	2.78
METHIONINE	1.00	0.94	0.90	0.87	0.90
CYSTINE	0.36	0.34	0.33	0.31	0.33
PHENYLALANINE	1.51	1.42	1.36	1.31	1.36
TYROSINE	1.20	1.13	1.09	1.04	1.09
THREONINE	1.67	1.58	1.51	1.45	1.51
TRYPTOPHAN	0.31	0.29	0.28	0.27	0.28
VALINE	1.73	1.63	1.56	1.50	1.56
CARBOHYDRATE (% max)	15	20	25	25	25
CRUDE FIBER (% max)	1	1.5	2	2.5	2.5
MAJOR MINERALS (%)					
CALCIUM (% max)	2.5	2.5	2	2	2
PHOSPHORUS (% min)	1.0	0.8	0.8	0.7	0.8
MAGNESIUM (% min)	0.08	0.07	0.07	0.06	0.07

TABLE 6-9 (cont.):

NUTRIENT LEVEL	FISH SIZE CLASS [2]				
	FRY	FINGERLING	JUVENILE	GROWER	BROOD FISH
ADDED DIETARY SUPPLEMENTS					
TRACE MINERALS (mg/kg min)					
IRON	60	50	40	30	60
ZINC	100	83	67	50	100
MANGANESE	50	42	33	25	50
COPPER	6	5	4	3	6
COBALT	I	0.84	0.67	0.5	
IODINE	6	5	4	3	6
CHROMIUM	0.5	0.42	0.33	0.25	0.5
SELENIUM	0.2	0.17	0.13	0.10	0.2
VITAMINS IU/kg min [5]					
VITAMIN A	3000 (6000)	2500 (5000)	2000 (4000)	1500 (3000)	3000 (6000)
VITAMIN D3	1500 (3000)	1250 (2500)	1000 (2000)	750 (1500)	1500 (3000)
VITAMINS mg/kg min					
VITAMIN E	150 (300)	125 (250)	100 (200)	75 (150)	150 (300)
VITAMIN K	10 (12)	8 (10)	6 (8)	5 (6)	10 (12)
THIAMINE	24 (48)	20 (40)	16 (32)	12 (24)	24 (48)
RIBOFLAVIN	30 (60)	25 (50)	20 (40)	15 (30)	30 (60)
PYRIDOXINE	24 (48)	20 (40)	16 (32)	12 (24)	24 (48)
PANTOTHENIC ACID	60 (180)	50 (150)	40 (120)	30 (90)	60 (180)
NICOTINIC ACID	120 (240)	100 (200)	80 (160)	60 (120)	120 (240)
BIOTIN	0.3 (0.6)	0.25 (0.5)	0.2 (0.4)	0.2 (0.4)	0.3 (0.6)
FOLIC ACID	6 (12)	5 (10)	4 (8)	3 (6)	6 (12)
VITAMIN B12	0.03 (0.06)	0.025 (0.05)	0.02 (0.04)	0.015 (0.03)	0.03 (0.06)
VITAMIN C [6]	400 (1200)	333 (1000)	266 (800)	200 (600)	400 (1200)
CHOLINE [6]	1500 (3000)	1250 (2500)	1000 (2000)	750 (1500)	1500 (3000)
INOSITOL	250 (500)	200 (400)	150 (300)	100 (200)	250 (500)

[1] Dietary nutrient levels recommended for clear-water 'intensive' aquaculture systems (ie. tanks, cages, raceways)

[2] Fish size class: fry 0.0.5 g; fingerling 0.5-10g; juvenile 10-50 g; grower 50+; brood fish 1kg+ (varies with species)

[3] For strict marine carnivorous fish species, plant lipid can be omitted from the formulation.

[4] Amino acid requirement based on carcass essential amino acid pattern of whole fish tissue

[5] Suggested minimum dietary vitamin levels required to prevent deficiency signs. Values in parentheses indicate suggested dietary vitamin levels taking into account processing, storage and leaching losses

[6] These vitamins should be added separately to the diet and not included in the form of a multivitamin premix; vitamin C should be added in fat coated form so as to minimize losses through diet processing and leaching

(Source: A.G.J. Tacon, *Standard Methods for the Nutrition of Farmed Fish and Shrimp*; 1990 - with permission of Argent Laboratories, Inc.)

TABLE 6-10: RECOMMENDED DIETARY NUTRIENT LEVELS FOR OMNIVOROUS SHRIMP SPECIES

NUTRIENT LEVEL	SHRIMP SIZE CLASS (2)					
	LARVAL	PL 1-25	PL25 - 1GM	JUVENILE	GROWER	BROOD STOCK
CRUDE LIPID (% min)	14	13	12	11	10	10
MARINE:PLANT LIPID [3]	5:1	5:1	5:1	5:1	5:1,5.1	
CHOLESTEROL (4)	2	1.5	1.5	1.0	1.0	2
CRUDE PROTEIN (% min)	55	50	45	40	35	45
AMINO ACIDS (% min) [5]						
ARGININE	2.98	2.71	2.44	2.17	1.90	2.44
HISTIDINE	0.85	0.77	0.69	0.62	0.54	0.69
ISOLEUCINE	1.31	1 19	1.07	0 95	0 83	1 07
LEUCINE	2.69	2.45	2.20	1.96	1.71	2.20
LYSINE	2.83	2.57	2.31	2.06	1.80	2.31
METHIONINE	1.04	0.95	0.85	0.76	0.66	0.85
CYSTINE	0.52	0.47	0.42	0.38	0.33	0.42
PHENYLALANINE	1.48	1.35	1.21	1.08	0.94	1.21
TYROSINE	1.50	1.37	1.23	1.09	0.96	1.23
THREONINE	1.85	1.68	1.51	1.34	1.18	1.51
TRYPTOPHAN	0.52	0.47	0.42	0.38	0.33	0.42
VALINE	1.64	1.49	1.34	1.19	1.04	1.34
CARBOHYDR. (% max)	15	20	25	30	35	25
CRUDE FIBRE (% max) 6/	1	1.5	2	2	3	2
MAJOR MINERALS (%)						
CALCIUM (% max)	3	3	2.5	2.5	2	2.5
PHOSPHORUS (% min)	1.8	1.6	1.4	1.2	1.2	1.4
POTASSIUM (% min)	1.1	1.0	0.9	0.8	0.7	0.9
MAGNESIUM (% min)	0.18	0.15	0.13	0.10	0.08	0.13
ADDED DIETARY SUPPLEMENTS						
TRACE MINERALS (mg/kg min)						
IRON	100	90	80	70	60	100
ZINC	120	110	100	90	80	120
MANGANESE	60	55	50	45	40	60
COPPER	12	11	10	9	8	12
COBALT	1.2	1.1	1.0	0.9	0.8	1.2
IODINE	6	5.5	5	4.5	4	
CHROMIUM	1.0	0.9	0.8	0.7	0.6	1.0
SELENIUM	0.25	0.23	0.21	0.19	0.17	0.25
VITAMINS (IU/kg min) [7]						
VITAMIN A	6000 (12000)	5500 (11000)	5000 (10000)	4500 (9000)	4000 (8000)	6000 (12000)
VITAMIN D3	2000 (4000)	1800 (3600)	1600 (3200)	1400 (2800)	1200 (2400)	2000 (4000)
VITAMINS (mg/kg min)						
VITAMIN E	200 (400)	180 (360)	160 (320)	140 (280)	120 (240)	200 (400)
VITAMIN K	12 (14)	11 (13)	10 (12)	9 (11)	8 (10)	12 (14)
THIAMINE	30 (90)	28 (84)	26 (78)	24 (72)	22 (66)	30 (90)

TABLE 6-10 (cont.):

NUTRIENT LEVEL	LARVAL	PL1-25	SHRIMP SIZE CLASS (2) PLP25-1g	JUVENILE	GROWER	BROOD
PYRIDOXINE	30 (90)	28 (84)	26 (78)	24 (72)	22 (66)	30 (90)
PENTOTHENIC ACID	75 (300)	70 (280)	65 (260)	60 (240)	55 (220)	75 (300)
NICOTINIC ACID	150 (450)	140 (420)	130 (390)	120 (360)	110 (330)	150 (450)
BIOTIN	0.25 (0.75)	0.23 (0.69)	0.21 (0.63)	0.19 (0.57)	0.17 (0.54)	0.25 (0.75)
FOLIC ACID	6 (18)	5.5 (16.5)	5 (15)	4.5 (13.5)	4 (12)	6 (18)
VITAMIN B_{12}	0.04 (0.12)	0.037 (0.111)	0.034 (0.102)	0.031 (0.093)	0.028 (0.084)	0.04 (0.12)
VITAMIN C	500 (2500)	450 (2250)	400 (2000)	350 (1750)	300 (1500)	500 (2500)
CHOLINE	1600 (3200)	1500 (3000)	1400 (2800)	1300 (2600)	1200 (2400)	1600 (3200)
INOSITOL	700 (2100)	650 (1950)	600 (1800)	550 (1650)	500 (1500)	700 (2100)

See comments after TABLE 6-11 for further explanation.

TABLE 6-11: RECOMMENDED DIETARY NUTRIENT LEVELS FOR CARNIVOROUS SHRIMP SPECIES [1]

NUTRIENT LEVEL	LARVAL	PL1-25	SHRIMP SIZE CLASS (2) PL25-1g	JUVENILE	GROWER	BROOD
CRUDE LIPID (% min)	16	13	12	11	10	10
MARINE:PLANT LIPID [3]	6:1	6:1	6:1	6:1	6:1	6:1
CHOLESTEROL (% min) [4]	2	1.5	1.5	I	1	2
CRUDE PROTEIN (% min)	65	62	59	55	53	55
AMINO ACIDS (% min) [5]						
ARGININE	3.53	3.36	3.20	2.98	2.87	2.98
HISTIDINE	1.00	0.95	0.91	0.85	0.82	0.85
ISOLEUCINE	1.55	1.48	1.40	1.31	1.26	1.31
ISOLEUCINE	3.18	3.04	2.89	2.69	2.60	2.69
LYSINE	3.34	3.19	3.03	2.83	2.73	2.83
METHIONINE	1.23	1.17	1.11	1.04	1.00	1.04
CYSTINE	0.61	0.59	0.56	0.52	0.50	0.52
PHENYLALANINE	1.75	1.67	1.59	1.48	1.43	1.48
TYROSINE	1.77	1.69	1.61	1.50	1.45	1.50
THREONINE	2.18	2.08	1.98	1.85	1.78	1.85
TRYPTOPHAN	0.61	0.59	0.56	0.52	0.50	0.52
VALINE	1.93	1.84	1.75	1-64	1-58	1-64
CARBOHYDRATE (%max)	5	10	15	15	20	15
CRUDE FIBRE (% max) [6]	1	1	1.5	1.5	2	1.5
MAJOR MINERALS (%)						
CALCIUM (% max)	3	3	2.5	2.5	2	2.5
PHOSPHORUS (% min)	1.8	1.6	1.4	1.2	1.2	1.4
POTASSIUM (% min)	1.1	1.0	0.9	0.8	0.7	0.9
MAGNESIUM (% min)	0.18	0.15	0.13	0.10	0.08	0.13
ADDED DIETARY SUPPLEMENTS						
TRACE MINERALS (mg/kg min)						
IRON	100	90	80	70	60	100
ZINC	120	110	100	90	80	120
MANGANESE	60	55	50	45	40	60

TABLE 6-11 (cont.):

NUTRIENT LEVEL	SHRIMP SIZE CLASS [2]					
	LARVAL	PL1-25	PL25-1g	JUVENILE	GROWER	BROOD
COPPER	12	11	10	9	8	12
COBALT	1.2	1.1	1.0	0.9	0.8	1.2
IODINE	6	5.5	5	4.5	4	
CHROMIUM	1.0	0.9	0.8	0.7	0.6	1.0
SELENIUM	0.25	0.23	0.21	0.19	0.17	0.25
VITAMINS (IU/kg min) [7]						
VITAMIN A	6000 (12000)	5500 (11000)	5000 (10000)	4500 (9000)	4000 (8000)	6000 (12000)
VITAMIN D$_3$	2000 (4000)	1800 (3600)	1600 (3200)	1400 (2800)	1200 (2400)	2000 (4000)
VITAMINS (mg/kg min)						
VITAMIN E	200 (400)	180 (360)	160 (320)	140 (280)	120 (240)	200 (400)
VITAMIN K	12 (14)	11 (13)	10 (12)	9 (11)	8 (10)	12 (14)
THIAMINE	30 (90)	28 (84)	26 (78)	24 (72)	22 (66)	30 (90)
PYROXIDINE	30 (90)	28 (84)	26 (78)	24 (72)	22 (66)	30 (90)
PENTOTHENIC ACID	75 (300)	70 (280)	65 (260)	60 (240)	55 (220)	75 (300)
NICOTINIC ACID	150 (450)	140 (420)	130 (390)	120 (360)	110 (330)	150 (450)
BIOTIN	0.25 (0.75)	0.23 (0.69)	0.21 (0.63)	0.19 (0.57)	0.17 (0.54)	0.25 (0.75)
FOLIC ACID	6 (18)	5.5 (16.5)	5 (15)	4.5 (13.5)	4 (12)	6 (18)
VITAMIN B$_{12}$	0.04 (0.12)	0.037 (0.111)	0.034 (0.102)	0.031 (0.093)	0.028 (0.084)	0.04 (0.12)
VITAMIN C	500 (2500)	450 (2250)	400 (2000)	350 (1750)	300 (1500)	500 (2500)
CHOLINE	1600 (3200)	1500 (3000)	1400 (2800)	1300 (2600)	1200 (2400)	1600 (3200)
INOSITOL	700 (2100)	650 (1950)	600 (1800)	550 (1650)	500 (1500)	700 (2100)

[1] Dietary nutrient levels recommended for clear-water 'intensive' aquculture systems (ie. tanks, cages and raceways)

[2] Shrimp size class: Larval - protozoea substage 1 to post-larval substage 1(PL1); PL1 to PL25 (25 days from PL1); PL25 to 1 g; juvenile - 1 g to 10 g; grower 10g to harvest; broodstock - 10 g+

[3] Marine lipid includes shrimp head oil, marine fish body oil, marine fish liver oil, or marine invertebrate oils.

[4] Cholesterol can be added either in purified form or by using dietary lipid sources naturally rich in cholesterol such as shrimp head oil.

[5] Amino acid requirement based on the carcass essential amino acid pattern of the short-necked clam.

[6] Maximum limit refers only to crude fiber of plant origin, and excludes crude fiber derived from shrimp meal (ie. chitin)

[7] Suggested minimum dietary vitamin levels required to prevent deficiency signs. Values in parentheses indicate suggested dietary vitamin levels taking into account processing, storage and leaching losses; the latter being 2 to 5 times greater than the recommended dietary requirements due to the extremely slow and extended feeding habits of marine shrimp, and to compensate for the considerable losses of vitamins which occur through leaching. However, the actual multiplication factor used should be adjusted in situ (ie. either higher or lower) depending on the water stability of the diet and the feeding response of the shrimp to the diet, and finally on the time period the feed remains in the water before total consumption.

(Source: A.G.J. Tacon, *Standard Methods for the Nutrition of Farmed Fish and Shrimp*; 1990 - with permission of Argent Laboratories, Inc.)

TABLE 6-12: AVERAGE PROXIMATE COMPOSITION OF THE MAJOR OILSEEDS AND THEIR BY-PRODUCTS.

All values are expressed as % by weight on an as-fed basis: Water-H₂O; Crude Protein-CP; Lipid or Ether Extract-EE; Crude Fiber-CF; Nitrogen-Free Extractives-NFE; Ash; Calcium-Ca; Phosphorus-P) [1]

OILSEED CROP	AVERAGE COMPOSITION (% BY WEIGHT)								
	H₂O	CP	EE	CF	NFE	Ash	Ca	P	NO. REF. [1]
ALMOND (*Prunus amygdalus/P. dulcis*)									
Seed (kernel)	4.9	18.9	55.3	2.2	15.8	2.9	0.26	0.50	(2)
Hulls	9.7	3.2	3.3	13.9	64.0	5.9	0.21	0.10	(3)
Oilcake (mechanically extr.; without hulls)	10.7	42.8	3.8	3.7	33.3	5.7			
BABASSU (*Orbignya speciosa/O. martiana*)									
Seed (kernel)	4.9	18.9	55.3	2.2	15.8	2.9	0.26	0.50	(2)
Oilcake (mechanically extracted)	9.5	20.2	6.1	14.5	44.8	4.9	0.16	0.85	(2)
Oilmeal (solvent extracted)	7.3	19.7	2.3	17.6	46.8	6.3	0.12	0.66	(1)
CASHEW (*Anacardium occidentale*)									
Seed (kernel)	5.0	21.4	46.8	1.6	21.6	3.6	0.04	0.86	(2)
Oimeal (solvent extracted)	7.0	41.9	1.3	2.8	41.0	6.0	0.06	1.68	(2)
CASTOR (*Ricinus communis*)									
Oilmeal (solvent extracted)	8.0	35.4	0.9	29.7	19.5	6.5	0.70	0.80	(2)
Hulls	8.1	26.5	2.0	28.8	28.2	6.4	0.50	0.60	(1)
COCOA (*Theobroma cacao*)									
Bean (seed kernel; fresh)	52.8	6.7	20.2	4.2	13.9	2.2	-	-	(1)
Bean (seed kernel; dried)	10.4	13.1	35.7	6.6	30.7	3.5	0.07	0.33	(2)
Shell (pericarp seed testa; dried)	9.3	18.8	7.0	13.5	43.5	7.9	0.15	0.21	(2)
Pods (without beans; fresh)	85.1	1.2	0.1	4.3	8.0	1.3	-	-	(3)
Pods (without beans; dried)	11.5	5.8	0.7	21.5	52.9	7.6	0.17	0.07	(4)
Oilcake (mechanically extracted)	11.4	23.1	5.3	8.9	46.0	5.3	0.14	0.68	(2)
COCONUT (*Cocus nucifera*)									
Kernel (endosperm; fresh)	47.9	4.2	34.0	2.6	9.8	1.5	0.01	0.13	(3)
Kernel (endosperm-meat-copra; dried)	4.0	7.2	64.6	3.8	18.5	1.9	0.03	0.19	(1)
Oilmeal (solvent extracted)	8.0	21.0	1.5	14.0	49.5	6.0	0.18	0.60	(5)
Oilcake (mechanically extracted)	8.5	20.8	6.3	12.0	45.4	7.0	0.19	0.60	(10)
Coir dust (husk processing dust)	12.9	2.0	0.6	29.8	48.1	6.6	-	-	(1)
CONOPHOR/AWUSA NUT (*Tetracarpidium conophorum*)									
Seed (kernel; whole)	4.9	22.7	56.0	3.7	9.1	3.6	0.45	0.33	(1)
Seed (kernel; without cotyledons)	4.2	20.8	57.7	3.5	10.3	3.5	0.53	0.39	(1)
Cotyledons	5.1	9.9	21.0	0.1	59.2	4.7	0.15	0.45	(1)
Oilcake (hydraulcally extracted; cooked)	3.5	31.5	27.2	5.9	27.2	4.7	-	-	(1)
Oilmeal (solvent extracted; cooked)	3.2	42.6	17.9	6.4	24.6	5.3	-	-	(1)
COTTON (*Gossypium* spp.)									
Seed (kernel; whole)	7.9	20.4	20.0	21.1	26.3	4.3	0.14	0.64	(8)
Hulls	9.6	4.2	1.9	44.5	37.3	2.5	0.14	0.09	(6)
Oicake (with hulls-undec.; mechan. extr.)	10.7	21.9	4.9	21.9	34.9	5.7	-	-	(3)
Oilcake (without hulls-dec.; mechan. extr.)	7.8	41.2	5.9	11.1	27.6	6.4	0.19	1.06	(12)
Oilmeal (dec.; solvent extr.; 41% protein)	9.8	41.7	1.5	11.3	28.8	6.9	0.16	1.09	(8)
Oilmeal (dec.; solvent extr.; 50% protein)	7.5	50.0	1.6	8.2	26.2	6.5	0.17	1.08	(2)
CRAMBE (*Crambe abyssinica*)									
Seed (kernel)	12.0	13.6	48.4	9.5	13.9	2.6	-	-	(1)
Oilmeal (solvent extracted)	12.0	29.9	1.1	20.9	30.0	6.1	-	--	(1)
GROUNDNUT/PEANUT (*Arachis hypogaea*)									
Seed (kernel)	12.0	13.6	48.4	9.5	13.9	2.6	-	-	(1)
Seed (kernel; with hull-undec.)	7.1	20.2	36.3	19.6	14.3	2.5			(1)
Seed (kernel; without hulls-dec.)	6.5	28.4	44.7	2.2	15.9	2.3	0.07	0.39	(5)
Hulls	11.4	6.2	1.6	54.3	21.4	5.1	1.10	0.91	(4)

TABLE 6-12 (cont.)

AVERAGE COMPOSITION (% BY WEIGHT)									
OILSEED CROP	H$_2$O	CP	EE	CF	NFE	Ash	Ca	P	NO. REF. [1]
Oilcake (undec.; mechanically extracted)	10.0	30.2	9.1	23.0	22.0	5.7	-	-	(2)
Oilmeal (undec.; solvent extracted)	7.8	31.7	1.9	25.2	29.1	4.3	-	-	(2)
Oilcake (dec.; mechanically extracted)	9.6	46.2	6.7	7.5	24.8	5.2	0.14	0.60	(12)
Oilmeal (dec.; solvent extracted)	8.7	48.7	1.1	7.7	27.8	6.0	0.25	0.59	(8)
HEMP (*Cannabis sativa*)									
Seed (kernel)	8.9	18.2	32.6	15.0	21.1	4.2	-	-	(2)
Oicake (mechanically extracted)	8.1	31.8	6.2	24.0	21.1	8.8	0.24	0.45	(2)
Oilmeal (solvent extracted)	11.3	34.8	1.7	26.6	16.3	9.3			(1)
KAPOK/SILK COTTON TREE (*Eriodendron anfractuosum/Ceiba pentandra*)									
Seed (kernel)	8.0	28.0	21.3	18.0	17.2	7.5	0.43	0.89	(1)
Oilcake (mechanically extracted)	11.9	27.5	6.7	23.7	23.3	6.9	-	-	(2)
LINSEED/FLAX (*Linum usitatissimum*)									
Seed (kernel)	6.7	25.6	34.7	5.4	22.9	4.7	0.20	0.54	(2)
Hulls	9.0	7.7	1.4	28.7	43.7	9.5	-	-	(1)
Oilcake (mechanically extracted)	9.0	16.2	9.6	12.5	46.0	6.7	0.35	0.43	(2)
Oilmeal (solvent extracted)	10.7	34.1	1.3	9.1	38.9	5.9	0.37	0.80	(6)
MUSTARD (*Brassica* spp.)									
Seed (kernel)	8.3	21.5	42.8	7.9	14.8	4.7	0.03	0.55	(1)
Oilcake (mechanically extracted)	9.5	36.3	8.9	8.1	29.0	8.2	1.24	1.13	(3)
NIGER (*Guizotia abyssinica*)									
Seed (kernel)	5.0	23.0	38.0	16.0	13.0	5.0	-	-	(1)
Oilcake (mechanically extracted)	10.2	31.4	5.7	16.9	25.6	10.2	0.08	0.73	(2)
Oilmeal (solvent extracted)	7.8	31.4	1.1	20.6	28.0	11.1	-	—	(1)
OLIVES (*Olea europaea*)									
Seed (kernel)	8.0	1.1	0.7	68.2	20.9	1.1	-	-	(1)
Pulp with seed (dried)	8.0	5.9	15.5	36.5	31.6	2.5	-	-	(1)
Pulp (dried)	5.0	13.9	27.4	19.3	31.0	3.4	-	-	(1)
Pulp (solvent extracted)	7.8	11.4	3.1	28.5	43.1	6.1	0.31	0 11	(2)
Oilcake (kernel plus pulp)	14.8	5.4	10.1	34.1	32.0	3.6	-	-	(1)
AFRICAN OIL PALM (*Elaeis guineesis*)									
Seed (kernel)	7.2	9.4	47.8	5.1	28.6	1.9	0.08	0.28	(2)
Oilcake (mechanically extracted)	10.5	17.7	9.7	14.7	43.5	3.9	0.20	0.49	(9)
Oilmeal (solvent extracted)	9.7	18.8	1.5	21.5	44.5	4.0	0.28	0.77	(3)
Press fiber bunch (fresh)	34.5	4.5	7.7	21.0	28.1	4.2	0.20	0.09	(2)
Press fiber bunch (dried)	13.8	4.8	18.1	31.4	24 2	7 7	0 27	0 11	(1)
Palm oil sludge (dried)	10.2	9.4	18.1	10.8	40 5	11 0	0 36	0 47	(2)
PARA RUBBER TREE (*Hevea brasiliensis*)									
Seed (kernel;decorticated)	7.6	21.7	39.0	2.8	25 8	3.1	-	-	(1)
Oilcake (undecor.; mechanically extracted)	8.1	13.2	4.4	42.8	29.0	2.5	-	-	(1)
Oilcake (decor.; mechanically extracted)	9.3	24.2	3.5	9.8	47.4	5.8	0.11	0.43	(1)
POPPY (*Papaver somniferum*)									
Seed (kernal)	7.2	20.9	32.5	8.6	23.3	7.5	-	-	(1)
Oilcake (mechanically extracted)	9.6	37.1	7.3	12.7	20.9	12.4	-	-	(3)
Oilmeal (solvent extracted)	11.3	36.0	1.0	16.0	22.5	13.2	-	-	(2)
RAPE (*Brassica napus, B campestris*)									
Seed (kernel)	8.2	19.2	42.0	25.5	1.4	3.7	0.30	0.60	(2)
Oilmeal (mechanically extracted)	8.1	34.1	7.9	12.8	30.6	6.5	0.75	1.07	(8)
Oilmeal (solvent extracted)	9.0	37.3	1.9	11.4	33.2	7.2	0.64	0.97	(8)
SAFFLOWER (*Carthamus tinctorius*)									
Seed (kernel)	7.0	17.1	31.1	27.6	14.3	2.9	0.24	0.62	(2)
Hulls	8.7	3.5	4.1	53.1	29.1	1.5	-	-	(2)

115

TABLE 6-12 (cont.)

AVERAGE COMPOSITION (% BY WEIGHT)									
OILSEED CROP	H$_2$O	CP	EE	CF	NFE	Ash	Ca	P	NO. REF. [1]
Oilmeal (undecor.; mechanically extracted)	8.1	21.7	5.9	30.7	29.3	4.3	0.22	0.64	(6)
Oilmeal (undecor.; solvent extracted)	8.7	22.5	1.0	32.8	30.2	4.8	0.34	0.80	(3)
Oilmeal (decor.; mechanically extracted)	8.6	41.1	6.4	13.6	23.0	7.3	0.45	1.00	(3)
Oilmeal (decor.; solvent extracted)	9.1	42.3	1.2	14.6	25.6	7.2	0.38	1.28	(3)
SESAME (*Sesamum orientale/S. radiatum*)									
Seed (kernel)	7.0	21.1	46.5	7.6	12.2	5.6	0.90	0.75	(3)
Oilcake (mechanically extracted)	8.0	40.4	10.6	6.4	24.2	10.4	2.10	1.20	(14)
SOYBEAN (*Glycine max*)									
Seed (kernel; with hulls undecor.)	8.8	24.1	10.0	17.3	33.2	6.6	-	-	(1)
Hulls	9.1	9.8	1.7	36.4	38.1	4.9	0.49	0.18	(3)
Seed (kernel; without hulls decor.)	9.1	37.8	17.8	4.9	25.6	4.8	0.25	0.59	(8)
Oilcake (undecor.; mechanically extracted)	11.0	41.6	5.3	5.9	30.1	6.1	0.20	0.61	(7)
Oilmeal (undecor.; solvent extracted)	11.6	44.4	1.2	6.1	30.6	6.1	0.28	0.62	(8)
Oilmeal (decor; solvent extracted)	10.4	49.0	0.8	3.0	30.9	5.9	0.25	0.63	(5)
Protein concentrate meal	8.0	84.3	0.5	0.1	3.6	3.5	0.11	0.68	(1)
Mill run	12.0	11.9	1.2	35.8	34.6	4.5	0.37	0.18	(2)
Mill feed (flour by-product)	10.3	12.9	1.7	32.5	37.9	4.7	0.41	0.18	(2)
SUNFLOWER (*Hellanthus annus*)									
Seed (kernel; with hulls-undecor.)	6.1	14.2	32.6	27.6	16.5	3.0	0.19	0.54	(2)
Seed (kernel; without hull-decor.)	5.0	25.7	44.2	5.0	16.3	3.8	0.16	0.88	(1)
Hulls	12.0	3.1	1.6	65.0	13.0	5.3	-	-	(1)
Sunflower (head with seed)	9.5	13.1	12.6	23.4	32.8	8.6	-	-	(2)
Sunflower (head without seed)	10.0	8.2	3.7	19.4	47.7	11.0	-	-	(2)
Oilcake (undecor.; mechanically extracted)	7.3	31.6	8.9	24.0	21.8	6.4	-	-	(2)
Oilmeal (undecor.; solvent extracted)	9.7	30.8	1.5	24.8	26.9	6.3	0.26	1.16	(4)
Oilcake (decor.; mechanically extracted)	7.8	37.1	9.3	12.3	27.2	6.3	0.36	1.08	(5)
Oilmeal (decor.; solvent extracted)	7.7	43.4	2.5	12.5	26.9	7.0	0.39	0.97	(4)

[1] The data presented represents the mean values from various sources, including Allen (1984); Bolton and Blair (1977); Branckaert, Tessema and Temple (1976); Capper, Wood and Jackson (1982); Cooley (1976); Devendra (1979); Fetuga Babatunde and Oyenuga (1975); Gohl (1981); Gogus (1975); Godin and Spensley (1971); Hastings (1973); Hickling (1971); Janseen (1935); Ling (1967); Miller (1976); NRC (1982, 1983); Oyenuga (1975); Platt (1962); Springhall (1969); and Stosic and Kaykay (1981).

(Source: A.G.J. Tacon, *Standard Methods for the Nutrition of Farmed Fish and Shrimp*; 1990 - with permission of Argent Laboratories, Inc.)

TABLE 6-13: AVERAGE ESSENTIAL AMINO ACID (EAA) COMPOSITION OF THE MAJOR OILSEEDS AND THEIR BY-PRODUCTS

[1] All values are expressed as % by weight on a as-fed basis: Arginine-ARG; Cystine-CYT; Methionine-MET; Threonine-THR; Isoleucine-ISO; Leucine-LEU; Lysine-LYS; Valine-VAL; Tyrosine-TYR; Tryptophan-TRY; Phenylalanine-PHE; Histidine-HIS

AVERAGE EAA COMPOSITION (%)												
OILSEED/BY-PRODUCT	ARG	CYT	MET	THR	ISO	LEU	LYS	VAL	TYR	TRY	PHE	HIS	REF. NO.
ALMOND (*P. amygdalus/P. dulcis*)												
Seed (kernel)	1.98	0.17	0.52	0.49	0.70	1.27	0.45	1.05	0.59	0.17	0.97	0.45	(1)
CASHEW (*A. occidentale*)												
Seed (kernel)	2.13	-	0.31	0.67	1.06	1.68	0.94	1.19	-	0.38	0.87	0.42	(1)
Oilmeal: extracted	4.30	0.72	0.58	1.30	1.60	2.75	1.65	2.30	1.08	0.56	1.60	0.84	(2)
COCONUT (*C. nucifera*)												
Kernel (endosperm): dried	1.03	0.09	0.15	0.26	0.31	0.52	0.27	0.42	0.21	0.08	0.35	0.16	(1)

TABLE 6-13 (cont.):

OILSEED/BY-PRODUCT	AVERAGE EAA COMPOSITION (%)												NO. REF.
	ARG	CYT	MET	THR	ISO	LEU	LYS	VAL	TYR	TRY	PHE	HIS	
Oilcake: mechanically extracted	2.31	0.26	0.32	0.65	0.80	1.33	0.58	1.00	0.50	0.19	0.82	0.35	(3)
Oilmeal: solvent extracted	2.41	0.25	0.32	0.66	0.83	1.44	0.60	1.04	0.57	0.20	0.86	0.38	(1)
CONOPHOR/AWUSA NUT (*T. conophorum*)													
Oilseed meal (Amino acid g/16gN)	9.45	2.71	2.62	5.30	4.41	6.79	3.74	4.94	4.84	4.45	3.03	2.17	(1)
COTTON (*Gossypium* spp.)													
Seed (kernel): whole	2.67	0.37	0.31	0.78	0.78	1.41	1.05	1.10	0.69	0.30	1.24	0.65	(1)
Oilcake: decor. mechanically extr.	4.1S	0.72	0.59	1.33	1.45	2.42	1.58	2.11	1.17	0.55	2.15	1.00	(5)
Oilmeal: decor. solvent extracted	4.57	0.77	0.59	1.42	1.41	2.35	1.71	1.93	0.92	0.54	2.33	1.13	(4)
GROUNDNUT/PEANUT (*A. hypogaea*)													
Oilcake: decor. mechanically extr.	4.79	0.70	0.47	1.29	1.65	3.17	1.50	2.22	1.69	0.44	2.42	1.09	(3)
Oilmeal: decor. solvent extracted	4.67	0.72	0.42	1.27	1.84	3.03	1.71	2.19	1.51	0.49	2.26	1.03	(3)
LINSEED/FLAX (*L. usitatissimum*)													
Seed (kernel)	2.03	0.41	0.42	0.81	0.92	1.30	0.81	1.15	0.58	0.33	1.02	0.44	(1)
Oilcake: mechanlcally extracted	2.86	0.49	0.54	1.18	1.65	1.95	1.17	1.67	0.85	0.52	1.44	0.62	(3)
Oilmeal: solvent extracted	2.82	0.59	0.51	1.21	1.74	2.01	1.13	1.67	1.09	0.50	1.48	0.69	(2)
MUSTARD (*Brassica* spp.)													
Oilcake: mechanically extracted	2.12	0.92	0.82	1.67	1.62	2.46	3.64	1.90	-	0.48	1.43	0.93	(2)
NIGER (*G. abyssinica*)													
Seed (kernel)	1.73	0.34	0.31	0.79	0.86	1.34	0.90	0.74	0.35	-	0.87	0.48	(1)
OIL PALM (*E. guineesis*)													
Seed (kernel)	1.16	0.15	0.20	0.27	0.30	0.52	0.30	0.47	0.23	0.08	0.32	0.18	(1)
Oilmeal (kernel): solvent extr.	2.36	0.28	0.33	0.61	0.64	1.19	0.54	0.82	0.47	0.20	0.79	0.32	(1)
PARA RUBBER (*H. brasiliensis*)													
Seed (kernel): decor. roasted	1.96	0.24	0.43	0.58	0.51	1.06	0.64	1.13	0.41	-	0.65	0.35	(1)
RAPE (*B. campestris/B. napus*)													
Oilcake: mechanically extracted	1.93	0.35	0.68	1.51	1.38	2.40	1.68	1.76	0.85	0.42	1.39	0.90	(2)
Oilmeal: solvent extracted	2.11	0.43	0.70	1.61	1.41	2.55	2.12	1.83	0.80	0.44	1.43	1.00	(4)
SAFFLOWER (*C. tinctorius*)													
Seed (kernel)	1.28	0.21	0.14	0.45	0.61	0.92	0.46	0.72	0.35	0.14	0.53	0.3S	(1)
Oilcake: undecor. mechanically extr.	1.29	0.65	0.40	0.52	0.42	1.16	0.69	1.05	-	0.30	1.08	0.46	(2)
Oilmeal: undecor. solvent extracted	1.93	0.36	0.34	0.51	0.28	1.20	0.71	1.00	-	0.27	1.00	0.50	(2)
Oilcake: decor. mechanically extr.	2.75	0.70	0.79	1.36	1.60	2.44	1.04	2.15	0.95	0.68	1.75	0.91	(1)
Oilmeal: decor. solvent extracted	3.68	0.70	0.69	1.33	1.63	2.48	1.29	2.33	1.07	0.60	1.80	1.04	(2)
SESAME (*S. orientale/S. radiatum*)													
Seed (kernel)	2.59	0.39	0.60	0.76	0.77	1.43	0.58	0.98	0.67	0.29	0.95	0.52	(1)
Oilcake: mechanically extracted	4.75	0.71	1.33	1.63	1.98	3.15	1.22	2.26	1.71	0.61	2.12	1.08	(3)
SOYBEAN (*G. max*)													
Seed: without hulls (decort.)	2.82	0.55	0.54	1.68	2.16	2.79	2.41	2.03	1.12	0.53	2.08	1.00	(3)
Hulls	0.59	0.07	0.12	1.29	0.30	0.71	0.64	0.36	0.39	0.07	0.41	0.25	(1)
Oilcake: undecor. mechanically extr.	3.14	0.59	0.63	1.71	2.72	3.71	2.75	2.24	1.55	0.63	2.15	1.12	(2)
Oilmeal: undecor. solvent extracted	3.48	0.71	0.59	1.62	2.14	3.12	2.76	2.27	1.33	0.61	2.00	1.12	(3)
Oilmeal: decor. solvent extracted	3.74	0.73	0.70	1.94	2.36	3.71	3.12	2.45	1.77	0.60	2.54	1.23	(3)
Protein concentrate meal	7.34	0.92	0.88	3.34	4.60	6.33	5.61	4.38	3.10	0.88	4.33	2.41	(1)
Mill feed (flour by-product)	0.75	0.14	0.13	0.30	0.41	0.58	0.65	0.38	0.23	0.13	0.38	0.18	(1)
SUNFLOWER (*H. annug*)													
Seed (kernel) with hulls (undecor.)	1.19	0.22	0.28	0.55	0.64	0.95	0.54	0.75	0.28	0.20	0.66	0.35	(1)
Oilcake: decor. mechanically extr.	4.00	0.78	1.24	1.57	2.06	2.67	1.79	2.29	1.12	0.56	2.17	1.08	(3)
Oilmeal: undecor. solvent extracted	2.57	0.53	0.55	1.14	1.13	1.79	1.17	1.51	-	0.33	1.25	0.99	(3)
Oilmeal: decor. solvent extracted	3.96	0.72	1.33	1.72	2.17	3.21	1.81	2.45	1.39	0.56	2.28	1.12	(2)

(continued on page 118)

117

(continued from page 117)

[1] The data presented represents the mean values from various sources, including: Allen (1984); Bolton and Blair (1977); Capper, Wood and Jackson (1982); Fetuga, Babatunde and Oyenuga (1975); FAO (1970); NRC (1982, 1983); Oyenuga (1975); and Stosic and Kaykay (1981).

(Source: A.G.J. Tacon, *Standard Methods for the Nutrition of Farmed Fish and Shrimp;* 1990 - with permission of Argent Laboratories, Inc.)

TABLE 6-14: AVERAGE FATTY ACID COMPOSITION OF THE MAJOR OILSEED AND PLANT BY-PRODUCT OILS

(Individual fatty acid values are expressed as a percentage of the total fatty acids present) [1]

PLANT LIPID	FATTY ACID	% TOTAL FATTY ACIDS
CASTOR OIL	Ricinoleic acid	91 - 95
	Linoleic acid	4 - 5
	Palmitic & stearic acid	1 - 2
COCOA BUTTER	Oleic acid	30 - 40
	Stearic acid	35
	Palmitic acid	25
	Linoleic acid	2 - 4
COCONUT OIL	Lauric acid	44 - 52
	Myristic acid	13 - 19
	Palmitic acid	7 - 10
	Caprylic acid	5 - 10
	Capric acid	4 - 10
	Oleic acid	5 - 8
	Stearic acid	1 - 3
	Linoleic acid	1 - 3
CONOPHOR SEED OIL	Linolenic acid	65 - 72
	Oleic acid	10 - 13
	Linoleic acid	10 - 12
	Stearic acid	3
	Palmitic acid	2 - 3
COTTON SEED OIL	Linoleic acid	40 - 55
	Palmitic acid	20 - 25
	Oleic acid	18 - 30
	Stearic acid	2 - 7
GROUNDNUT OIL	Oleic acid	42 - 72
	Linoleic acid	13 - 28
	Palmitic acid	6 - 12
	Arachidic & higher sat. acids.	5 - 7
	Stearic acid	2 - 4
HEMPSEED OIL	Linoleic acid	45 - 65
	Linolenic acid	15 - 30
	Oleic acid	14 - 16
	Saturated acids	4 - 10
KAPOK SEED OIL	Linoleic acid	33
	Oleic acid	30
	Palmitic acid	23
	Stearic acid	1
	Cyclopropene bearing fatty acids	13
LINSEED OIL	Linolenic acid	30 - 60
	Oleic acid	13 - 36
	Linoleic acid	10 - 25
	Stearic & palmitic acids	6 - 16
MAIZE (CORN) OIL	Linoleic acid	34 - 62
	Oleic acid	19 - 49
	Palmitic acid	8 - 12
	Stearic acid	2 - 5
NEEM OIL	Oleic acid	49 - 62
	Stearic acid	14 - 19
	Palmitic acid	14 - 15
	Linoleic acid	8 - 16

TABLE 6-14 (cont.)

PLANT LIPID	FATTY ACID	% TOTAL FATTY ACIDS
NIGER SEED OIL	Linoleic acid	51 - 55
	Oleic acid	31 - 39
	Saturated acids	9 - 17
PALM OIL	Oleic acid	40 - 53
	Palmitic acid	32 - 47
	Linoleic acid	2 - 11
	Stearic acid	1 - 9
	Myristic acid	1 - 3
PALM KERNEL OIL	Lauric acid	46 - 52
	Myristic acid	14 - 17
	Oleic acid	13 - 19
	Palmitic acid	6 - 9
	Capric acid	3 - 7
	Caprylic acid	3 - 4
	Stearic acid	1 - 3
	Linoleic acid	0.5 - 2
OLIVE OIL	Oleic acid	65 - 86
	Palmitic acid	7 - 20
	Linoleic acid	5 - 15
	Stearic acid	0.5 - 3
POPPY SEED OIL	Linoleic acid	65
	Oleic acid	25
	Saturated acids	6 - 10
RAPESEED OIL	Erucic acid	20 - 45
	Eicosenoic acid	9 - 15
	Linoleic acid	9 - 15
	Linolenic acid	2 - 7
	Saturated fatty acids	3 - 6
RICE BRAN OIL	Oleic acid	40 - 50
	Linoleic acid	29 - 42
	Palmitic acid	13 - 18
	Stearic acid	1 - 3
	Linoleic acid	0.5 - 1
SAFFLOWER OIL	Linoleic acid	78
	Oleic acid	13.5
	Palmitic acid	6.6
	Stearic acid	1.8
SESAME OIL	Oleic acid	37 - 50
	Linoleic acid	37 - 47
	Palmitic acid	7 - 9
	Stearic acid	4 - 5
SOYBEAN OIL	Linoleic acid	52 - 60
	Oleic acid	23 - 34
	Palmitic acid	7 - 14
	Stearic acid	2 - 6
	Linolenic acid	2 - 8
SUNFLOWER OIL	Linoleic acid	55 - 70
	Oleic acid	15 - 30
	Palmitic & Stearic acids	5 - 15

Based on the data of Godin and Spensley (1971) and Oyenuga (1975)

(Source: A.G.J. Tacon, *Standard Methods for the Nutrition of Farmed Fish and Shrimp*; 1990 - with permission of Argent Laboratories, Inc.)

TABLE 6-15: AVERAGE PROXIMATE COMPOSITION OF GRASS, GREEN FODDER CROPS, AND SOME MISCELLANEOUS PLANT FEEDSTUFFS

All values are expressed as % by weight on a as-fed basis; water-H2O; crude protein-CP; lipid or ether extract- EE; crude fiber-CF; nitrogen free extractives-NFE; ASH; calcium-CA; phosporus-P

GRASS/GREEN CROP	AVERAGE COMPOSITION (% BY WEIGHT)								
	H2O	CP	EE	CF	NFE	Ash	Ca	P	NO REF.[1]
FRESH GREEN PASTURE GRASS [2]									
Very leafy	82.0	4.0	0.6	3.6	7.5	2.3	-	-	(1)
Leafy	81.0	3.3	0.5	4.5	8.5	2.2	-	-	(1)
Early flowering	79.0	3.0	0.7	5.4	9.8	2.1	-	-	(1)
Flowering	77.0	2.4	0.5	6.2	11.7	2.2	-	-	(1)
Seed set	75.0	2.1	0.6	7.4	13.1	1.8	-	-	(1)
FRESH GREEN FODDER CROPS									
ALFALFA/LUCERNE (Medicago sativa)									
Late vegetative	79.0	4.3	0.6	4.9	9.1	2.1	-	-	(1)
Early bloom (flowering)	77.0	4.4	0.7	5.8	9.9	2.2	-	-	(1)
Mid bloom	76.0	h.5	0.6	6.8	10.0	2.1	-	-	(1)
Full bloom	75.0	3.5	0.7	7.7	11.0	2.1	-	-	(1)
RED CLOVER (Trifollum pratense)									
Early vegetative	81.0	6.3	0.9	2.0	8.3	1.5	0.30	0.07	(2)
Late vegetative	78.0	5.5	0.6	3.5	10.7	1.7	-	-	(1)
Early bloom	79.0	3.9	0.8	5.2	9.2	1.9	0.33	0.05	(3)
Mid bloom	74.6	3.9	0.8	6.6	12.1	2.0	-	-	(2)
WHITE CLOVER (Trifolium repens)									
Early vegetative	81.0	5.3	0.5	2.7	7.9	2.6	-	-	(1)
Early bloom	81.0	4.4	0.8	4.3	7.4	2.1	-	-	(1)
VETCHES (Vicia spp.)									
Mid bloom	82.0	3.2	0.5	5.1	7.7	1.5	-	-	(1)
TREFOIL (Lotus corniculatus)									
Aerial part	77.4	4.0	0.8	5.8	9.9	2.1	0.34	0.05	(2)
SESBANIA (Sesbania spp.)									
Leaves	77.0	6.6	0.5	2.6	11.0	2.3	0.52	0.10	(3)
LEADTREE/Ipil-Ipil (L. glauca)									
Leaves	68.4	8.8	1.0	3.3	17.4	1.1	0.17	0.10	(1)
SAMAN (S. saman)									
Leaves	60.9	8.7	2.7	11.5	13.9	2.3	0.55	0.08	(1)
KALE (Brassica oleracea)									
Aerial crop	85.9	2.4	0.5	2.2	7.2	1.8	0.18	0.08	(4)
MANGOLD (B. vulgaris)									
Leaves and crowns (tops)	87.4	2.1	0.5	1.4	6.2	2.4	-	-	(1)
SUGAR BEET (B. vulgaris)									
Leaves and crowns (tops)	85.2	2.2	0.4	1.8	6.8	3.0	0.17	0.03	(3)
Leaves	85.3	2.6	0.4	1.8	7.1	2.8	0.26	0.05	(2)
Cassava (M. esculenta)									
Leaves and stem	76.9	4.5	1.2	3.9	11.8	1.7	-	-	(1)
Leaves	74.4	7.7	1.3	7.7	7.1	1.8	0.17	0.10	(6)
SWEET POTATO (I. batatas)									
Leaves	89.2	2.1	0.4	1.1	4.4	2.8	-	-	(1)
Vines	89.0	2.2	0.3	1.9	5.0	1.6	0.16	0.02	(2)
GRASS/GREEN CROP									
Leaves and vines	84.9	2.5	0.5	2.5	7.2	2.4	-	-	(3)
TARO (C. esculenta)									
Leaves	89.8	2.2	0.8	1.2	4.7	1.3	0.14	0.05	(2)

TABLE 6-15 (cont.):

GRASS/GREEN CROP	AVERAGE COMPOSITION (% BY WEIGHT)								
	H₂O	CP	EE	CF	NFE	Ash	Ca	P	NO. REF.[1]
FRESH GREEN PASTURE GRASS (2)									
Aerial crop	83.3	3.7	1.2	1.9	7.6	2.3	0.01	0.01	(1)
SWEDE (*B. napus*)									
Leaves and crowns (tops)	88.0	2.3	0.5	1.5	5.5	2.2	-	-	(1)
TURNIP (*B. rapa rapa*)									
Leaves and crowns (tops)	87.0	2.8	0.3	1.3	6.4	2.2	0.38	0.07	(1)
CARROT (*D. carota*)									
Leaves and crowns (tops)	84.0	2.1	0.6	2.9	8.0	2.4	0.31	0.03	(1)
CABBAGE (*Brassica oleracea var. capitata*)									
Leaves	87.4	2.1	0.5	2.0	6.4	1.6	0.07	0.03	(4)
BRUSSEL SPROUTS (*Brassica oleracea var.*)									
Leaves	85.0	4.9	0.4	1.6	6.9	1.2	0.04	0.08	(1)
BROCCOLI (*Brassica oleracea var.italica*)									
Leaves and stems	89.0	3.6	0.3	1.5	4.5	1.1	0.10	0.08	(1)
LETTUCE (*Lactuca sativa*)									
Leaves	95.0	1.1	0.2	0.6	2.3	0.8	0.04	0.02	(1)
RAMIE (*Boehmeria nivea*)									
Leaves	84.9	2.0	0.6	4.2	5.9	2.4	0.60	0.05	(2)
JERUSALEM ARTICHOKE (*H. tuberosus*)									
Aerial parts (tops)	70.2	2.4	0.7	5.2	17.9	3.6	0.44	0.09	(2)
Leaves	78.3	4.5	0.5	2.7	10.5	3.5	0.44	0.08	(1)
ELEPHANT YAM (*Amorphophallus* spp.)									
Leaves	86.5	3.0	0.5	5.6	-	-	-	-	(1)
QUEENSLAND ARROWROOT (*C. edulis*)									
Aerial part (early vegetatlve)	83.5	1.7	0.8	3.2	8.1	2.7	-	-	(1)
RADISH (*R. sativus*)									
Leaves	87.4	2.2	0.4	1.5	6.1	2.4	0.40	0.30	(1)
SUGARCANE (*Saccharum officinarum*)									
Stems	85.0	1.2	0.1	4.2, 8.6	0.9	-	-		(1)
Cane tops	75.0	1.3	0.4	8.5	11.8	3.0	-	-	(1)
GUAVA (*Psidium guajava*)									
Leaves 62.5	3.8	2.8	7.2	20.8	2.9	0.41	0.10	(2)	
PAPAYA (*C. papaya*)									
Leaves	77.4	5.3	1.1	3.6	9.7	2.9	0.31	0.06	(3)
BREADNUT TREE (*B. alicastrum*)									
Leaves	61.1	5.4	1.3	10.3	18.9	3.0	-	-	(1)
PINEAPPLE (*A. comosus*)									
Leaves	79.4	1.9	0.3	4.9	12.5	1.0	-	-	(1)
Green tops	83.0	1.5	0.4	4.2	9.5	1.4	0.05	0.01	(2)
PUMPKIN (*Cucurbita* spp.)									
Vine	82.5	1.5	0.9	5.6	6.6	2.9	-	-	(1)
ALOCASIA (*A. macrorrhiza*)									
Leaves	90.4	2.3	0.6	1.2	3.8	1.7	0.15	0.01	(1)
BANANA (*Musa* spp.)									
Leaves	75.0	2.4	1.3	6.1	-	-	0.32	0.04	(1)
Pseudostem (trunk)	95.0	0.15	0.1	1.1	2.9	0.7	0.05	0.01	(2)
JACKFRUIT (*Artocarpus heterophyllus*)									
Leaves	60.3	6.8	1.7	8.3	18.2	4.7	0.74	0.08	(3)
GRASS/GREEN CROP									
NEEM TREE (*Azadirachta indica*)									
Leaves	60.3	6.8	1.7	8.3	18.2	4.7	0.74	0.08	(3)

TABLE 6-15 (cont.):

GRASS/GREEN CROP	AVERAGE COMPOSITION (% BY WEIGHT)								
	H$_2$O	CP	EE	CF	NFE	Ash	Ca	P	NO. REF. [1]
Leaves	64.2	4.8	2.2	5.3	19.8	3.7	0.69	0.06	(1)
GROUNDNUT (*A. hypogaea*)									
Leaves	73.1	4.7	0.6	5.4	13.9	2.3	0.25	0.05	(1)
PIGEON PEA (*C. caian*)									
Aerial part (forage)	71.0	5.9	1.5	8.1	11.7	1.8	0.22	0.06	(4)
JACK/SWORD BEAN (*C. ensiformis*)									
Aerial part (forage)	76.8	5.2	0.5	6.4	8.4	2.7	-	-	(1)
CAROB (*C. siliqua*)									
Leaves and stem	75.7	5.4	0.6	7.5	8.5	2.3	-	-	(1)
CHICKPEA (*C. arietinum*)									
Young shoots	60.6	8.2	0.5	-	-	3.5	0.31	0.21	(1)
CLUSTER BEAN (*C. tetragonoloba*)									
Aerial part (fodder)	80.8	3.1	0.4	4.4	8.0	3.3	0.61	0.07	(1)
EGYPTIAN BEAN / HYACINTH BEAN (*L. purpureus*)									
Aerial part	81.6	2.5	0.9	5.8	6.9	2.3	0.30	0.06	(1)
SOYBEAN (*G. max*)									
Aerial part	74.0	3.7	1.1	10.5	8.3	2.4	0.35	0.07	(4)
GRASS PEA (*L. sativus*)									
Aerial part (late vegetative)	82.6	3.6	0.6	4.0	7.3	1.9	0.23	0.06	(1)
Aerial part (early bloom)	78.2	3.2	0.6	6.5	8.4	3.1	0.34	0.08	(1)
Aerial part (mid-bloom)	72.0	5.0	0.7	8.5	10.3	3.5	0.28	0.06	(1)
LUPIN (*Lupinus* spp.)									
Aerial part	88.3	3.1	0.3	2.2	4.5	1.6	0.15	0.03	(1)
VELVET BEAN (*M. pruriens*)									
Aerial part (vegetative)	82.9	3.9	0.4	4.7	6.1	2.0	-	-	(2)
Aerial part (mid-bloom)	81.5	3.2	0.7	6.2	7.3	1.1	0.23	0.02	(2)
BROAD BEAN (*V. faba*)									
Stems and leaves	85.0	2.5	0.4	4.9	5.4	1.8	0.22	0.04	(1)
WINGED BEAN (*P. tetragonolobus*)									
Stems and leaves	78.9	6.3	1.0	4.1	7-9	1.8	0.37	0.12	(1)
PEA (*P. sativum*)									
Aerial part (late vegetative)	86.6	2.3	0.4	3.6	5.4	1.7	0.25	0.05	(1)
Aerial part (mid bloom)	84.8	2.2	0.4	4.3	6.3	2.0	0.28	0.06	(1)
URD/BLACK GRAM (*V. mungo*)									
Aerial part	84.0	3.1	0.4	4.3	5.6	2.6	0.32	0.04	(1)
HORSE GRAM (*V. unguiculata*)									
Aerial part	81.8	3.2	0.4	3.9	9.4	1.3	0.10	0.05	(1)
RED BEAN (*V. umbellata*)									
Aerial part (mid-bloom)	68.0	5.4	0.6	9.8	13.7	2.5	0.33	0.08	(1)
SILAGES [3]									
Grass (leafy)	80.0	3.5	1.0	5.0	8.7	1.8	-	-	(1)
Grass (early bloom)	75.0	3.2	0.9	7.0	11.6	2.3	-	-	(1)
Grass (full-bloom)	75.0	2.9	0.7	7.9	10.8	2.7	-	-	(1)
Alfalfa/lucerne(*M. sativa*)	79.2	4.3	1.8	5.0	7.1	2.6	0.40	0.10	(2)
Red clover (*T. pratense*)	74.6	3.2	1.3	6.6	11.3	3.0	0.43	0.06	(2)
Maize/corn (*Z. mays*)	75.0	2.4	1.1	6.1	13.9	1.5	0.09	0.08	(2)
Oats (*A. sativa*)	76.0	2.5	0.7	7.9	11.2	1.7	0.10	0.07	(2)
Rye (*S. cereale*)	68.0	4.1	1.1	10.9	13.4	2.5	0.13	0.10	(1)
Sorghum (*S. bicolor*)	70.0	2.2	0.9	8.2	16.1	2.6	0.10	0.06	(1)
Wheat (*Triticum* spp.)	72.5	2.8	0.7	8.0	13.8	2.2	0.08	0.08	(2)
Soybean (*G. max*)	73.0	4.8	0.7	7.8	11.0	2.7	0.37	0.13	(1)

TABLE 6-15 (cont.):

AVERAGE COMPOSITION (% BY WEIGHT)									
GRASS/GREEN CROP	H₂O	CP	EE	CF	NFE	Ash	Ca	P	NO. REF.[1]
Pea (*P. sativum*) (vines only)	76.0	3.1	0.8	7.2	10.8	2.1	0.31	0.06	(2)
Urd (*V. mungo*)	72.7	3.8	1.3	5.2	9.8	7.2	-	-	(1)
Sugar beet (*B. vulgaris*) (crowns with tops)	79.0	2.8	0.6	2.9	7.6	7.1	0.38	0.05	(3)
Pineapple (*A. comosus*) (leaves)	80.9	1.1	0.5	4.4	11.2	1.9	-	-	(1)
HAY (Sun-cured)									
Meadow grass (leafy)	15.0	13.7	3.0	19.5	41.0	7.8	-	-	(1)
Meadow grass (early-bloom)	15.0	10.0	1.6	26.6	40.0	6.8	-	-	(1)
Meadow grass (full-bloom)	15.0	7.6	1.5	28.7	40.8	6.4	-	-	(1)
Meadow grass (seed set)	15.0	4.8	1.2	30.6	43.1	5.3	-	-	(1)
Bahia grass (*Paspalum notatum*)	9.2	4.3	1.5	20.5	59.1	5.4	0.41	0.17	(1)
Orchard grass (*Dactylis glomerata*) (early bloom)	11.0	13.4	2.5	27.6	37.7	7.8	-	-	(1)
late bloom	9.0	7.6	3.1	33.6	37.5	9.2	-	-	(1)
Pangola grass (*Digitaria decumbers*)	12.0	6.7	1.5	27.4	40.7	11.7	0.40	0.20	(1)
Rye grass (*Lollum perenne*)	14.0	7.4	1.9	26.1	40.7	9.9	0.56	0.28	(1)
Bermuda grass (*Cynodon dactylon*)	9.7	7.3	1.9	29.0	44.9	7.2	0.43	0.16	(3)
Alfalfa/lucerne (*M. sativa*)	9.5	12.3	1.4	31.6	38.2	7.0	1.33	0.24	(2)
Red clover (*T. pratense*)	11.9	12.9	2.8	26.2	39.6	6.6	1.11	0.17	(3)
White clover (*T. repens*)	12.5	17.6	2.9	20.9	38.1	8.0	1.21	0.28	(2)
Crimson clover (*T. incarnatum*)	13.0	16.1	2.1	26.3	32.9	9.6	1.22	0.19	(1)
Trefoil (*L. corniculatus*)	8.0	15.0	2.3	28.3	39.9	6.5	1.57	0.25	(1)
Barley (*H. vulgare*)	13.0	7.6	1.9	24.1	46.8	6.6	0.20	0.23	(1)
Oats (*A. sativa*)	12.0	8.2	2.5	27.6	42.8	6.9	0.22	0.20	(2)
Sorghum (*S. bicolor*) (early vegetative)	8.0	14.7	3.0	25.8	36.5	12.0	0.46	0.17	(1)
Sorghum (late vegetative)	8.0	11.0	2.4	30.4	38.1	10.1	0.37	0.18	(1)
Sorghum (early bloom)	7.0	7.0	1.9	35.3	40.4	8.4	0.28	0.13	(1)
Wheat (*Triticum spp.*)	12.0	7.4	1.9	24.6	47.9	6.2	0.13	0.17	(1)
Groundnut (*A. hypogaea*)	9.2	9.2	2.7	30.3	40.2	8.4	0.97	0.13	(4)
Pigeon pea (*C. cajan*)	11.2	14.8	1.7	28.9	39.9	3.5	-	-	(1)
Cluster bean (*C. tetragonoloba*)	9.3	16.5	1.3	19.3	41.2	12.4	-	-	(1)
Soybean (*G. max*)	10.0	15.3	3.9	27.0	36.8	7.0	1.10	0.22	(3)
Grass pea (*L. sativus*)	12.8	13.0	2.4	31.8	32.0	8.0	-	-	(2)
Lentil (*L. esculenta*)	10.2	4.4	1.8	21.4	50.0	12.2	-	-	(1)
Lupin (*Lupinus spp.*)	5.9	15.4	3.1	23.3	44.6	7.7	0.99	0.19	(1)
Velvet bean (*M. pruriens*)	9.4	13.4	2.4	27.8	38.9	8.1	-	-	(1)
Pea (*P. sativum*)	11.3	12.6	2.3	27.4	39.4	7.0	1.20	0.25	(2)
Mung bean (*V. radiata*)	9.7	9.8	2.2	24.0	46.6	7.7	-	-	(1)
Cowpea (*V. unguiculata*)	9.7	17.4	2.7	23.9	35.6	10.7	1.33	0.32	(3)
Pineapple (*A. comosus*) (aerial part)	11.0	6.9	2.5	26.3	47.8	5.5	0.35	0.21	(1)
Sweet potato (*I. batatas*) (vines)	13.4	14.2	4.5	23.7	33.3	10.9	-	P	(1)
STRAWS AND CHAFF									
Barley (*H. vulgare*)	11.8	3.4	1.6	37.5	40.4	5.3	0.30	0.07	(5)
Maize/corn (*Z. mays*)	10.0	5.3	1.2	33.4	44.9	5.2	0.44	0.08	(1)
Oats (*A. sativa*) (straw)	10.5	3.7	2.1	35.8	41.5	6.4	0.25	0.08	(4)
Oats (*A. sativa*) (chaff)	14.0	6.0	2.1	22.8	44.8	10.3	-	-	(1)
Rice (*0. sativa*)	8.0	3.9	1.0	31.8	40.5	14.8	0.24	0.08	(4)
Rye (*S. cereale*)	11.7	2.9	1.5	37.8	42.3	3.8	0.23	0.08	(3)
Wheat (*Triticum spp.*)	10.4	2.7	1.4	36.4	41.4	7.7	0.15	0.06	(4)
Soybean (*G. max*)	12.0	4.6	1.3	38.9	37.6	5.6	1.40	0.05	(1)
Chickpea (*C. arletinum*)	9.4	5.4	0.4	40.2	32.6	12.0	0.31	0.11	(1)

TABLE 6-15 (cont.):

GRASS/GREEN CROP	H_2O	CP	EE	CF	NFE	Ash	Ca	P	NO. REF.[1]
AVERAGE COMPOSITION (% by weight)									
Lima bean (*P. lunatus*)	10.0	6.8	1.6	27.9	46.3	7.4	0.09	0.37	(1)
Kidney bean (*P. vulgaris*)	11.0	6.0	1.5	40.0	34.0	7.5	1.70	0.10	(1)
Pea (*P. sativum*)	13.0	8.1	1.5	34.4	37.0	6.0	0.85	0.09	(3)
Broad bean (*V. faba*)	12.0	5.3	1.1	41.5	34.1	6.0	1.70	0.13	(2)
ARTIFICIALLY DEHYDRATED LEAVES									
Grass (very leafy)	10.0	18.7	3.o	17.7	40.6	10.0	-	-	(1)
Grass (leafy)	10.0	15.0	2.6	20.9	40.7	10.8	-	-	(1)
Grass (early bloom)	10.0	12.1	2.2	24.4	42.3	9.0	-	-	(1)
Alfalfa lucerne (*M. sativa*)	7.8	17.3	2.7	24.4	38.3	9.5	1.37	0.23	(7)
Cassava (*M. esculenta*)	10.0	27.0	4.6	27.1	25.0	6.3	0.60	0.35	(1)
Papaya (*C. papaya*)	7.5	21.7	3.9	9.8	45.7	11.4	-	-	(1)
Bambarra groundnut (*V. subterranea*)	9.8	14.3	1.6	28.6	38.9	6.8	-	-	(1)
Cabbage (*B. oleracea var. capitata*)	11.7	14.9	3.9	8.4	54.1	7.0	-	-	(1)
Banana (*Musa* spp.)	5.9	9.3	11.1	22.6	42.8	8.3	0.71	0.23	(1)
Cluster bean (*C. tetragonoloba*)	14.2	19.3	3.0	8.3	42.8	12.4	-	-	(1)
MISCELLANEOUS PLANT-BASED FEEDSTUFFS									
Bakery waste (dried)	8.8	10.0	12.4	0.8	64.8	3.2	0.10	0.23	(2)
Bread (dried)	8.0	12.2	2.9	0.9	74.2	1.8	0.08	0.15	(4)
Leaf (rye grass) protein concentrate (dried)	5.4	57.7	20.6	1.7	10.9	3.7	0.27	0.35	(1)
Potato protein concentrate (dried)	8.5	78.1	0.4	0.1	11.6	1.3	0.07	0.25	(1)
Pyrethrum (*Chrysanthemum cinerariifolium*) (fresh)	22.7	11.9	0.5	20.6	38.2	6.1	0.41	0.19	(4)
Pyrethrum (*C. cinerariifolium*) (sun cured)	14.5	12.6	0.5	20.6	45.4	6.4	0.34	0.26	(2)
Sago palm (*Metroxylon sagu*) meal	14.0	1.8	1.1	4.9	74.4	3.8	0.05	0.04	(2)
Sago palm refuse (starch extracted)	22.9	2.0	0.2	7.7	51.0	16.2	0.29	0.02	(2)
Sugar cane (*S. officinarum*) bagasse (dried)	9.6	1.5	0.8	40.3	42.7	5.1	-	-	(4)
Sugar can bagasse (fresh)	45.0	0.8	0.2	26.9	24.1	3.0	0.19	0.15	(1)
Sugar cane filter press mud (fresh)	75.5	2.7	2.6	3.0	10.9	5.3	0.64	0.27	(2)
Sugar cane molasses (final)	25.0	3.0	trace	trace	63.5	8.5	0.70	0.10	(12)
Sugar cane strippings (fresh)	55.0	1.6	0.4	20.3	18.1	4.6	0.16	0.12	
Seaweed (*Laminariales/Fucales* spp.)									
Kelp meal (dehydrated)	8.9	6.5	0.5	6.6	42.3	35.	2.50	0.26	(3)

[1] The data presented represents the mean values from various sources. The number in parenthesis indicates the number of sources used to calculate the mean. These include: Allen (1984); Bath *et.al.*, (1984); Branckaert, Tessema and Temple (1976); Cooley (1976); Devendra (1979); Gohl (1981); Hickling (1971); Kay (1973, 1979); Ling (1967); MacDonald, Edwards and Greenhalgh (1977); Miller (1976); NRC (1982); Ogino, Cowey and Chlou (1978); Tacon (1983; 1986).

[2] Data presented for pasture grass was obtained from McDonald, Edwards and Greenhalgh (1977); for specific species composition see Gohl (1981) and NRC (1982).

(Source: A.G.J. Tacon, *Standard Methods for the Nutrition of Farmed Fish and Shrimp*; 1990 - with permission of Argent Laboratories, Inc.)

TABLE 6-16: AVERAGE ESSENTIAL AMINO ACID (EAA) COMPOSITION OF THE MAJOR GRASS AND GREEN FODDER CROPS AND SOME SELECTED MISCELLANEOUS PLANT FEEDSTUFF

All values are expressed as % by weight on a as-fed basis: Arginine-ARG; Cystine-CYT; Methionine-MET; Threonine-THR; Isoleucine-ISO; Leucine-LEU; Lysine-LYS; Valine-VAL; Tyrosine-TYR; Tryptophan-TRY; Phenylalanine-PHE; Histidine-HIS [1]

INGREDIENT	AVERAGE EAA COMPOSITION (%)												NO. REF. [1]
	ARG	CYT	MET	THR	ISO	LEU	LYS	VAL	TYR	TRY	PHE	HIS	
DEHYDRATED/DRY CROPS													
Pasture grass (dehydrated)	0.75	0.19	0.31	0.62	0.56	1.21	0.71	0.70	0.45	0.31	0.71	0.27	(1)
Alfalfa (*M.sativa*) (meal dehyd. 15% prot.)	0.56	0.21	0.21	0.63	0.82	0.91	0.60	0.77	0.33	0.33	0.62	0.26	(3)
(meal dehyd. 17% prot.)	0.76	0.23	0.27	0.73	0.83	1.29	0.76	0.97	0.54	0.40	0.85	0.34	(2)
(meal dehyd. 20% prot.)	0.97	0.27	0.32	0.84	0.94	1.45	0.89	1.12	0.62	0.43	0.99	0.40	(2)
(meal dehyd. 22% prot.)	0.96	0.30	0.34	0.97	1.06	1.63	0.97	1.29	0.64	0.49	1.13	0.44	(1)
(meal dehyd. 25% prot.)	0.58	0.17	0.20	0.60	0.60	1.10	0.60	0.60	-	0.38	0.58	0.22	(1)
White clover (*I.repens*) (hay; sun cured)	0.99	0.36	0.27	1.17	1.08	1.88	1.08	1.17	0.63	0.45	1.08	0.45	(1)
Cassava (*M. esculenta*) (leaves; dehyd.)	1.41	0.32	0.39	1.27	1.11	2.72	2.78	1.67	1.07	0.29	1.05	0.67	(1)
(stems; dehy.) 0.79	0.20	0.37	1.16	0.89	1.78	1.20	1.71	0.84	0.25	0.89	0.47	-	(1)
Cowpea (*V.unguiculata*) (hay; sun cured)	1.11	-	0.51	1.06	1.26	2.00	1.08	1.43	-	0.52	1.25	0.45	(1)
FRESH CROPS													
Alfalfa (*M. sativa*) (leaves)	0.239	0.052	0.064	0.194	0.194	0.331	0.247	0.239	0.155	0.072	0.206	0.093	(1)
Sugar beet (*B. vulgaris*)	0.060	0.020	0.017	0.062	0.040	0.093	0.060	0.060	0.050	0.021	0.054	0.033	(1)
Cabbage (*B. oleracea capitata*) (leaves)	0.135	0.018	0.017	0.061	0.050	0.086	0.050	0.068	0.030	0.017	0.049	0.041	(1)
Cassava (*M. esculenta*) (leaves)	0.381	0.077	0.118	0.327	0.339	0.900	0.437	0.401	0.274	0.102	0.386	0.157	(1)
Groundnut (*A. hypogaea*) (leaves)	0.276	0.057	0.062	0.171	0.175	0.307	0.223	0.219	0.162	-	0.228	0.105	(1)
Lead tree (*L.leucocephala*) (leaves)	0.135	0.040	0.046	0.106	0.259	0.216	0.144	0.155	0.121	-	0.135	0.058	(1)
Lupin (*Lupinus* spp.) (leaves)	0.075	0.018	0.019	0.098	0.105	0.153	0.108	0.087	0.056	-	0.128	0.042	(1)
Mulberry (*Morus* spp.) (leaves)	0.401	-	0.133	0.232	0.301	0.534	0.276	0.382	0.259	-	0.371	0.152	(1)
Pumpkin (*Cucurbita* spp.) (leaves)	0.292	0.038	0.080	0.204	0.218	0.400	0.254	0.250	0.196	0.052	0.250	0.090	(1)
Sesame (*Sesamum* spp.) (leaves)	0.238	0.061	0.058	0.218	0.214	0.364	0.203	0.251	0.167	-	0.236	0.089	(1)
Spinach (*Spinacia oleracea*) (leaves)	0.139	0.036	0.046	0.116	0.106	0.208	0.159	0.133	0.110	0.034	0.133	0.056	(1)
Tamarind (*Tamarindus indica*) (leaves)	0.184	0.028	0.022	0.144	0.166	0.291	0.184	0.181	0.110	-	0.198	0.072	(1)
Turnip (*B. rapa rapa*) (leaves)	0.118	0.024	0.032	0.127	0.106	0.210	0.157	0.136	0.086	0.042	0.142	0.057	(1)
MISCELLANEOUS PLANT FEEDSTUFFS													
Bakery waste (dried)	0.43	0.16	0.16	0.38	0.40	0.77	0.30	0.41	0.41	0.09	0.40	0.17	(2)
Leaf (rye grass) protein concentrate (dried)	3.75	0.52	1.04	2.88	2.88	5.48	3.81	3.75	2.37	0.86	3.63	1.32	(1)
Potato protein concentrate (dried)	3.70	1.40	2.10	4.20	4.70	7.60	6.00	4.50	4.80	0.95	5.00	1.40	(1)

[1] The data presented represents the mean values from various sources, including: Allen(1984); Bolton and Blair (1977); FAO (1970); NRC (1982,1983); Ogino, Cowey and Chiou (1978), and Tacon (1983)

(Source: A.G.J. Tacon, *Standard Methods for the Nutrition of Farmed Fish and Shrimp*; 1990 - with permission of Argent Laboratories, Inc.)

TABLE 6-17: AVERAGE PROXIMATE COMPOSITION OF SELECTED INVERTEBRATE FOOD ORGANISMS

All Values Are Expressed As % By Weight On An As-fed Basis: Water-H_2O; Crude Protein-CP, Lipid Or Ether Extract-EE; Crude Fiber-CF; Nitrogen Free Extractives-NFE; Ash; Calcium-Ca; Phosphorus-P [1]

| INVERTEBRATE | AVERAGE COMPOSITION (% by weight) | | | | | | | | NO. REF. [1] |
	H_2O	CP	EE	CF	NFE	Ash	Ca	P	
ROTIFERS [2]									
Brachionus plicatilis (wet basis)									
Cultured on bakers yeast	90.7	6.2	1.8	-	-	0.7	0.015	0.127	(1)
Cultured on bakers yeast + marine chlorella	88.7	7.7	2.4	-	-	0.5	0.016	0.138	(1)
Cultured on marine chlorella	86.9	7.9	3.9	-	-	0.7	0.016	0.142	(1)
Cultured on marine oil enriched bakers yeast	86.8	7.2	4.6	-	-	0.5	0.015	0.130	(1)
BRINE SHRIMP [2]									
Artemia salina									
Eggs (dry basis)	-	51.1	7.2	-	-	11.5	0.24	0.74	(1)
Larvae (nauplii); Just after hatching (wet basis)	89.0	6.7	2.1	-	-	1.1	0.03	0.14	(1)
Larvae (nauplii); Just after hatching (dry basis) [3]	-	52.2	18.9	14.8		9.7	-	-	(1)
Juveniles and adults; cultured (dry basis)	-	54.6	13.2	-	-	16.6	-	-	(1)
Adults; wild (dry basis) [3]	-	58.4	11.1	12.1		17.8	-	-	(1)
Brine shrimp meal; Adults (dried)	18.7	44.3	4.0	17.5		15.6	-	-	(1)
MARINE COPEPODS [2]									
Tigriopus japonicus (wet basis)									
Cultured under natural conditions	88.6	8.1	2.6	-	-	0.5	0.01	0.09	(1)
Cultured on bakers yeast	87.2	8.9	2.6	-	-	0.6	0.02	0.12	(1)
Cultured on bakers yeast + marine Chlorella	86.6	9.0	3.2	-	-	0.5	0.02	0.13	(1)
Cultured on bakers + dry prawn diet	86.3	9.8	3.1	-	-	0.5	0.02	0.14	(1)
Cultured on marine oil enriched bakers yeast	87.2	8.7	2.6	-	-	0.6	0.04	0.14	(1)
Cultured on soy sauce cake	86.3	9.4	2.3	-	-	0.6	0.04	0.14	(1)
Acartia clausi	87.6	8.8	1.3	-	-	2.1	0.05	0.15	(1)
FRESHWATER COPEPODS [4]									
Moina spp. (wet basis)									
Cultured on bakers yeast	87.2	8.8	2.9	-	-	-	0.01	0.18	(1)
Cultured on bakers yeast + poultry manure	89.0	8.6	1.3	-	-	-	0.01	0.12	(1)
Cultured on poultry manure	87.9	8.2	3.3	-	-	-	0.02	0.16	(1)
Daphnia pulex (wet basis)	94.0	3.0	1.0	0.4	-	1.2	-	-	(1)
Daphnia spp. (wet basis)	89.3	7.5	1.4	-	-	0.7	0.02	0.15	(1)
Diaptomus spp. (wet basis)	92.4	4.4	1.9	0.5	-	0.4	-	-	(1)
MISCELLANEOUS FRESHWATER INVERTEBRATES (wet basis)									
Amphipod (*Gammarus lacustris*)	85.9	5.7	1.5	1.0	-	4.0	-	-	(2)
Damselfly nymph (*Enallagma* spp.)	86.5	7.9	1.8	1.3	-	0.8	-	-	(1)
Dragonfly nymph (*Aeshna* spp.)	86.4	4.7	2.9	1.0	-	0.6	-	-	(1)
Water boatmen (Corixidae)	78.9	12.2	5.7	2.5	-	0.7	-	-	(1)
Chironomid larvae (*Chironomidae* spp.)	83.9	9.1	13.6	-	-	7.1	-	-	(1)
Blood worm (*Tubifex* tubifex)	87.1	8.1	2.0	-	1.9	0.9	-	-	(1)
Riversnail (*Lymnea* spp.)									
Whole snail	36.8	5.7	0.7	-	2.0	54.8	-	-	(1)
Snail meat	78.4	12.2	1.4	-	4.3	3.7	-	-	(1)
Freshwater mussel	79.6	18.4	0.8	-	-	1.2	-	-	(1)
MISCELLANEOUS MARINE INVERTEBRATES									
Short necked clam (*Venerupis phillippinarum*) (flesh)	81.8	12.6	0.6	-	-	2.5	-	-	(1)
Squid (*Ommastrephes pacifica*) (meal; dried)	8.1	74.8	8.8	0.0	4.9	3.4	-	-	(1)

TABLE 6-17 (cont.):

INVERTEBRATE	AVERAGE COMPOSITION (% by weight)								NO. REF. (1)
	H₂O	CP	EE	CF	NFE	Ash	Ca	P	
KRILL (*Euphausia pacifica*) (whole;dried)	82.0	6.0	5.0	-	-	5.0	0.46	0.29	(1)
CRAB (*Cancer/Carcinus/Callinectes* spp.)									
Process residue (meal; dried)	6.5	31.0	2.1	10.6	13.7	36.1	16.28	1.54	(2)
Protein concentrate (dried)	10.0	60.5	0.4	-	-	6.8	0.09	0.60	(1)
Mysid shrimp meal (dried)	10.4	68.2	2.4	5.0		14.0	-	-	(1)
Sergestid shrimp (*Acetes* spp.) (whole; sun dried)	14.0	46.9	3.2	4.2	-	13.1	2.96	1.07	(1)
CRAWFISH (*Procambarus clarkii*) (by-product meal)	-	36.0	3.7	13.6	-	42.2	15.80	0.95	(1)
Shrimp meal (process residue; dehydrated)	10.0	40.6	2.6	14.2	2.6	30.0	9.70	1.57	(6)
Shrimp heads (dried) (5)	-	58.2	8.9	11.1	-	22.6	7.20	1.68	(1)
Shrimp shells (exoskeleton/hull; dried) (5)	-	45.9	0.4	27.2	-	31.7	11.10	3.16	(1)
Shrimp head silage (fresh)	81.0	14.1	1.4	-	-	3.5	1.08	0.30	(1)
Shrimp head silage (dried)	7.0	69.0	6.8	-	-	17.1	5.29	1.47	(1)
TERRESTRIAL INVERTEBRATES									
AFRICAN GIANT SNAIL (*Achatina fulica*)									
Snail meat meal (dried)	11.1	45.6	2.4	-	-	7.0	0.73	0.48	(1)
Snail meal (without shell; dry matter basis)	0.0	60.9	6.1	4.5	18.9	9.6	2.0	0.84	(1)
Snail shell (dry matter basis)	0.0	2.8	1.0	-	-	54.5	36.1	0.14	(1)
Whole snail (including shell; dry matter basis)	0.0	16.1	2.0	-	-	46.0	31.1	0.32	(1)
EUROPEAN SNAIL (*Helix* spp.)									
Snail meat (*H. aspersa*) (fresh)	78.5	14.6	0.7	-	-	1.4	-	-	(1)
Snail meat (*H. lucorum*) (fresh)	80.3	12.9	0.6	-	-	1.8	-	-	(1)
Snail meat (*Helix* spp.) (dried)	5.7	62.7	7.5	-	-	7.8	-	-	(1)
SILKWORM (*Bombyx mori/Antheraea mylittapaphia*)									
Pupae (fresh)	74.9	13.7	8.3	1.1	0.9	1.1	0.03	0.18	(3)
Pupae (dry)	10.0	55.9	24.5	-	-	1.9	-	-	(1)
Pupae (solvent extracted; dried)	7.9	72.0	1.2	6.7	6.0	6.2	0.14	1.06	(3)
LOCUST (*Schistocerca gregaria*)									
Whole fresh	68.2	22.1	3.0	4.0	0.3	2.4	-	-	(2)
Whole dried	10.5	46.2	9.7	12.0	-	-	-	-	(1)
SOLDIER FLY (*Hermetia illucens*)									
Larvae meal (dried) (6)	7.9	42.1	34.8	7.0	-	14.6	5.0	1.5	(1)
TERRESTRIAL OLIGOCHAETE WORMS									
Eisenia foetida (fresh)	83.3	9.8	1.5	-	-	2.9	-	-	(1)
Eisenia foetida (meal; dried)	7.4	56.4	7.8	1.6	18.0	8.8	0.48	0.87	(1)
Eudrilus eugenige (fresh)	85.3	8.9	1.8	-	-	1.5	0.22	0.13	(1)
Eudrilus eugenige (dry matter basis)	0.0	60.4	12.0	-	-	10.5	1.49	0.89	(1)
Dendrodrilus subrubicundus (dried)	9.1	65.1	9.6	-	-	13.0	0.18	-	(1)
Allolobophora longa (fresh)	78.3	10.9	0.3	-	-	7.6	-	-	(1)
Lumbricus terrestris (fresh)	81.1	10.6	0.4	-	-	5.4	-	-	(1)

(1) The data presented represents the mean values from various sources, including: Allen (1984); Choubert and Luquet (1983); Creswell and Kompiang (1981); Deshimaru and Shigeno (1972); Deshimaru *et.al.*, (1985); Elmslie (1982); Gallagher and Brown (1975); Gohl (1981); Hilton (1983); Imada *et.al.*, (1979); Ling (1967); Mathias *et.al.*, (1982); Meyers (1986, 1987); Newton *et.al.*, (1977), NRC (1983); Seidel et.al., (1980); Simpson, Klein-MacPhee and Beck (1982); Stafford and Tacon (1984, 1985); Tacon (1986a); Tacon, Stafford and Edwards (1983); Watanabe, Kitajima and Fujita (1983); Yoshida and Hoshii (1978), Yurkowski and Tabachek (1978), and Leger *et.al*, (1986).

(2) Data obtained from Watanabe, Kitajima and Fujita (1983).

(3) Data obtained from Leger *et.al.*, (1986); no information provided on moisture, crude fiber or NFE content - the carbohydrate content being represented by a single value.

(4) Data compiled from Watanabe, Kitajima and Fujita (1983) and Yurkowski and Tabachek (1978).

(continued on page 128)

(continued from page 127)

[5] Data obtained from Meyers (1986); crude protein (N x 6.25) values do not correspond to the corrected true protein value of 53.5% and 22.8% for shrimp heads and shrimp hulls respectively (ie. values corrected for chitin content)

[6] Data obtained from Newton et al., (1977); no value is presented in this table for NFE, as the existing values reported by these authors total 106.4%.

(Source: A.G.J. Tacon, *Standard Methods for the Nutrition of Farmed Fish and Shrimp*; 1990 - with permission of Argent Laboratories, Inc.)

TABLE 6-18: AVERAGE ESSENTIAL AMINO ACID (EAA) COMPOSITION OF SELECTED INVERTEBRATE FOODS

All values are expressed as % by weight on a as-fed basis: Arginine-ARG; Cystine-CYT; Methionine-MET; Threonine-THR; Isoleucine-ISO; Leucine-LEU; Lysine-LYS; Valine-VAL; Tyrosine-TYR; Tryptophan-TRY; Phenylalanine-PHE; Histidine-HIS [1]

INVERTEBRATE	AVERAGE EAA COMPOSITION (% dry meal or % total recovered amino acids)												NO. REF. [1]
	ARG	CYT	MET	THR	ISO	LEU	LYS	VAL	TYR	TRY	PHE	HIS	
Rotifer (*B. plicatilis*)[2] (% total AA)	6.3	1.1	1.2	4.7	4.8	8.2	8.2	5.5	4.2	1.6	5.3	2.1	(1)
Brine shrimp (*A. salina*)													
Nauplii (newly hatched) (% total AA)	7.3	0.6	1.3	2.5	3.8	8.9	8.9	4.7	5.4	1.5	4.7	1.9	(1)
Nauplii (3-day old) (% total AA)	6.5	1.1	2.3	4.8	4.8	7.6	8.0	5.1	7.8	-	5.7	3.6	(1)
Adults (wild) (% total AA)	6.5	2.2	2.7	4.6	5.3	8.0	7.6	5.4	4.5	1.0	4.7	1.8	(1)
Brine shrimp meal [3] (% total AA)	6.8	1.3	2.3	4.9	5.1	8.6	7.4	5.3	4.6	-	5.3	2.2	(1)
Copepod (*A. clausi*) (% total AA)	5.6	1.0	2.0	5.5	4.6	7.2	7.1	5.9	4.7	1.4	4.8	2.5	(1)
Copepod (*T. japonicus*) (% total AA)	6.9	0.9	1.5	5.0	3.3	6.6	7.5	4.3	5.3	1.5	4.6	2.1	(1)
Copepod (*Moina* spp.) (% total AA)	7.0	0.8	1.4	5.2	3.4	8.3	8.0	4.4	4.5	1.6	4.9	2.2	(1)
Amphipod(*G. lacustris*) (% dry matter)	2.5	0.4	0.8	2.0	1.7	3.0	2.8	2.2	3.2	-	1.9	1.1	(1)
Soldier fly (*H.illucens*) (larvae) (% dry matter)	2.2	<0.1	0.9	0.5	2.0	3.5	3.4	3.4	2.5	0.2	2.2	1.9	(1)
Snail (*A. fulica*) (meal) (% dry matter)	4.9	0.6	1.0	2.8	2.6	4.6	4.3	3.1	2.4	-	2.6	1.4	(2)
Crab process residue (meal) (% by weight as fed)	1.7	0.2	0.5	1.1	1.2	1.6	1.4	1.5	1.2	0.3	1.2	0.5	(2)
Crab protein concentrate (meal) (by wt. as fed)	5.5	<0.1	0.8	3.5	3.4	5.3	3.6	5.0	4.8	-	5.1	2.3	(1)
Squid (*0. pacifica*) (meal) (% total AA)	7.2	0.7	2.9	5.1	4.9	7.7	8.0	4.4	3.8	-	5.6	2.1	(1)
Mysid shrimp meal (% total AA)	6.5	1.2	3.1	5.6	0.5	7.3	8.6	5.3	4.5	-	5.0	2.5	(1)
Short-necked clam (*V. phillippinarum*) (%tot. AA flesh)	7.7	1.3	2.6	4.8	3.4	6.9	7.3	4.2	3.9	1.3	3.8	2.2	(1)
Shrimp meal (processing residue; dehyd.) (% by wt.as fed)	2.5	0.6	0.8	1.4	1.7	2.7	2.2	1.8	1.3	0.4	1.6	1.0	(1)
Shrimp meal (sun dried) (% total AA)	6.9	1.7	3.1	4.7	3.6	8.3	6.7	4.8	4.0	1.4	5.0	2.1	(1)
Shrimp head meal (% total AA)	6.8	2.4	1.7	4.3	6.3	6.8	9.3	6.9	3.7	0.6	4.7	2.3	(1)
Shrimp (*Acetes* spp.)[4] (whole; dried; % by wt. as fed)	4.6	0.2	1.6	2.3	2.5	4.9	4.4	2.7	2.0	-	2.5	1.0	(1)
Shrimp (*Acetes* spp.)[4] (whole; dried; % total AA)	8.2	0.4	3.0	4.1	4.5	8.8	8.0	4.8	3.6	-	4.6	1.8	(1)
TERRESTRIAL OLIGOCHAETE WORMS													+
E. eugenige (% dry matter basis)	1.73	0.23	0.77	1.37	0.99	-	1.83	1.15	1.01	-	1.19	0.40	(1)
E. foetida (% dry meal)	2.73	0.34	1.36	2.72	2.01	4.03	3.17	2.26	1.68	0.35	1.93	1.44	(1)
A. longa (% dry meal)	3.15	0.30	0.50	2.11	2.24	3.57	3.43	2.46	1.99	-	2.65	1.01	(1)
D. subricunda (% dry meal)	3.39	0.35	1.29	2.50	1.72	3.86	3.25	1.89	1.79	0.57	2.15	1.39	(1)
L. rubellus (% dry meal)	3.68	0.39	1.31	2.77	1.97	4.17	3.86	2.26	1.95	0.46	1.88	1.29	(1)
L. terrestris (% dry meal)	3.17	0.32	1.11	2.48	2.20	4.11	3.52	2.30	1.78	0.44	2.02	1.38	(1)

[1] The data presented represence the mean values from various sources, including: Allen (1984); Cresswell and Komplang(1981); Deshimaru and Shigeno (1972); Deshimaru et al., (1985); Gallagher and Brown (1975); Hilton (1983), Mathias et al. (1982); Meyers (1986); Newton et al., (1977); NRC (1983); Seidel et al., (1980); Stafford (1984); Tacon, Stafford and Edwards (1983); and Watanabe, Kitajima and Fujita (1983).
[2] Mean of the eight amino acid analyses presented by Watanabe, Kitajima and Fujita (1983)

[3] 0rigin of meal not specified (Deshimaru and Shigeno, 1972).

(continued on page 129)

(continued from page 128)

(4) Tacon, A.G.J. (unpublished data).

(Source: A.G.J. Tacon, *Standard Methods for the Nutrition of Farmed Fish and Shrimp*; 1990 - with permission of Argent Laboratories, Inc.)

TABLE 6-19: AVERAGE PROXIMATE COMPOSITION OF THE MAJOR ROOT CROPS AND THEIR BY-PRODUCTS

All Values Are Expressed As % By Weight On An As-fed Basis: Water-H_2O; Crude Protein-CP, Lipid Or Ether Extract-EE; Crude Fiber-CF; Nitrogen Free Extractives-NFE; Ash; Calcium-Ca; Phosphorus-P [1]

ROOT CROP	AVERAGE COMPOSITION (% by weight)								NO. REF. [1]
	H_2O	CP	EE	CF	NFE	Ash	Ca	P	
GIANT TARO/ALOCASIA (*Alocasia macrorrhiza*)									
Fresh tuber	81.2	0.6	0.1	-	-	-	0.15	0.05	(1)
ELEPHANT YAM (*Amorphophallus spp.*)									
Fresh tuber	74.2	5.1	0.4	0.6	18.0	1.7	0.05	0.02	(2)
MANGOLD/MANGEL (*Beta vulgaris macrorhiza*)									
Fresh root	88.5	1.2	0.1	0.8	8.5	0.9	0.02	0.03	(6)
SUGAR BEET (*Beta vulgaris altissima*)									
Fresh root	83.6	1.4	0.1	0.9	13.0	1.0	0.04	0.04	(4)
Beet crowns	82.0	3.0	0.3	1.9	9.2	3.6	-	-	(1)
Beet pulp (sugar extracted; wet)	87.0	1.4	0.2	3.1	7.8	0.5	0.10	0.01	(2)
Beet pulp (sugar extracted; dry)	9.6	8.4	0.5	19.3	58.3	3.9	0.64	0.10	(5)
Pulp with molasses (dry)	8.0	9.2	0.6	15.4	61.0	5.8	0.56	0.10	(2)
Beet molasses	20.4	7.3	0.1	0.0	62.6	9.6	0.10	0.02	(4)
SWEDE (*Brassica napus*)									
Fresh root	91.0	0.8	0.1	1.1	6.3	0.7	0.06	0.02	(2)
TURNIP (*Brassica rapa rapa*)									
Fresh root	91.0	1.2	0.2	1.0	5.8	0.8	0.06	0.02	(3)
QUEENSLAND ARROWROOT (*Canna edulis*)									
Fresh tuber	70.9	1.1	0.2	0.8	25.1	1.9	-	-	(2)
TARO/OLD COCOYAM/DASHEEN (*Colocasia esculenta*)									
Fresh tuber	74.0	1.7	0.2	0.7	22.4	1.0	0.06	0.60	(5)
Fresh tuber (peeled)	67.6	1.9	0.1	0.6	28.7	1.1	-	-	(2)
Fresh peelings	81.2	0.9	0.2	1.7	14.7	1.3	-	-	(1)
SWAMP TARO (*Cyrtosperma chamissonis*)									
Fresh tuber	60.0	1.0	0.5	1.0	36.5	1.0	-	-	(1)
CHUFA/TIGER NUT (*Cyperus esculentus*)									
Tuber	19.8	5.3	24.2	10.0	38.9	1.8	-	-	(1)
CARROT (*Daucus carota*)									
Fresh tuber	86.8	1.5	0.2	1.3	9.0	1.2	-	-	(2)
Pulp fresh	86.0	0.9	1.1	2.6	8.2	1.	-	-	(1)
GREATER YAM/WATER YAM/WINGED YAM (*Dioscorea alata*)									
Fresh tuber	70.0	2.0	0.2	1.0	25.4	1.4	0.04	0.06	(3)
Fresh tuber (peeled)	73.8	1.9	0.2	0.6	22.1	1.4	-	-	(1)
Fresh peelings	74.1	3.0	0.3	1.7	18.4	2 5	-	-	(1)
POTATO YAM (*Dioscorea bulbifera*)									
Fresh tuber	65.0	1.3	<0.1	0.7	31.6	1.3	-	-	(1)
YELLOW YAM (*Dioscorea cavenensis*)									
Fresh tuber	83.0	1.0	<0.1	0.4	15.0	0.5	-	-	(3)
Fresh tuber (peeled)	83.9	0.9	<0.1	0.1	14.6	0.5	-	-	(3)
Fresh peelings	78.3	1.6	0.2	1.6	16.7	1.6	-	-	(1)

TABLE 6-19 (cont.):

ROOT CROP	AVERAGE COMPOSITION (% by weight)								NO. REF. (1)
	H₂O	CP	EE	CF	NFE	Ash	Ca	P	
LESSER YAM (*Dioscorea esculenta*)									
Fresh tuber	74.0	1.6	0.2	0.8	22.5	0.9	0.03	0.03	(3)
Fresh tuber (peeled)	81.4	1.4	<0.1	0.2	16.4	0.5	-	-	(1)
Fresh peelings	93.0	0.7	<0.1	0.5	5.3	0.4	-	-	(1)
BITTER YAM (*Dioscorea dumetorum*)									
Fresh tuber	79.0	2.8	0.3	0.3	16.9	0.7	0.09	-	(1)
INTOXICATING YAM (*Dioscorea hispida*)									
Fresh tuber	78.0	1.8	0.2	0.9	18.4	0.7	-	-	(1)
WHITE YAM (*Dioscorea rotundata*)									
Fresh tuber	65.5	1.5	0.1	0.6	30.7	1.6	-	-	(1)
Fresh tuber (peeled)	75.9	1.1	<0.1	0.4	21.8	0.7	-	-	(1)
Fresh peelings	82.3	2.0	0.2	1.7	12.1	1.7	-	-	(1)
JERUSALEM ARTICHORE (*Helianthus tuberosus*)									
Fresh tuber	77.9	2.1	0.2	1.2	16.4	1.7	0.02	0.10	(5)
SWEET POTATO/SPANISH POTATO (*Ipomoea batatas*)									
Fresh tuber 70.9	1.5	0.3	0.8	25.6	0.9	0.05	0.06	(8)	
Dried tuber meal	12.6	4.2	0.7	4.2	74.9	3.4	0.09	0.13	(5)
Fresh peelings	88.3	0.7	0.2	<0.1	10.2	0.5	-	-	(1)
CASSAVA/TAPIOCA/MANIOC/MANIHOT (*Manihot esculenta*)									
Fresh tuber	65.9	0.9	0.2	1.0	30.9	1.0	0.03	0.05	(8)
Tuber (dehydrated)	13.5	2.1	0.5	3.8	77.9	2.2	0.17	0.11	(8)
Fresh tuber (peeled)	68.8	0.9	0.2	0.5	28.6	1.0	0.03	0.01	(3)
Fresh peelings	72.1	1.6	0.4	4.4	20.1	1.4	-	-	(2)
Cassava meal (starch extracted)	14.8	1.3	0.6	13.5	67.5	2.3	0.50	0.03	(2)
ARROWROOT (*Maranta arundinacea*)									
Rhizome (root; fresh)	70.5	1.6	0.1	1.0	25.4	1.4	-	-	(2)
Rhizome (starch extracted; dry)	12.2	3.0	0.3	14.4	67.7	2.4	0.30	0.15	(2)
OCA (*Oxalis tuberosa*)									
Fresh tuber	84.0	1.1	0.8	1.0	12.3	0.8	-	-	(2)
YAM BEAN/POTATO BEAN (*Pachyrrhizus erosus*)									+
Fresh tuber	82.4	1.5	0.1	0.6	14.9	0.5	0.02	-	(1)
Young green pods	86.4	2.6	0.3	2.9	7.1	0.7	0.12	0.04	(1)
Seed (mature; dry)	6.7	26.2	27.3	7.0	29.2	3.6	-	-	(1)
YACON STRAWBERRY (*Polymnia sonchifolia/ P. edulis*)									
Fresh tuber	75.2	1.4	0.1	1.5	20.2	1.6	-	-	(3)
KUDZU (*Pueraria lobata*)									
Fresh root (peeled)	68.6	2.1	0.1	0.7	27.1	1.4	0.02	0.02	(1)
RADISH (*Raphanus sativus*)									
Fresh root	92.4	0.8	0.1	0.7	5.2	0.8	0.04	0.02	(1)
POTATO/IRISH POTATO (*Solanum tuberosum*)									
Fresh tuber	76.7	2.3	0.1	0.7	19.1	1.01	0.02	0.02	(7)
Tuber (dry meal)	9.9	7.9	0.3	1.7	75.5	4.7	0.07	0.20	(7)
Fresh peelings	78.8	2.1	0.1	0.7	17.0	1.3	-	-	(1)
Pulp residue (starch extracted)	11.6	7.9	0.3	5.9	70.7	3.6	0.14	0.23	(2)
HAUSA POTATO (*Solenostemon rotundifolius*)									
Fresh tuber	75.2	1.4	0.4	0.6	21.4	1.0	0.02	-	(1)
AFRICAN YAM BEAN (*Sphenostylis stenocarpa*)									
Fresh tuber	64.7	3.7	0.1	0.4	30.4	0.7	0.01	-	(1)
NEW COCOYAM/TANNIA (*Xanthosoma sagittifolium*)									
Fresh tuber	70.0	2.1	0.2	0.9	25.8	1.0	0.04	0.06	(3)

TABLE 6-19 (cont.):

ROOT CROP	AVERAGE COMPOSITION (% by weight)								NO. REF. (1)
	H_2O	CP	EE	CF	NFE	Ash	Ca	P	
Fresh tuber (peeled)	75.9	1.4	<0.1	0.3	21.1	1.2	-	-	(1)
Fresh peelings	70.5	2.4	0.4	3.4	20.8	2.5	-	-	(1)

[1] The data presented represents the mean values from various sources, including: Allen (1984; Bath et al., (1984); Bolton and Blair (1977); Branckaert, Tessema and Temple (1976); Cooledy (1976); Devendra (1979); Gohl (1981; Hastings (1973); Hickling (1971); Kay (1973); MacDonald, Edwards and Greenhalgh (1977; Miller (1976); NRC (1982); and Platt (1962)

(Source: A.G.J. Tacon, *Standard Methods for the Nutrition of Farmed Fish and Shrimp;* 1990 - with permission of Argent Laboratories, Inc.)

TABLE 6-20: AVERAGE ESSENTIAL AMINO ACID (EAA) COMPOSITION OF THE MAJOR ROOT CROPS AND THEIR BY-PRODUCTS

All values are expressed as % by weight on a as-fed basis: Arginine-ARG; Cystine-CYT; Methionine-MET; Threonine-THR; Isoleucine-ISO; Leucine-LEU; Lysine-LYS; Valine-VAL; Tyrosine-TYR; Tryptophan-TRY; Phenylalanine-PHE; Histidine-HIS [1]

ROOT CROP	AVERAGE EAA COMPOSITION (% dry meal or % total recovered amino acids)												NO. REF. (1)
	ARG	CYT	MET	THR	ISO	LEU	LYS	VAL	TYR	TRY	PHE	HIS	
ELEPHANT YAM (*Amorphophallus spp.*)													
Fresh tuber	0.247	0.055	0.033	0.093	0.083	0.135	0.107	0.140	0.083	-	0.112	0.055	(1)
SUGAR BEET (*B. vugaris*)													
Beet pulp (dehydrated)	0.30	0.01	0.01	0.40	0.30	0.60	0.60	0.40	0.40	0.10	0.30	0.20	(2)
TARO (*C. esculenta*)													
Fresh tuber	0.162	0.047	0.024	0.074	0.064	0.133	0.070	0.111	0.066	0.026	0.092	0.032	(1)
CHUFA/TIGER NUT (*C. esculentus*)													
Fresh tuber	0.532	0.042	0.035	0.133	0.105	0.199	0.175	0.182	0.073	0.035	0.119	0.059	(1)
YAM (*Dioscorea spp.*)													
Fresh tuber	0.181	0.027	0.038	0.86	0.089	0.154	0.097	0.110	0.076	0.030	0.114	0.045	(1)
SWEET POTATO (*I. batatas*)													
Fresh tuber	0.064	0.014	0.022	0.050	0.048	0.071	0.045	0.059	0.031	0.022	0.051	0.018	(1)
CASSAVA (*M. esculenta*)													
Tuber (meal; dehydrated)	0.178	0.023	0.022	0.043	0.046	0.064	0.067	0.054	0.026	0.019	0.041	0.034	(1)
POTATO (*S. tuberosum*)													
Fresh tuber	0.100	0.012	0.026	0.075	0.076	0.121	0.096	0.093	0.055	0.033	0.080	0.030	(1)
Tuber meal (dehydrated)	0.32	0.11	0.12	0.39	0.28	0.79	0.42	0.39	0.06	0.10	0.43	0.15	(2)
HAUSA POTATO (*S. rotundifolius*)													
Fresh tuber	0.106	0.009	0.030	0.075	0.060	0.076	0.075	0.081	0.052	0.013	0.069	0.029	(1)
NEW COCOYAM (*Xanthosoma spp.*)													
Fresh tuber	0.153	0.061	0.018	0.065	0.059	0.124	0.070	0.112	0.062	0.030	0.096	0.032	(1)

[1] The data presented represents the mean values from various sources including, Allen (1984); FAO (1970); and NRC (1982)

(Source: A.G.J. Tacon, *Standard Methods for the Nutrition of Farmed Fish and Shrimp;* 1990 - with permission of Argent Laboratories, Inc.)

TABLE 6-21: AVERAGE PROXIMATE COMPOSITION OF THE MAJOR EDIBLE FRUITS AND THEIR BY-PRODUCTS

All values are expressed as % by weight on a as-fed basis; Water-H_2O; Crude Protein-CP; Lipid or Ether Extract-EE: Crude Fiber-CF; Nitrogen Free Extractives-NFE; Ash; Calcium-Ca; Phosphorus-P [1]

| FRUIT | AVERAGE COMPOSITION (% by weight) | | | | | | | | NO. REF. (1) |
	H_2O	CP	EE	CF	NFE	Ash	Ca	P	
PINEAPPLE (*Ananas comosus*)									
Fruit (ripe; fresh)	85.3	0.4	0.2	0.4	13.3	0.4	0.02	0.01	(1)
Stump meal (fresh)	54.0	1.4	0.4	10.1	33.2	0.9	0.13	0.04	(1)
Juice presscake	79.0	1.1	0.2	5.5	13.6	0.6	0.06	0.02	(1)
Cannery residue (pulp/bran; dehyd.)	11.7	3.6	1.1	15.9	64.2	3.5	0.21	0.12	(4)
BREADFRUIT (*Artocarpus altilis*)									
Fruit (ripe; fresh)	70.2	1.7	0.3	1.5	24.3	2.0	0.04	0.04	(1)
Fruit (ripe; cooked & peeled)	68.2	1.5	0.3	1.4	27.6	1.0	-	-	(1)
Fruit meal (dehydrated)	15.1	2.7	0.8	4.7	74.1	2.6	0.07	0.14	(1)
BREADNUT TREE (*Brosimum alicastrum*)									
Fruit pulp (fresh)	84.9	3.1	0.8	1.8	8.4	1.0	-	-	(1)
Fruit seeds (fresh)	63.0	4.7	1.6	3.4	25.9	1.4	-	-	(2)
Fruit fiber and skin (fresh)	86.6	0.9	0.6	2.4	8.0	1.5	-	-	(1)
PAPAYA/PAWPAW (*Carica papaya*)									
Fruit (ripe; fresh)	88.0	0.8	0.1	1.0	4.4	0.8	0.02	0.01	(2)
Fruit (immature; fresh)	92.8	0.8		0.9	4.9	0.5	-	-	(1)
WATERMELON (*Citrullus lanatus*)									
Fruit (ripe; fresh)	95.9	0.5	0.1	0.9	2.3	0.3	-	-	(1)
Seeds (dry)	8.5	8.3	17.5	42.9	20.6	2.2	-	-	(1)
LIME (*Citrus aurantiifolia*)									
Whole fruit (ripe; fresh)	68.1	3.6	3.0	13.5	9.7	2.1	-	-	(1)
Fruit skin (peel) and rag (fiber)	81.7	1.4	0.9	3.1	12.2	0.7	-	-	(1)
Seed (fresh)	70.9	6.4	3.6	3.8	14.7	0.6	0.05	0.09	(1)
Silage of skins (peels)	77.0	2.4	1.5	4.8	12.1	2.2	-	-	(1)
Fruit pulp (dehydrated)	15.0	7.7	2.9	15.2	-	-	-	-	(1)
LEMON (*Citrus limon*)									
Fruit pulp (dehydrated)	7.0	6.4	1.4	14.8	65.1	5.3	-	-	(1)
GRAPE FRUIT (*Citrus paradisi*)									
Whole fruit (ripe; fresh)	86.6	1.0	0.5	1.3	10.1	0.5	0.09	0.02	(2)
Fruit pulp (wet)	79.7	1.3	<0.1	2.1	16.0	0.8	0.12	-	(1)
Fruit pulp (dehydrated)	9.0	6.1	1.4	12.6	65.4	5.5	1.30	0.16	(1)
Fruit skin (peels; fresh)	82.1	1.2	0.3	1.9	13.8	0.7	-	-	(1)
Silage of fruit peels (fresh)	80.8	1.4	0.4	2.5	14.1	0.8	-	-	(1)
TANGERINE (*Citrus reticulata*)									
Fruit pulp (dehydrated)	13.0	7.0	4.9	9.6	61.1	4.4	1.40	0.12	(1)
SWEET ORANGE (*Citrus sinensis*)									
Whole fruit (ripe; fresh)	87.2	1.0	0.2	1.3	9.7	0.6	0.07	0.02	(2)
Fruit skin (peels; fresh)	83.9	1.1	0.3	1.0	13.1	0.6	0.21	0.02	(1)
Silage of fruit peels (fresh)	80.4	1.5	0.5	2.8	13.8	1.0	0.27	0.02	(1)
Fruit pulp (wet)	75.0	2.2	0.4	3.3	18.2	0.9	0.05	0.07	(1)
Fruit pulp (silage)	88.7	1.0	0.2	2.0	7.5	0.6	-	-	(1)
Fruit pulp (dehydrated)	11.1	7.5	2.0	10.2	65.8	3.4	0.63	0.09	(3)
CITRUS PULP (*Citrus spp.*)									
Citrus pulp (fresh)	81.7	1.2	0.6	2.3	12.8	1.4	-	-	(1)
Citrus pulp (silage)	80.0	1.5	2.1	3.2	12.1	1.1	0.42	.03	(2)
Citrus pulp (dehydrated)	9.1	6.3	3.3	12.4	62.9	6.0	1.80	0.11	(6)
MOLASSES (*Citrus spp.*)									
Citrus molasses (fresh)	32.0	5.6	0.2	0.0	57.3	4.9	1.12	0.09	(4)
COFFEE (*Coffea arabica/robusta*)									
Fruit pulp (fresh)	76.8	2.4	0.5	4.6	13.8	1.9	0.13	0.03	(3)
Fruit pulp (sun dried)	11.4	10.9	2.3	22.9	44.8	7.7	0.53	0.11	(7)

TABLE 6-21 (cont.):

FRUIT	AVERAGE COMPOSITION (% by weight)								NO. REF. (1)
	H₂O	CP	EE	CF	NFE	Ash	Ca	P	
Seed hulls (dried)	8.8	2.3	0.6	68.9	18.9	0.5	-	-	(2)
PUMPKIN/SQUASH/GOURD (*Cucurbita* spp.)									
Fruit (ripe; fresh)	91.5	1.2	0.4	1.0	5.2	0.7	0.02	0.04	(3)
MANGO (*Mangifera indica*)									
Fruit pulp (immature fruit; fresh)	82.3	6.2	<0.1	0.5	10.6	0.3	0.04	0.02	(1)
Fruit pulp (mature fruit; fresh)	82.7	1.0	0.1	0.4	15.4	0.4	-	-	(1)
Fruit kernel (seed; fresh)	50.0	4.2	4.4	1.4	37.3	2.7	-	-	(1)
Fruit silage (wet)	84.0	0.8	1.0	2.7	10.0	1.5	0.03	0.01	(1)
TOMATO (*Lycopersicon esculentum*)									
Whole fruit (ripe; fresh)	93.8	1.0	0.2	0.6	3.7	0.7	0.01	0.03	(1)
Pomace (pulp; dehydrated)	8.1	21.4	10.3	24.8	30.1	5.3	0.36	0.56	(4)
Pomace (pulp; silage; wet)	70.5	5.7	4.3	13.2	5.0	1.3	0.15	0.14	(1)
Fruit skins with juice (dried)	10.4	18.5	2.2	17.1	43.0	8.8	0.55	0.41	(1)
APPLE (*Malus sylvestris*)									
Whole fruit (ripe; fresh)	83.0	0.5	0.4	1.2	14.5	0.4	0.01	0.01	(1)
Fruit pomace (pulp; dried)	11.0	4.4	4.5	15.1	63.0	2.0	0.12	0.11	(1)
Fruit pomace (pulp; silage; wet)	78.6	1.7	1.3	4.4	13.0	1.0	0.02	0.02	(1)
BANANA/PLANTAIN (*Musa sapientum/M. paradisiaca*)									
Banana fruit (immature/green; fresh)	80.6	0.9	0.5	0.6	16.5	0.9	-	-	(2)
Banana fruit (ripe; fresh)	76.0	1.3	0.3	0.7	20.7	1.0	0.01	0.03	(3)
Peeled fruit (immature; fresh)	74.9	0.9	0.4	0.2	22.8	0.8	-	-	(1)
Peeled fruit (ripe; fresh)	69.5	1.3	0.2	<0.1	27.5	1.4	-	-	(1)
Green fruit with peel (meal)	12.0	4.3	2.8	3.0	73.6	4.3	-	-	(1)
Ripe fruit (dried)	14.0	3.5	0.5	1.0	78.4	2.6	0.03	0.09	(1)
Fruit skins (peels; ripe; fresh)	85.9	1.1	1.6	1.1	8.4	1.9	-	-	(1)
Fruit skins (peels; ripe; dried)	12.0	6.8	7.1	7.6	57.3	9.2	-	-	(1)
Fruit skins (peels; immature; dried)	10.0	6.9	5.4	11.7	51.2	14.8	-	-	(1)
Plantain fruit (ripe; fresh)	68.8	1.1	0.2	0.3	30.5	1.1	0.22	0.08	(2)
Plantain fruit (green with peel; meal)	10.0	4.3	1.0	6.2	74.0	4.5	-	-	(1)
Plantain peels (mature; fresh)	81.6	1.7	1.0	1.2	11.3	3.2	-	-	(1)
AVOCADO (*Persea americana*)									
Avocado seeds (fresh)	59.0	2.0	1.6	2.4	-	-	0.02	0.08	(1)
Avocado skins (fresh)	76.0	1.7	8.4	5.9	-	-	0.03	0.04	(1)
Avocado oil meal	9.0	18.5	1.1	17.6	42.5	11.3	-	-	(1)
DATE PALM (*Phoenix dactylifera*)									
Whole fruit date (dried)	25.7	2.2	0.7	4.8	62.4	4.2	-	-	(1)
Fruit seeds (ground; dried)	9.8	5.9	8.1	14.1	59.2	2.9	-	-	(2)
Fruit pulp (sugar extracted)	11.8	4.8	0.3	10.4	70.3	2.4	-	-	(1)
PRUNE (*Prunus* spp.)									
Fruit with seeds	18.0	4.3	2.4	10.7	-	-	0.11	0.09	(1)
Fruit without seeds	20.0	3.3	1.1	1.8	-	-	0.03	0.09	(1)
Fruit (prune) mix	18.2	5.2	2.0	16.2	53.3	5.1	-	-	(1)
POMEGRANATE (*Punica granatum*)									
Fruit pulp (wet)	74.0	2.2	1.3	4.3	17.2	1.0	-	-	(1)
PEAR (*Pyrus communis*)									
Fruit pulp (ground; dried)	8.5	5.6	1.9	21.8	58.5	3.7	2.20	0.11	(1)
Fruit cannery residue (wet)	84.8	0.6	0.2	2.6	11.5	0.3	-	-	(1)
RAISIN (*Vitis* spp.)									
Fruit pulp (dried)	11.0	9.5	7.7	16.0	50.4	5.4	-	-	(1)
GRAPE (*Vitis vinifera*)									
Seeds	11.0	9.4	9.8	41.3	25.2	3.3	0.58	0.20	(2)
Fruit pomace (dried)	9.0	11.6	6.9	30.0	37.5	5.0	0.46	0.36	(1)
Winery pomace (stalk; skin; seed; wet)	59.4	4.7	4.0	10.4	18.4	3.1	-	-	(1)
Winery pomace (skin; seed; wet)	53.5	6.4	3.2	11.0	19.9	6.0	0.38	0.09	(1)
Winery pomace (stalk; skin; dried)	11.2	13.2	4.4	31.8	31.5	7.9	-	-	(1)
Winery pomace (skin; dried)	11.1	16.3	5.7	28.4	31.4	7.1	1.45	0.29	(1)

(continued on page 134)

(continued from page 133)

[1] The data presented represents the mean values from various sources, including: Allen (1984); Bath *et al.*, (1984): Branckaert, Tessema and Temple (1976): Cooley (1976); Gohl (1981); Janssen (1985): Ling (1967); Miller (1976); NRC (1982) Platt (1962); Springhall (1969); and Tacon (1986).

(Source: A.G.J. Tacon, *Standard Methods for the Nutrition of Farmed Fish and Shrimp*; 1990 - with permission of Argent Laboratories, Inc.)

TABLE 6-22: AVERAGE PROXIMATE COMPOSITION OF THE MAJOR AQUATIC MACROPHYTES USED FOR AQUACULTURE FEEDING

All values are expressed as % by weight on a as-fed basis: Water-H_2O; Crude Protein-CP; Lipid or Ether Extract-EE; Crude Fiber-CF; Nitrogen-Free Extractives-NFE; Ash; Calcium-Ca; Phosphorus-P[1]

MACROPHYTE	AVERAGE COMPOSITION (% by weight)								NO. REF. [1]
	H_2O	CP	EE	CF	NFE	Ash	Ca	P	
ALLIGATOR WEED (*Alternanthera philoxeroides*)									
Whole plant (fresh)	84.1	2.4	0.4	2.4	7.5	3.2	-	-	(4)
Whole plant (dry matter basis)	0	15.1	2.5	15.1	47.2	20.1			
AQUATIC FERN (*Azolla* spp.)									
Whole plant (fresh)	93.5	1.7	0.3	0.6	3.2	0.9	0.07	0.03	(2)
Whole plant (dry matter basis) [2]	0	25.3	3.8	9.3	49.1	12.5	1.16	0.59	(4)
COONTAIL (*Ceratophyllum demersum*)									
Whole plant (fresh)	93.1	1.3	0.3	1.7	2.0	1.6	0.06	0.04	(4)
Whole plant (dry matter basis)	0	17.9	3.8	18.3	40.5	19.5	1.30	0.32	(7)
CHARA (*Chara vulgaris/Chara* spp.)									
Whole plant (fresh)	93.1	1.3	0.3	1.7	2.0	1.6	0.06	0.04	(4)
Whole plant (fresh)	91.6	1.5	0.1	2.0	2.0	2.7	-	-	(2)
Whole plant (dry matter basis) [3]	0	8.8	0.8	14.0	48.1	28.3	-	-	(4)
WATER HYACINTH (*Eichhornia crassipes*)									
Whole plant (fresh)	91.5	1.2	0.3	1.9	3.8	1.3	0.18	0.09	(14)
Whole plant (dried)	10.9	14.8	2.9	22.9	26.4	22.1	1.69	0.37	(5)
Whole plant compost (dried)	10.5	14.2	1.3	9.4	20.0	44.6	-	-	(1)
Whole plant (silage fresh)	89.9	1.0	0.1	2.0	5.1	1.9	-	-	(1)
CANADIAN PONDWEED (*Elodea canadensis*)									
Whole plant (fresh)	91.1	1.9	0.3	2.0	3.1	1.6	0.19	0.04	(6)
Whole plant (dry matter basis)	0	18.0	2.9	14.7	44.7	19.7	1.75	0.36	(7)
HYDRILLA (*Hydrilla verticillata*)									
Whole plant (fresh)	91.7	1.8	0.3	2.6	1.5	2.0	-	-	(2)
Whole plant (dry matter basis)	0	23.1	4.1	30.2	15.6	27.0	4.40	0.28	(1)
KANGKONG/WATER BIND-WEED (*Ipomoea aquatica/I. reptans*)									
Leaves and stem (fresh)	92.5	2.1	0.2	0.9	2.9	1.4	0.09	0.03	(1)
Leaves and stem (dry matter basis)	0	28.0	2.7	12.0	38.6	18.7	1.20	0.40	(1)
WATER WILLOW (*Justicia americana*)									
Whole plant (fresh)	85.0	3.4	0.5	3.9	4.6	2.6	-	-	(1)
Whole plant (dry matter basis)	0	17.6	3.5	24.0	38.8	16.1	0.82	0.12	(3)
DUCKWEED (*Lemna minor*)									
Whole plant (fresh)	91.9	1.7	0.5	0.9	4.0	0.9	-	-	(3)
Whole plant (dry matter basis)	0	20.9	4.1	13.2	48.2	13.6	1.75	0.17	(5)
MILFOIL (*Myriophyllum* spp.)									
Whole plant (fresh)	88.7	2.0	0.3	1.9	5.3	1.9	-	-	(5)
Whole plant (dry matter basis)	0	20.3	2.5	13.9	45.1	18.2	2.82	0.41	(11)
NAJAS (*Najas guadalupensis*)									
Whole plant (fresh)	90.4	2.3	0.4	2.9	2.6	1.4	-	-	(2)
Whole plant (dry matter basis)	0	23.9	4.2	30.2	27.1	14.6	0.98	0.15	(2)

TABLE 6-22 (cont.):

MACROPHYTE	AVERAGE COMPOSITION (% by weight)								NO. REF. (1)
	H$_2$O	CP	EE	CF	NFE	Ash	Ca	P	
WATER LETTUCE (*Pistia stratiotes*)									
Whole plant (fresh)	93.6	1.2	0.3	1.0	2.3	1.6	-	-	(6)
Whole plant (dry matter basis)	0	15.9	4.2	20.8	36.1	23.0	2.35	0.30	(2)
POND WEED (*Potamogeton* spp.)									
Whole plant (fresh)	85.0	2.0	0.4	3.1	7.1	2.4	-	-	(4)
Whole plant (dry matter basis)	0	13.1	2.1	20.0	46.1	18.7	1.68	0.24	(13)
SAGITTARIA (*Sagittaria* spp.)									
Whole plant (fresh)	85.0	2.6	1.0	4.1	5.8	1.5	-	-	(1)
Whole plant (dry matter basis)	0	18.2	6.6	23.9	42.4	8.9	0.83	0.35	(5)
SALVINIA (*Salvinia auriculata/S. molesta*)									
Whole plant (fresh) (4)	77.2	1.8	0.6	7.7	11.2	1.5	-	-	(1)
Whole plant (dry matter basis)	0	7.9	2.6	33.8	49.1	6.6	-	-	(1)
BURREED (*Sparganium americanum*)									
Whole plant (fresh)	89.1	2.6	0.9	2.2	4.0	1.2	-	-	(1)
Whole plant (dry matter basis)	0	23.8	8.3	20.2	36.7	11.0	-	-	(1)
REED-MACE (*Typha latifolia*)									
Whole plant(fresh)	77.1	2.4	0.9	7.6	10.4	1.6	-	-	(1)
Whole plant (dry matter basis)	0	10.7	3.9	30.3	47,0	8.1	0.64	0.17	(3)
WOLFFIA (*Wolffia spp.*)									
Whole plant (fresh) (4)	96.4	1.0	0.3	-	1.0	0.6	-	-	(1)
Whole plant (dry matter basis)	0	27.8	8.3	-	47.2	16.7	-	-	(1)
MARINE MULTICELLULAR ALGAE									
Chaetomorpha spp. (fresh)	90.4	3.1	0.6	1.1	2.5	2.3	-	-	(1)
Enteromorpha intestinalis (fresh)	81.4	3.7	0.5	-	-	6.0	-	-	(1)

(1) The data presented represents the mean values from various sources including: Boyd (1968, 1969, 70); Carraro (1983); Edwards, Kamai and Wee (1985); Gohl (1981); Ling (1967); Linn (1975); Little (1979); Little and Henson (1967); Pullin and Almazan (1983); and Siriwardene, Ranawana and Piyasena (1970).

(2) Carraro (1982) reports composition of *Azolla* sp. as 23.4% crude protein, 22.1% cellulose, 11.3% hemicellulose, 23.0% lignin and 14.5% ash on a dry matter basis.

(3) Composition reported in the literature reviewed was highly variable, crude protein ranging from 4.5 to 17.5% and ash ranging from 5.6 to 41.2% on a dry weight basis.

(4) Data obtained from the study of Siriwardene, Ranawana and Piyasena (1970), although Little and Herson (1967) report the water content of the fresh whole plant as 92.8%.

(5) Data for crude fiber and nitrogen-free extractives reported together (Ling, 1967).

(Source: A.G.J. Tacon, *Standard Methods for the Nutrition of Farmed Fish and Shrimp*; 1990 - with permission of Argent Laboratories, Inc.)

TABLE 6-23: AVERAGE ESSENTIAL AMINO ACID (EAA) COMPOSITION OF SOME AQUATIC MACROPHYTES

All values are expressed as % by weight on a as-fed basis: Arginine-ARG; Cystine-CYT; Methionine-MET; Threonine-THR; Isoleucine-ISO; Leucine-LEU; Lysine-LYS; Valine-VAL; Tyrosine-TYR; Tryptophan-TRY; Phenylalanine-PHE; Histidine-HIS [1]

| MACROPHYTE | AVERAGE EAA COMPOSITION (%) | | | | | | | | | | | | NO. REF. [1] |
	ARG	CYT	MET	THR	ISO	LEU	LYS	VAL	TYR	TRY	PHE	HIS	
ALLIGATOR WEED (*A. philoxeroides*)	2.10	-	0.60	1.60	1.50	1.90	1.50	1.80	-	-	Trace	1.10	(1)
AQUATIC FERN (*Azolla* spp.)	0.78	0.07	1.24	1.04		3.15	0.44	1.26	0.66	1.60	1.18	0.19	(1)
WATER HYACINTH (*E. crassipes*)	1.18	0.05	0.35	0.97	1.00	1.76	1.21	1.16	0.76	-	1.06	0.42	(1)
HYDRILLA (*H. verticillata*)	0.73	0.05	0.28	0.66	0.68	1.25	0.72	0.82	0.62	-	0.80	0.25	(1)
WATER WILLOW (*J. americana*)	3.00	-	0.90	2.30	2.50	4.30	2.80	2.90	-	-	2.80	1.10	(1)
WATER LETTUCE (*P. stratiotes*)	0.83	0.06	0.31	0.88	0.92	1.62	1.21	1.10	0.73	-	1.02	0.39	(1)
SAGITTARIA (*Sagittaria* spp.)	1.10	-	0.20	1.00	0.90	1.70	1.60	1.40	-	-	Trace	0.60	(1)

[1] The data presented represents the mean values from various sources, including: Boyd (1968, 1969) and Carraro (1983).

(Source: A.G.J. Tacon, *Standard Methods for the Nutrition of Farmed Fish and Shrimp*; 1990 - with permission of Argent Laboratories, Inc.)

TABLE 6-24: AVERAGE PROXIMATE COMPOSITION OF THE MAJOR CEREAL GRAINS AND THEIR BY-PRODUCTS

All values are expressed as % by weight on a as-fed basis: Water-H_2O; Crude Protein-CP; Lipid or Ether Extract-EE; Crude Fiber-CF; Nitrogen-Free Extractives-NFE; Ash; Calcium-Ca; Phosphorus-P [1]

| CEREAL | AVERAGE COMPOSITION (% by weight) | | | | | | | | NO. REF. [1] |
	H_2O	CP	EE	CF	NFE	Ash	Ca	P	
BARLEY (*Hordeum vulgare/H. distichum*)									
Grain	12.4	10.5	1.8	5.6	67.1	2.6	0.05	0.37	(11)
Bran	10.0	11.6	3.4	14.6	55.4	5.0	0.36	0.70	(2)
Middlings	11.0	14.5	4.5	9.3	56.4	4.3	-	-	(1)
Mill run	10.0	10.5	2.5	14.1	58.8	4.1	-	-	(2)
Grain screenings (sweepings)	11.0	11.2	2.3	8.4	66.0	3.1	0.33	0.29	(3)
Pearl by-product	10.4	13.2	3.5	10.7	57.1	5.1	0.04	0.41	(1)
Malt (dehydrated)	9.0	i4.1	1.7	2.6	70.4	2.2	0.07	0.46	(2)
Malt sprouts (culms; fresh)	86.8	2.4	0.7	2.6	7.0	0.5			
Malt sprouts (culms; dehydrated)	8.4	25.4	1.7	14.4	43.8	6.3	0.22	0.73	(5)
Brewers grains (fresh)	74.9	6.2	2.0	4.0	11.7	1.2	0.06	0.13	(3)
Brewers grains (dehydrated)	9.4	20.8	5.7	15.3	45.3	5.1	0.29	0.54	(11)
Brewers grains (silage)	74.5	6.0	1.9	4.7	10.8	1.6			
Distillers dried grains	8.0	27.7	11.6	10.1	40.8	1.8			
Distillers grains (fresh)	76.1	4.7	2.0	4.1	12.3	0.8			
Distillers dried solubles	5.1	26.8	0.2	-	-	17.2	4.60	-	(1)
Distillers dried grains with solubles	9.9	26.0	6.9	12.9	39.7	4.6	0.20	0.79	(2)
CORN/MAIZE (*Zea mays*)									
Grain (ground)	12.2	9.6	3.9	2.0	70.8	1.5	0.02	0.28	(22)
Grain (flaked)	11.0	10.0	3.6	1.2	73.2	1.0	-	-	(3)
Gluten feed	10.3	23.7	2.4	7.1	50.7	5.8	0.30	0.64	(5)
Gluten meal	9.9	45.d	2.7	3.7	34.7	3.2	0.10	0.43	(5)
Hominy feed	9.7	10.7	5.8	5.0	66.2	2.6	0.05	0.50	(5)
Germ oil meal (oilcake)	8.9	17.5	14.1	8.2	41.1	3.2	0.18	0.45	(5)
Feed meal	12.5	9.0	4.5	3.5	68.0	2.5	0.05	0.40	(1)

136

TABLE 6-24 (cont):

CEREAL	AVERAGE COMPOSITION (% by weight)								NO. REF. (1)
	H₂O	CP	EE	CF	NFE	Ash	Ca	P	
Corn-and-cob meal (corn ears)	12.8	7.8	3.1	8.6	66.2	1.5	0.05	0.22	(3)
Cobs (ground meal)	9.7	2.5	0.6	34.5	51.2	1.5	0.10	0.06	(4)
Cannery process residue (waste; fresh)	77.0	2.0	0.6	5.1	13.9	1.4	0.18	0.14	(1)
Cannery process residue (waste; silage)	69.5	2.5	1.2	9.5	15.7	1.6	0.10	0.26	(2)
Distillers dried grains	7.0	27.2	8.9	12.0	42.4	2.5	0.09	0.39	(4)
Distillers dried solubles	6.8	26.7	9.0	4.3	45.4	7.8	0.32	1.19	(3)
Distillers dried grains with solubles	8.4	26.7	9.9	8.7	41.6	4.7	0.24	0.76	(3)
MILLET (pearl or bulrush millet-*Pennisetum typhoideum*; foxtail; Italian millet-*Setaria italica*; Japanese millet-*Echinochloa crusgalli*; broom corn millet-*Panicum miliaceum*; finger millet-*Eleusine coracana*; Scrobic millet-*Paspalum scrobiculatum*)									
Grain	10.7	11.2	3.9	6.3	64.6	3.3	0.06	0.30	(15)
Hulls	8.7	4.8	1.3	38.3	41.2	5.7	0.60	0.30	(3)
OATS (*Avena sativa*)									
Grain	11.5	10.4	4.8	11.5	58.4	3.4	0.10	0.32	(12)
Dehulled grain (naked oats; groats)	10.9	13.6	6.4	2.8	64.0	2.3	0.09	0.39	(5)
Oatmeal/middlings (feeding)	9.5	15.9	5.7	2.9	63.8	2.2	0.06	0.43	(2)
Oat-mill feed	7.7	5.0	1.6	28.5	51.1	6.1	0.12	0.12	(4)
Oat shorts	9.0	12.8	5.6	13.5	54.3	4.8	-	-	(1)
Oat sprouts (fresh)	86.8	2.4	0.7	2.6	7.0	0.5	-	-	(1)
RICE (*Oryza sativa*)									
Rough (paddy) rice	11.2	8.3	1.6	9.4	65.1	4.4	0.07	0.26	(5)
Brown (cargo) rice (dehulled)	9.0	9.1	1.6	1.0	78.2	1.0	0.07	0.65	(3)
Broken (brewers) rice (rice meal)	11.3	7.5	0.6	0.3	79.7	0.6	0.19	0.13	(3)
Polished (milled) rice	11.8	7.1	0.3	0.3	79.7	0.8	0.06	0.18	(5)
Hulls (husk. chaff)	9.4	3.7	1.0	36.9	32.6	16.4	0.09	0.07	(8)
Bran	10.0	12.2	11.8	12.3	40.6	13.1	0.12	1.38	(13)
Bran (solvent extracted)	10.5	12.3	2.1	14.6	47.9	12.6	0.20	1.33	(7)
Polishings	10.0	12.1	11.5	4.7	52.9	8.8	0.05	1.26	(8)
Pollards (mixture of bran/polishings)	11.1	12.8	11.7	7.6	48.0	8.8	0.05	1.41	(3)
Rice-mill feed (mixture of hulls/bran)	8.3	6.6	5.3	29.4	36.1	14.3	0.10	0.45	(4)
Distillers spent rice (fresh)	62.0	3.4	1.3	0.8	28.5	4.0	-	-	(1)
RYE (*Secale cereale*)									
Grain	13.0	11.2	1.5	2.3	70.3	1.7	0.06	0.34	(6)
Bran	11.1	15.9	2.9	6.3	59.3	4.5	0.10	0.74	(3)
Middlings	10.5	16.4	3.3	5.0	61.2	3.6	0.06	0.62	(2)
Mill run	10.0	16.7	3.3	4.6	61.6	3.8	0.07	0.64	(1)
Distillers dried grains	8.0	20.9	7.4	12.7	48.6	2.4	0.14	0.43	(2)
Distillers dried solubles	5.6	35.1	1.2	3.4	47.5	7.2	0.35	1.20	(1)
Distillers dried grains with solubles	9.5	27.2	4.1	8.1	44.7	6.4	-	-	(1)
SORGHUM (*Sorghum bicolor/S. vulgare*)									
Grain	11.2	10.6	3.0	1.9	71.4	1.9	0.08	0.27	(19)
Bran	12.0	7.8	4.8	7.6	65.7	2.1	-	-	(1)
Gluten feed	9.6	23.7	3.6	8.4	46.7	8.0	0.13	0.63	(4)
Gluten meal	9.2	42.0	4.9	3.9	37.6	2.4	0.05	0.40	(3)
Hominy feed	11.0	10.0	5.8	3.4	67.4	2.4	-	-	(1)
Distillers dried grains	6.0	31.8	8.7	12.1	37.7	3.7	0.14	0.64	(2)
Distillers dried solubles	7.0	26.5	5.5	3.9	48.6	8.5	0.37	0.61	(1)
Distillers dried grains with solubles	5.0	33.2	9.4	10.2	38.0	4.2	0.17	0.92	(1)
WHEAT (*Triticum aestivum/T. vulgare/T. Sativum/T. durum*)									
Grain	12.1	12.0	1.7	2.5	70.0	1.7	0.05	0.36	(16)
Bran	12.1	14.7	4.0	9.9	53.5	5.8	0.12	1.28	(17)

137

TABLE 6-24 (cont.):

MACROPHYTE	AVERAGE COMPOSITION (% by weight)								NO. REF. (1)
	H₂O	CP	EE	CF	NFE	Ash	Ca	P	
Germ meal	11.1	25.0	8.0	3.3	47.9	4.7	0.05	0.98	(5)
Mill run	11.5	15.2	4.1	8.5	57.0	5.4	0.10	1.10	(4)
Grain screenings	9.5	13.2	3.7	9.1	58.9	5.6	0.18	0.35	(3)
Shorts (fine bran/feed flour mixture)	11.8	16.3	4.3	6.1	56.7	4.8	0.10	0.70	(5)
Middlings (pollard)	10.5	17.4	4.3	7.5	55.4	4.9	0.14	0.91	(9)
Feed flour	12.0	11.7	.12	1.3	73.3	0.5	0.03	0.18	(1)

(1) The data presented represents the mean values from various sources, including: Allen (1984); Bath *et al.*, (1984); Bolton and Blair (1977), Branckaert, Tessema and Temple (1976); Cooley (1976); Devendra (1979); Gohl (1981); Hastings (1973); Hickling (1971); Janssen (1985); Ling (1967); McDonald, Edwards and Greenhalgh (1977); Miller (1976); NRC (1982, 1983, 1983a).

(Source: A.G.J. Tacon, *Standard Methods for the Nutrition of Farmed Fish and Shrimp;* 1990 - with permission of Argent Laboratories, Inc.)

TABLE 6-25: AVERAGE ESSENTIAL AMINO ACID (EAA) COMPOSITION OF THE MAJOR CEREAL GRAINS AND THEIR BY-PRODUCTS

All values are expressed as % by weight on a as-fed basis: Arginine-ARG; Cystine-CYT; Methionine-MET; Threonine-THR; Isoleucine-ISO; Leucine-LEU; Lysine-LYS; Valine-VAL; Tyrosine-TYR; Tryptophan-TRY; Phenylalanine-PHE; Histidine-HIS
(1)

CEREAL	AVERAGE EAA COMPOSITION (%)												NO. REF. (1)
	ARG	CYT	MET	THR	ISO	LEU	LYS	VAL	TYR	TRY	PHE	HIS	
BARLEY (*Hordeum vulgare/H. distichum*)													
Grain	0.53	0.23	0.18	0.36	0.44	0.73	0.40	0.54	0.35	0.15	0.55	0.23	(6)
Naked grain	0.64	0.21	0.10	0.42	0.42	0.83	0.48	0.59	0.48	0.17	0.67	0.26	(1)
Malt sprouts (culms; dehydrated)	1.12	0.24	0.33	0.95	1.04	1.56	1.18	1.38	0.59	0.40	0.87	0.50	(3)
Brewers grains (dehydrated)	1.28	0.35	0.46	0.99	1.62	2.73	0.95	1.62	1.38	0.36	1.55	0.54	(3)
CORN/MAIZE (*Z. mays*)													
Grain	0.47	0.17	0.17	0.35	0.35	1.10	0.23	0.43	0.38	0.08	0.45	0.23	(6)
Grain (opaque; high lysine)	0.66	0.20	0.17	0.37	0.35	0.99	0.42	0.50	0.40	0.11	0.43	0.35	(1)
Gluten feed	0.94	0.51	0.49	0.85	0.75	2.21	0.63	1.15	0.80	0.18	0.86	0.68	(4)
Gluten meal - 41% protein	1.36	0.72	1.00	1.45	2.09	6.70	0.77	2.10	1.33	0.23	2.84	0.90	(3)
Gluten meal - 60% protein	1.99	1.04	1.84	2.11	2.42	9.81	1.00	2.89	3.19	0.30	3.90	1.30	(2)
Germ meal	1.20	0.50	0.58	1.05	0.68	1.52	0.83	1.16	0.54	0.21	0.79	0.68	(3)
Hominy feed	0.53	0.14	0.19	0.41	0.38	0.87	0.43	0.54	0.50	0.12	0.37	0.27	(2)
Corn-and-cob meal (corn ears)	0.36	0.12	0.14	0.28	0.35	0.86	0.17	0.31	0.32	0.07	0.39	0.16	(1)
Distillers dried grains	0.99	0.28	0.43	0.40	0.96	2.81	0.84	1.19	0.84	0.21	0.74	0.61	(2)
Distillers dried solubles	1.05	0.52	0.57	1.01	1.27	2.23	0.90	1.58	0.87	0.22	1.50	0.64	(3)
Distillers dried grains with solubles	1.03	0.40	0.52	1.00	1.44	2.42	0.70	1.55	0.70	0.19	1.55	0.68	(3)
MILLET (pearl or bulrush millet-*Pennisetum typhoideum*; foxtail; Italian millet-*Setaria italica*; Japanese millet-*Echinochloa crusgalli*; broom corn millet-*Panicum miliaceum*; finger millet-*Eleusine coracana*; Scrobic millet-*Paspalum scrobiculatum*)													
Grain	0.35	0.14	0.29	0.40	0.46	1.16	0.25	0.58	0.27	0.17	0.57	0.21	(3)
OATS (*A. sativa*)													
Grain	0.70	0.20	0.16	0.33	0.43	0.77	0.39	0.55	0.48	0.13	0.51	0.18	(6)
Dehulled grain (naked; groats)	0.86	0.24	0.19	0.45	0.52	0.94	0.46	0.68	0.50	0.17	0.64	0.24	(1)
Hulls	0.16	0.07	0.10	0.19	0.18	0.28	0.19	0.15	0.18	0.10	0.19	0.10	(2)
RICE (*0. sativa*)													
Rough (paddy) rice	0.60	0.11	0.16	0.26	0.33	0.56	0.27	0.47	0.47	0.11	0.35	0.16	(3)

TABLE 6-25 (cont.):

CEREAL	AVERAGE EAA COMPOSITION (%)												NO. REF. (1)
	ARG	CYT	MET	THR	ISO	LEU	LYS	VAL	TYR	TRY	PHE	HIS	
Brown (cargo rice; dehulled)	1.30	0.18	0.35	0.54	0.70	1.26	0.70	1.00	0.54	0.24	0.76	0.35	(1)
Broken (brewers) rice	0.49	0.08	0.12	0.24	0.33	0.68	0.27	0.47	0.41	0.10	0.39	0.18	(1)
Polished (milled) rice	0.44	0.09	0.25	0.36	0.45	0.71	0.28	0.53	0.62	0.09	0.53	0.18	(1)
Polishings	0.55	0.12	0.19	0.34	0.34	0.86	0.50	0.70	0.43	0.12	0.37	0.18	(2)
Mill run	0.34	0.10	0.10	0.21	0.20	0.36	0.24	0.30	0.16	0.06	0.25	0.12	(1)
Bran	0.69	0.10	0.21	0.42	0.43	1.04	0.50	0.65	0.60	0.10	0.43	0.24	(2)
RYE (*S. cereale*)													
Grain	0.55	0.17	0.15	0.37	0.46	0.68	0.39	0.56	0.29	0.13	0.59	0.28	(3)
SORGHUM (*S. bicolor/S. vulgare*)													
Grain	0.35	0.16	0.14	0.33	0.44	1.28	0.23	0.52	0.37	0.12	0.48	0.22	(7)
Gluten meal	1.40	0.80	0.75	1.40	2.30	7.40	0.80	2.50	-	0.40	2.60	1.40	(1)
Gluten feed	0.80	0.45	0.40	0.80	1.00	2.50	0.90	1.30	-	0.20	1.00	0.80	(1)
WHEAT (*T. aestivum/T. vulgare/T. sativum/T. durum*)													
Grain	0.54	0.28	0.18	0.34	0.48	0.81	0.35	0.53	0.39	0.15	0.58	0.24	(10)
Grain (durum)	0.60	0.13	0.15	0.38	0.50	1.35	0.95	0.57	0.31	0.26	0.58	0.28	(1)
Bran	0.86	0.28	0.20	0.45	0.57	0.95	0.50	0.67	0.29	0.26	0.59	0.33	(3)
Germ meal	1.84	0.43	0.41	0.96	0.85	1.37	1.51	1.18	0.73	0.29	0.95	0.61	(3)
Middlings	1.01	0.29	0.22	0.52	0.61	1.01	0.65	0.78	0.46	0.22	0.61	0.42	(5)
Grain screenings	0.60	0.14	0.15	0.33	0.44	0.78	0.39	0.55	0.23	0.11	0.52	0.24	(4)
Feed flour	0.43	0.30	0.18	0.33	0.47	0.87	0.25	0.50	0.34	0.12	0.60	0.25	(1)
Shorts	1.07	0.31	0.23	0.55	0.64	1.05	0.75	0.79	0.46	0.22	0.69	0.39	(2)

[1] The data presented represents the mean values from various sources, including: Allen (1984); Bolton and Blair (1977); FAO (1970); Gohl (1981); McDonald, Edwards and Greenhalgh (1977); and NRC (1982, 1983).

(Source: A.G.J. Tacon, *Standard Methods for the Nutrition of Farmed Fish and Shrimp*; 1990 - with permission of Argent Laboratories, Inc.)

TABLE 6-26: AVERAGE PROXIMATE COMPOSITION OF THE SINGLE-CELL PROTEINS

All values are expressed as % by weight on a as-fed basis: Water-H_2O; Crude Protein-CP; Lipid or Ether Extract-EE; Crude Fiber-CF; Nitrogen-Free Extractives-NFE; Ash; Calcium-Ca; Phosphorus-P [1]

SINGLE- CELLED PROTEIN	SUBSTRATE USED FOR CULTURE	AVERAGE COMPOSITION (% by weight)								NO. REF. (1)
		H_2O	CP	EE	CF	NFE	Ash	Ca	P	
BACTERIAL SCP										
Pseudomonas/Methylophllus spp.	METHANOL	6.4	73.1	5.7	0.4	2.7	11.7	0.54	2.33	(10)
FUNGAL SCP										
Brewers yeast (*Saccharomyces cerevisiae*) (dried)	MALT	8.6	45.0	1.2	3.9	34.3	7.0	0.17	1.45	(8)
Yeast (*S. cerevisiae*) (dried)	MOLASSES	9.2	46.8	5.7	1.6	30.5	6.2	-	-	(1)
Bakers yeast (*S. cerevisiae*) (fresh)		68.2	16.2	2.3	-	-	1.9	<0.01	0.16	(1)
W-yeast (*S.cerevisiae*) (fresh) [2]	MARINE OILS (2)	62.1	14.6	12.7	-	-	2.2	<0.01	0.25	(1)
Torula yeast (*Torulopsis utilis*) (dried)	-	7.0	48.0	2.7	2.1	32.2	8.0	0.49	1.52	(4)

TABLE 6-26 (cont.):

SINGLE- CELLED PROTEIN	SUBSTRATE USED FOR CULTURE	AVERAGE COMPOSITION (% by weight)								NO. REF. (1)
		H₂O	CP	EE	CF	NFE	Ash	Ca	P	
Candida utilis (dried)	SULPHITE LIQUOR	8.3	47.3	5.2	1.1	30.8	7.3	-	-	(2)
Candida boldinii (dried)	METHANOL	6.2	36.4	7.2	10.0	34.5	5.7	-	-	(1)
Candida lipolytica (dried)	N-PARAFFIN	6.0	58.8	7.2	3.9	16.4	7.7	0.01	0.80	(5)
Candida lipolytica (dried)	GAS-OIL	9.0	53.3	7.1	3.8	19.1	7.7	-	-	(1)
Candida pseudotrophus (dried)	WHEY	10.0	57.6	5.0	4.5	13.9	9.0			
Candida spp. (dried)	CITRUS MOLASSES	7.6	43.3	0.2	8.1	33.7	7.1	0.20	1.42	(1)
Pichia guillerm	N-ALKANES	2.9	48.6	11.8	7.4	23.6	5.7	-	-	(1)
Aspergillus oryzae	SOYBEAN WASTE	6.3	44.1	3.5	13.2	25.0	7.9	0.34	1.63	(1)
Aspergillus tomarii	FISH WASTE WATER	8.5	44.4	9.4	16.9	16.1	4.7	0.10	0.95	(1)
Mixed fungal SCP culture (dried) (3)	WHISKY SPENT WASH	3.1	53.7	4.5	1.8	31.4	5.5	-	-	(2)
ALGAL SCP - FRESHWATER										
Chlorella vulgaris (dried)		5.7	47.2	7.4	8.3	20.8	10.6	-	-	(2)
Spirulina maxima		6.7	58.6	4.8	0.5	22.7	6.7	-	-	(5)
Scenedesmus obliquus (dried)		6.0	52.6	13.0	6.5	13.5	8.0	0.16	1.76	(1)
Scenedesmus acutus (dried)		8.1	43.6	10.5	6.0	24.4	7.4	0.59	3.66	(2)
Cladophora glomerata (dried)		1.6	31.0	5.2	1.0	28.0	23.2	-	-	(1)
ALGAL SCP - MARINE										
Filamentous bluegreen algae (mixed; fresh) (4)		90.1	2.3	0.2	0.7	0.6	5.1	-	-	(1)
Oscillatoria/Phormidium spp. (fresh) (4)		82.9	1.6	0.4	1.5	1.4	12.2	-	-	(1)
Diatoms (mixed; fresh) (4)		87.1	2.9	0.9	0.3	2.3	6.5	-	-	(1)
Phytoflagellates (mixed; fresh) (4)		88.9	3.9	1.3	0.4	4.8	0.7	-	-	(1)
Marine chlorella (C. vulgaris) (fresh)		75.8	12.2	5.4	-	-	2.3	0.03	0.61	(1)
Tetraselmis maculata (dry matter basis) (5)		0.0	52.0	2.9	15.0		23.8	-	-	(1)
Dunaniella salina (dry matter basis) (5)		0.0	57.0	6.4	31.6		7.6	-	-	(1)
Monochrysis lutteri (dry matter basis) (5)		0.0	49.0	11.6	31.4		6.4	-	-	(1)
Syracosphaera carterae (dry matter basis) (5)		0.0	56.0	4.6	17.8		36.5	-	-	(1)
Chaetoceros spp. (dry matter basis) (5)		0.0	35.0	6.9	6.6		28.0	-	-	(1)
S. costatum (dry matter basis) (5)		0.0	37.0	4.7	20.8		39.0	-	-	(1)
Coscinodiscus spp. (dry matter basis) (5)		0.0	17.0	1.8	4.1		57.0	-	-	(1)
Phaeodactulum tricornotum (dry matter basis) (5)		0.0	33.0	6.6	24.0		7.6	-	-	(1)
Amphidinium carteri (dry matter basis) (5)		0.0	28.0	18.0	30.5		14.1	-	-	(1)
Exuviella spp. (dry matter basis) (5)		0.0	31.0	15.0	37.0		8.3	-	-	(1)
Agmenellum quadruplicatum (dry matter basis) (4)		0.0	36.0	12.8	31.5		10.7	-	-	(1)
MIXED SCP CULTURES										
Activated sludge (domestic sewage; dried)		5.6	39.6	2.6	11.3	19.8	21.1	1.84	1.65	(1)
Activated sludge (brewery processing waste; dried)		5.0	44.4	8.0	-	-	12.6	-	-	(1)
Activated sludge (paper processing waste; dried)		3.0	42.3	0.4	10.6	16.0	27.7	11.4	2.3	(1)

(continued on page 141)

(continued from page 140)

[1] The data presented represents the mean values from various sources, including: Allen (1984); Appler and Jauncey (1983); Bath *et al.*, (1984); Cooley (1976); Gohl (1981); Imada *et al.*,(1979); Janssen (1985); Raushik and Luquet (1980); Ling(1967); Matty and Smith (1978); Murray and Marchant (1986); NRC (1983); Orme and Lemm (1973); Parsons, Stephens and Stickland (1961); Schultz and Oslage (1976); Smith and Palmer (1976); Smith *et al.*, (1975); Soeder (1981); Tacon and Ferns (1979); Tacon *et al.*, (1983); Verkataraman, Becker and Shamala (1977); Windell, Armstrong and Clineball (1974); Yoshida and Hoshii (1980); Zimmerman and Tegbe (1977).

[2] W-Yeast is bakers yeast which has been grown in a culture medium supplemented with fish oil or cuttle fish oil (Imada *et al.*, 1979).

[3] Mixed fungal culture of *Hansenula anomala*, *Candida kruzei* and *Geotrichum candidum* (Murray and Marchant, 1986).

[4] Data as presented by Ling (1967).

[5] Data only available on a dry matter basis (Parsons, Stephens and Strickland, 1961).

(Source: A.G.J. Tacon, *Standard Methods for the Nutrition of Farmed Fish and Shrimp*; 1990 - with permission of Argent Laboratories, Inc.)

TABLE 6-27: AVERAGE PROXIMATE COMPOSITION OF SELECTED ANIMAL BY-PRODUCTS

All values are expressed as % by weight on a as-fed basis; Water-H_2O; Crude Protein-CP; Lipid or Ether Extract-EE; Crude Fiber-CF; Nitrogen Free Extractives-NFE; Ash; Calcium-Ca; Phosphorus-P [1]

ANIMAL BY-PRODUCT	AVERAGE COMPOSITION (% by weight)								NO. REF. [1]
	H_2O	CP	EE	CF	NFE	Ash	Ca	P	
POULTRY									
Chicken (*Gallus domesticus*) (Eggs)									
Whole egg (excluding shell; fresh)	74.4	12.4	11.0	0.0	1.3	0.9	0.06	0.18	(2)
Whole egg (excluding shell; dried)	4.0	46.5	41.6	0.0	4.3	3.6	0.20	0.74	(2)
Egg white (Albumen; fresh)	87.1	11.4	0.1	0.0	0.8	0.6	-	-	(1)
Egg white (Albumen; dried)	9.0	77.4	0.0	0.0	9.3	4.3	0.08	0.08	(1)
Egg yolk fresh	49.1	16.2	33.0	0.0	0.6	1.1	-	-	(1)
Egg shells (dried meal)	1.5	14.0	0.1	4.9	0.0	86.8	31.25	0.07	(1)
Egg-processing waste (dry matter basis)	0.0	60.9	22.8	-	6.1	10.2	-	-	(1)
Poultry by-product meal	6.5	57.5	15.0	2.3	3.1	15.6	3.40	1.90	(6)
Hydrolysed feather meal	8.1	84.2	2.8	1.0	0.5	3.4	0.25	0.66	(5)
Day-old chickens (culled; dried)	4.9	55.4	32.0	0.3	0.0	7.6	1.36	1.24	(1)
Poultry viscera (raw)	73.7	13.9	11.2	0.0	0.0	1.2	-	-	(1)
SLAUGHTERHOUSE									
Meat meal with blood (tankage)	7.3	60.0	8.7	2.2	0.6	21.2	5.95	3.62	(5)
Meat meal	6.9	53.0	4.8	2.4	11.7	21.2	8.22	4.22	(4)
Meat and bone meal (solvent extracted)	8.1	50.0	1.8	2.5	5.9	31.7	10.25	5.25	(3)
Meat and bone meal (rendered)	7.4	49.1	10.3	2.6	0.7	29.9	9.50	4.98	(9)
Blood (cattle; fresh)	79.6	19.7	0.1	0.0	0.0	0.6	0.18	0.05	(3)
Blood meal	10.4	81.5	1.0	0.7	1.6	4.8	0.32	0.25	(10)
Liver (cows; fresh)	73.1	20.2	4.5	0.1	0.6	1.5	0.01	0.23	(3)
Liver (pig; fresh)	70.0	20.5	5.0	0.1	2.8	1.6	0.01	0.37	(1)
Liver meal	8.0	66.7	12.2	0.7	4.4	8.0	0.56	1.26	(2)
Liver and lung meal	6.7	65.0	14.8	2.0	5.5	6.0	0.50	0.95	(2)
Rumen contents (fresh)	57.5	4.6	0.6	15.4	19.6	2.3	-	-	(2)
Rumen contents (solid part; hung)	34.0	8.4	1.1	21.9	26.8	7.8	0.44	0.30	(2)
Rumen contents (liquid part)	91.3	2.1	2.7	1.9	0.9	1.1	0.17	0.09	(1)
Blood meal/rumen contents (6:4,w/w; dry matter basis)	0.0	68.5	5.2	12.2	7.9	6.2	-	-	(1)
MILK BY-PRODUCTS									
Whole cows milk (fresh)	87.6	3.3	3.6	0.0	4.8	0.7	0.12	0.09	(5)
Whole cows milk (dried)	4.6	25.4	26.3	0.1	38.2	5.4	0.88	0.70	(3)
Skim milk (fresh)	90.0	3.2	0.3	0.0	5.8	0.7	0.12	0.09	(5)

TABLE 6-27 (cont.):

ANIMAL BY-PRODUCT	AVERAGE COMPOSITION (% by weight)								NO. REF. (1)
	H₂O	CP	EE	CF	NFE	Ash	Ca	P	
Skim milk (dried)	6.9	33.5	0.9	0.1	50.6	8.0	1.20	1.00	(8)
Buttermilk (fresh)	90.3	3.9	0.5	0.0	4.6	0.7	0.13	0.09	(1)
Buttermilk (dried)	8.7	31.8	5.0	0.3	44.5	9.7	1.21	0.84	(3)
Whey (fresh)	93.0	1.0	0.3	0.0	5.5	0.2	0.07	0.06	(1)
Whey (dried)	7.0	12.5	0.7	0.1	70.9	8.8	0.88	0.76	(5)
Whey (delactose; dried)	6.7	16.5	1.1	0.1	59.0	16.6	1.56	1.08	(3)
Dairy-processing waste (dried) (2)	5.0	29.9	3.8	-	34.1	26.5	4.25	2.67	(1)
FISH PRODUCTS									
RAW FISH (3)									
Group A - <5% lipid; <15% protein	83.0	13.3	1.3	-	-	1.9	-	-	
Group B - <5% lipid; 15-20% protein	81.5	17.9	0.6	-	-	1.6	-	-	
Group C - <5% lipid; >20% protein	72.4	26.2	0.7	-	-	1.5	-	-	
Group D - 5-15% lipid; 15-20% protein	67.5	18.0	13.0	-	-	1.5	-	-	
Group E - >15% lipid; <15% protein	52.5	11.3	36.0	-	-	0.5	-	-	
Fish processing waste									
Catfish (Ictalurus punctatus)(dried) (4)	-	42.0	35.0	-	-	16.0	5.40	2.80	(1)
FISH MEAL									
Anchovy (Engraulis ringens)	8.2	65.3	7.1	1.0	3.4	15.0	4.03	2.61	(6)
Herring (Clupea harengus)	7.9	72.7	8.5	0.8	-	10.1	2.04	1.42	(5)
Sardine/Pilchard	8.5	65.0	6.7	1.0	3.5	15.3	4.44	2.72	(7)
Tuna (Thumus spp.) (mixed)	7.0	59.0	6.9	0.8	4.4	21.9	7.86	4.21	(1)
Menhaden (Brevoortia tyrannus)	7.8	61.3	9.3	1.0	1.4	19.2	5.11	2.92	(5)
Red fish	8.0	57.0	8.0	1.0	-	26.0	7.70	3.80	(1)
White fish (5)	9.1	63.2	4.2	0.9	0.8	21.8	7.17	3.80	(8)
Freshwater (various species) (6)	9.0	66.7	9.1	1.0	-	14.9	5.40	2.90	(1)
FISH SOLUBLES (7)									
Fish solubles (condensed)	49.5	32.0	5.7	0.5	2.6	9.7	0.14	0.61	(5)
Fish solubles (dehydrated)	6.8	56.0	7.8	2.5	13.6	13.3	1.00	1.46	(3)
Fish protein concentrate (dried) (8)	6.4	78.5	0.2	0.0	4.1	10.8	2.56	2.11	(1)
Acid preserved silaged (fresh) (9)									
Tilapia (Oreochromis niloticus) (whole)	71.9	15.6	4.2	-	-	5.0	-	-	(1)
Sprat (Sprattus sprattus) (whole)	74.3	16.7	6.4	-	-	2.7	-	-	(1)
Winter sprat (S. spratus) (whole) (10)	65.7	15.6	13.9	-	-	3.3	-	-	(1)
Herring (C. harengus) (whole)	77.7	15.5	3.4	-	-	2.1	-	-	(1)
Herring (C. harengus) (offal)	68.1	14.5	16.3	-	-	2.6	-	-	(1)
Sandeels (Ammodytes toblanus) (whole)	77.7	15.4	3.4	-	-	2.4	-	-	(1)
White fish offal (excluding viscera)	78.9	15.0	0.5	-	-	4.2	-	-	(1)
Mackerel (Scromber scrombus) (whole)	70.2	16.9	12.0	-	-	2.1	-	-	(1)
Whiting (Merlangius merlangus) (whole) (11)	78.3	15.4	0.5	-	-	2.6	-	-	(1)
FERMENTED SILAGES (fresh) (12)									
Tilapia/molasses (80:20 w/w)	67.8	13.9	3.1	-	-	4.8	-	-	(1)
Tilapia/cassava starch(80:20 w/w)	64.5	12.9	3.0	-	-	3.6	-	-	(1)

(1) The data presented represents the mean values from various sources, including: Allen (1984); Barlow and Windsor (1984); Bath et.al., (1984); Bolton and Blair (1977); Cooley (1976); Davies, Rumsey & Nickum (1976); Gohl (1981); Hastings (1974); Jackson, Kerr and Cowey (1984); Ling (1967); Lovell (1979); McDonald, Edwards and Greenhalgh (1977); NRC (1982, 1983); Reece et.al., (1975); Rumsey *et al.*, (1981); Tacon (1982); Tacon (1986); Tacon, Webster & Martinez (1984); Tatterson and Windsor (1974); Wee, Kerdchuen and Edwards (1987); Wilson, Freeman and Poe (1984) and Wood, Capper and Nicolaides (1985).

(continued on page 143)

(continued from page 142)

(2) Dairy processing wastes, including the processing of cheeses; carbohydrate concent expressed as 34.1% lactose (Rumsey *et al.*, 1981)

(3) Due to the wide variation in proximate composition of fish (depending on species, time of year, growth, nutritional history, and spawning period, etc), the 5 categories of Gohl (1981) are presented; Group A - oysters and clams; Group B - Carp, cod, flounder, haddock, hake, mullet, oceanperch, pollack, rockfish, whiting, crab, scallop and shrimp; Group C - Halibut and tuna; Group D - Anchovy, herring, mackerel, salmon, sardine; and Group E - Siscowet lake trout (*Cristivomer namacush*)

(4) Catfish processing wastes includes head, skin and viscera: mean water content of waste reported as 67% (Lovell, 1979).

(5) Includes various marine species such as Gadidae/Lophiidae/Rajidae, which have a white flesh and low lipid content.

(6) Mean of various freshwater fish species as reported by Allen (1984).

(7) Condensed fish solubles is a product (press liquor) resulting from the pressing of fish during the fish meal manufacturing process (Barlow and Windsor, 1984).

(8) Fish protein concentrate is the product arising from the solvent extraction of fish meal.

(9) Acid preserved silages are produced by the external addition of mineral or organic acids to the macerated whole fish or wet fish by-product.

(10) Composition presented is for a high lipid sprat silage, after a 2-week storage period at 20°C with added ethoxyquin antioxidant (Jackson, Kerr and Cowey, 1984).

(Source: A.G.J. Tacon, *Standard Methods for the Nutrition of Farmed Fish and Shrimp*; 1990 - with permission of Argent Laboratories, Inc.)

TABLE 6-28: AVERAGE ESSENTIAL AMINO ACID (EAA) COMPOSITION OF SELECTED ANIMAL BY-PRODUCTS

All values are expressed as % by weight on a as-fed basis: Arginine-ARG; Cystine-CYT; Methionine-MET; Threonine-THR; Isoleucine-ISO; Leucine-LEU; Lysine-LYS; Valine-VAL; Tyrosine-TYR; Tryptophan-TRY; Phenylalanine-PHE; Histidine-HIS [1]

BY-PRODUCT	AVERAGE EAA COMPOSITION (%)												NO. REF. [1]
	ARG	CYT	MET	THR	ISO	LEU	LYS	VAL	TYR	TRY	PHE	HIS	
POULTRY													
CHICKEN (*G. domesticus*) EGGS													
Whole egg (excl. shell; fresh)	0.76	0.29	0.40	0.61	0.76	1.07	0.83	0.86	0.50	0.19	0.70	0.30	(2)
Whole egg (excl. shell; dried)	2.94	1.09	1.48	2.26	2.87	4.03	3.10	3.30	1.91	0.73	2.59	1.10	(1)
Egg white (albumen; fresh)	0.63	0.27	0.44	0.53	0.57	0.92	0.74	0.54	0.39	0.18	0.66	0.26	(1)
Egg white (albumen; dried)	4.51	1.91	3.00	3.44	4.71	6.73	4.77	5.79	3.10	1.19	4.89	1.76	(1)
Egg shells (dried)	0.67	0.89	0.23	0.44	0.33	0.45	0.32	0.51	0.22	-	0.20	0.28	(1)
Poultry by-product meal	3.80	0.96	1.05	1.98	2.35	4.20	2.73	2.75	0.94	0.50	1.82	1.30	(2)
Hydrolysed feather meal	5.90	3.08	0.53	3.76	3.76	7.32	1.72	5.96	2.33	0.65	3.31	0.61	(3)
SLAUGHTERHOUSE													
Meat meal with blood (tankage)	3.59	0.48	0.76	2.36	1.92	5.11	3.76	3.98	1.29	0.65	3.82	1.86	(2)
Meat meal	3.65	0.67	0.73	1.72	1.82	3.35	3.11	2.56	0.96	0.35	1.86	1.03	(2)
Meat and bone meal (solvent extr.)	3.72	0.20	0.69	1.49	1.21	2.65	2.66	1.99	1.24	0.30	1.49	0.91	(2)
Meat and bone meal (rendered)	3.49	0.39	0.64	1.72	1.63	3.21	2.72	2.39	1.13	0.29	1.79	1.10	(5)
Blood meal (dried)	3.18	1.21	1.00	3.88	0.89	10.77	6.11	7.01	2.18	1.05	5.74	4.21	(5)
Liver meal (dried)	4.04	0.94	1.22	2.49	3.10	5.31	5.21	4.15	1.70	0.69	2.92	1.48	(1)
Liver and lung meal (dried)	3.52	1.05	1.09	2.33	3.02	5.23	4.36	3.53	1.62	0.53	3.00	1.44	(1)
MILK BY-PRODUCTS													
Whole cows milk (dried)	0.92	0.40	0.62	1.03	1.33	2.571	2.26	1.74	1.33	0.41	1.33	0.72	(2)

TABLE 6-28 (cont.):

BY-PRODUCT	ARG	CYT	MET	THR	ISO	LEU	LYS	VAL	TYR	TRY	PHE	HIS	NO. REF. (1)
Skim milk (dried)	1.08	0.37	0.87	1.57	2.23	3.23	2.51	2.22	0.93	0.41	1.60	0.92	(3)
Casein (dried)	3.40	0.30	2.70	3.80	5.70	8.70	7.00	6.80	-	1.00	4.60	2.50	(1)
Buttermilk (dried)	1.09	0.38	0.71	1.56	2.56	3.31	2.34	2.69	1.00	0.50	1.48	0.87	(2)
Whey (dried)	0.37	0.30	0.20	0.85	0.85	1.19	1.02	0.69	0.25	0.19	0.38	0.19	(2)
Whey (delactose; dried)	0.59	0.50	0.49	0.91	1.01	1.62	1.44	0.91	0.46	0.31	0.63	0.30	(2)
Dairy processing waste (dried)	1.23	0.46	0.32	1.43	1.04	1.78	1.04	1.42	0.73	-	1.11	0.39	(1)
FISH PRODUCTS													
Fresh fish (all types) [2]	1.06	0.22	0.53	0.86	0.90	1.44	1.71	1.15	0.68	0.21	0.73	0.66	(1)
Anguilliformes	1.13	0.22	0.52	0.89	0.86	1.61	1.50	1.05	0.78	0.20	0.75	0.41	(1)
Beloniformes	0.96	0.27	0.44	0.83	0.76	1.25	1.67	0.85	0.58	0.22	0.58	0.81	(1)
Clupeiformes-clupeoidei	1.26	0.24	0.62	1.02	1.05	1.76	1.80	1.2Z	0.77	0.21	0.92	0.61	(1)
Clupeiformes-salmonoidei	1.01	0.18	0.46	0.78	0.81	1.25	1.60	0.95	0.54	0.19	0.67	0.54	(1)
Cypriniformes	1.16	0.22	0.66	0.78	0.83	1.43	1.59	1.16	0.59	0.18	0.64	0.46	(1)
Gadiformes	1.12	0.19	0.57	0.87	0.79	1.45	1.70	0.88	0.66	-	0.86	0.50	(1)
Galeiformes	1.37	0.18	0.57	0.82	1.38	1.70	1.93	1.10	0.73	0.22	0.82	0.44	(1)
Mugiliformes	1.11	0.19	0.47	0.91	0.91	1.44	1.81	1.05	0.54	0.18	0.70	0.55	(1)
Perciformes-scombroidei	1.37	0.29	0.65	1.06	1.19	1.83	2.32	1.78	0.96	0.32	0.91	1.34	(1)
Perciformes	0.96	0.20	0.48	0.71	0.79	1.21	1.60	1.10	0.63	0.18	0.63	0.60	(1)
Pleuronectiformes	1.03	0.13	0.39	0.79	0.82	1.34	1.63	0.91	0.62	0.22	0.65	0.46	(1)
Rajiformes	1.57	-	0.63	0.72	1.49	1.67	2.18	1.08	1.08	0.28	0.77	0.38	(1)
FISH MEAL													
Anchovy (*E. ringens*)	3.67	0.61	1.94	2.78	2.99	4.98	5.08	3.52	2.17	0.76	2.63	1.52	(4)
Herring (*C. harengus*)	4.61	0.71	2.14	3.01	3.21	5.30	5.66	4.37	2.23	0.80	2.77	1.71	(4)
Sardine/Pilchard	3.25	0.76	1.95	2.70	3.09	4.42	5.55	3.64	2.29	0.58	2.34	1.88	(4)
Tuna (*Thunnus* spp.) (offal)	3.42	0.44	1.46	2.31	2.41	3.81	4.04	2.80	1.72	0.56	2.16	1.78	(2)
Menhaden (*B. tyrannus*)	3.58	0.57	1.77	2.43	2.81	4.64	4.70	3.27	1.97	0.68	2.40	1.44	(3)
Whitefish	4.16	0.67	1.72	2.56	2.71	4.38	4.56	3.05	1.86	0.64	2.30	1.45	(4)
Redfish	4.10	0.40	1.80	2.60	3.50	4.90	6.60	3.30	-	0.60	2.50	1.30	(1)
Freshwater (various species)	4.62	0.47	1.92	3.25	3.27	4.87	5.89	3.50	-	0.62	2.92	2.03	(3)
Catfish (*I. punctatus*) (offal)	3.92	-	1.23	2.43	1.94	3.47	3.20	2.21	1.62	-	2.22	1.15	(1)
Catfish (*I. punctatus*) (bone)	2.75	-	0.72	1.19	0.99	1.59	1.70	1.21	0.64	-	0.97	0.58	(1)
FISH SOLUBLES													
Fish solubles (condensed)	1.25	0.19	0.62	0.75	0.79	1.62	1.51	1.10	0.32	0.19	0.74	1.26	(3)
Fish solubles (dehydrated)	2.42	0.56	0.91	2.22	1.62	2.80	3.10	1.85	0.85	1.44	1.41	1.50	(2)
PROTEIN HYDROLYSATES (dried)													
Catfish (*I. punctatus*) (offal)	3.49	-	1.23	1.94	1.50	2.82	3.19	2.21	1.09	-	1.72	0.81	(1)
ACID PRESERVED SILAGES (% dry matter basis)													
Catfish (*I. punctatus*) (offal)	5.40	-	1.49	2.86	2.51	4.38	4.66	3.33	2.30	-	2.74	1.44	(1)
Winter sprat (*S.sprattus*) (whole) [3]	2.69		1.27	2.16	1.90	3.54	3.86	2.65	-	0.25	2.60	1.22	(1)
Whiting (*M. merlangus*) (whole; % total AA) [4]	7.34	0.61	3.67	3.67	4.65	7.96	8.93	5.75	3.55	-	3.55	1.96	(1)

(1) Data presented represents the mean values from various sources, including: Allen (1984); Barlow and Windsor (1984); Bolton and Blair (1977); FAO (1970); Jackson, Kerr and Cowey (1984); NRC (1982/1983); Tacon (1982); Wilson, Freeman and Poe (1984); Wood, Capper and Nicolaides (1985).

(continued on page 145)

(continued from page 144)

[2] Values for fresh fish obtained from FAO (1970).

[3] Composition presented is for a silage after a 8-week storage period at 20°C with added ethoxyquin: Authors report a 54.52% loss in tryptophan from an initial concentration of 0.55% after a 8-week storage period; Methionine and Cystine reported as total sulphur amino acids (Jackson, Kerr and Cowey, 1984).

[4] Composition presented is for a silage after a 6-month storage period at 18-22°C; values are expressed as % of total recovered amino acids (Wood, Capper and Nicolaides, 1985).

(Source: A.G.J. Tacon, *Standard Methods for the Nutrition of Farmed Fish and Shrimp*; 1990 - with permission of Argent Laboratories, Inc.)

TABLE 6-29: BINDING AGENTS USED IN STEAM PELLETED AQUACULTURE FEEDS

COMPOUND	AMOUNT USED (%)	COMMENTS
CARBOXYMETHYLCELLULOSE	0.5-2.0	Good; expensive
ALGINATES	0.8-3.0	Good in moist feeds; must combine with di-or polyvalent cation
POLYMETHYLOCARBAMIDE	0.5-0.8	Very good; not FDA approved for use in United States; unpalatable to some fish
GUAR GUM	1-2	Good
HEMICELLULOSE	2-3	Fair
LIGNIN SULFONATE	2-4	Good
NA AND CA BENTONITE	2-3	Inferior to organic binders
MOLASSES	2-3	Fair; has nutritional value
WHEY	1-3	Fair; has nutritional value
GELATINIZED STARCHES (CORN; POTATO; SORGHUM; RICE; CASSAVA)	10-20	Good; large amount required; has nutritional value
WHEAT GLUTIN	2-4	Good; expensive

(Source: T. Lovell, *Nutrition and Feeding in Fish*; 1989)

TABLE 6-30: FOOD AND DRUG ADMINISTRATION ACTION LEVELS FOR TOXIC OR DELETERIOUS SUBSTANCES IN FINISHED ANIMAL FEEDS

SUBSTANCE	QUANTITY mg/kg
AFLATOXIN	0.02
ALDRIN AND DIELDRIN	0.03
BENZENE HEXACHLORIDE (BHC)	0.10
CHLORODANE	0.10
DIBROMOCHLOROPROPANE (DBCP)	0.05
DDT; DDE; TDE	0.50
ENDRIN	0.03
HEPTACHLOR AND HEPTACHLOR EPOXIDE	0.03
KELTHANE	0.50
LINDANE	0.10
POLYBROMINATED BIPHENYLS (PBB's)	0.05
TOXAPHENE	0.60

All substances except PBB's: Food and Drug Administration Compliance Policy Guides, Guide 7126.27, Animal Feeds: Chapter 26, 1980.
For PBB's: Congressman William M. Broadhead's petition to reduce FDA action levels for PBB's in food, July 27,1977.

(Source: T. Lovell, *Nutrition and Feeding in Fish*; 1989)

TABLE 6-31: PROXIMATE ANALYSES OF ORGANISMS SERVING AS FOOD FOR POND FISHES

ORGANISM GROUP	DRY MATTER (n)[2] %	PROTEIN (n) %	DRY MATTER COMPOSITION[1] CARBOHYDRATE (n) %	LIPID (n) %	ASH (n) %	ENERGY (n) kcal/kg	REF.
BACTERIA					(2) 5.4	(2)4,710	5
ALGAE							
CYANOPHYTA (blue green)	-	(1) 31.3			(5) 46.7	(8)2,213	5
CHLOROPHYTA (greens)	(13) 16.8	(8) 17.6		(7) 3.7	(21) 26.9	(29) 3,773	2;5;11;14;21
PHAEOPHYTA (brown algae)	(15) 14.1				(24) 32.3	(25) 3,056	5
BACILLARIACAEA (diatoms)		(4) 30.7		(3) 9.9	(5) 38.3	(4) 3,654	5;10;15;19
RHODOPHYTA (red algae)	(21) 21.7				(39) 32.1	(39) 3,170	5
AQUATIC MACRO-VEGETATION	(51) 15.8	(64) 14.6		(29) 4.5	(62) 13.9	(107) 3,906	2;3;5
PROTOZOA						(1) 5,938	
ROTIFERS	(15) 11.2	(11) 64.3		(13) 20.3	(5) 6.2	(4) 4866	13;17
OLIGOCHAETES	(4) 7.3	(2) 49.3		(1) 19.0	(1) 5.8	(9) 5,569	5;10;1
LEECHES	(1) 24.0	(2) 61.0			(2) 5.1	(3) 5,432	5;7;12
CRUSTACEA							
ANOSTRACA (*Artemia*)	(6) 11.0	(6) 61.6		(6) 19.5	(6) 10.1	(2) 5,835	5;17
CLADOCERA	(18) 9.8	(30) 56.5	(20) 28.2	(28) 19.3	(57) 7.7	(75) 4,800	5;6;8;9;10;11;12;13;16;17;20
COPEPODA	(10) 10.3	(35) 52.3	(36) 9.2	(29) 26.4	(29) 7.1	(39) 5,445	5;6;9;12;13;14;16;18;21
OSTRACODA	(2) 35.0	(1) 41.5				(2) 5,683	5;14
MALACOSTRACA	(32) 24.6	(16) 49.9	(7) 18.4	(10) 20.3	(50) 19.6	(112) 5,537	4;5;6;7;8;9;12;13;14;18;20;21
INSECTS	(1) 23.2	(1) 55.94	(1) 20.1	(1) 18.6	(6) 4.9	(6) 5,075	9;20
PLECOPTERA (stone flies)						(1) 4,900	1
EPHEMERIDAE (mayflies)	(3) 17.6	(5) 50.2			(8) 3.7	(19) 5646	1;5;7;12;14;15;20
ODONATA (dragon flies)	(6) 21.1	(4) 51.9			(12) 5.8	(14) 4,985	5;7;12;14
HEMIPTERA (water bugs)	(2) 26.0	(2) 68.8				(2) 5,150	7
TRICHOPTERA (caddisflies)	(2) 14.8	(3) 34.7			(6) 11.8	(11) 5,019	1;5;12;14
DIPTERA	(1) 16.0	(1) 55.3			(1) 6.9	(3) 5,177	12
CHIRONOMIDS (larvae)	(5) 19.1	(4) 59.0	(1) 22.5	(2) 4.9	(17) 5.8	(31) 5,034	1;5;7;9;10;12;14;21
MOLLUSCS	(15) 32.2	(19) 39.5	(15) 7.5	(15) 7.8	(26) 32.9	(84) 3,889	1;4;5;14
AQUATIC DETRITUS	(59) 91.5				(68) 12.4	(67) 4,701	5

[1] Since the values are averages of figures collected from different sources they do not necessarily add up to 100%.

[2] The numbers in brackets are those of data collected from the literature. Each may be an average of a number of analyses.

REFERENCES:

1. Alimov & Shadrin, 1977
2. Boyd, 1968,
3. Boyd & Goodyear, 1971
4. Conover, 1978
5. Cummins & Wuycheck, 1971
6. Dabrowski & Rusiecki, 1981
7. Driver *et al.*, 1974
8. Farkas, 1958
9. Lieder, 1965a
10. Ogino, 1963
11. Richman, 1958
12. Salonen *et al.*, 1976
13. Schindler *et al.*, 1971
14. Stiaramaiah, 1967
15. Trama, 1957
16. Vijverberg & Frank, 1965
17. Watanabe *et al.*, 1978a
18. Watanabe *et al.*, 1978b
19. Wikfors, 1986
20. Wissing & Hasler 1968
21. Wissing & Hasler, 1971

(Source: B. Hepher, *Nutrition of Pond Fishes*; copyright © 1988 - with permission of Cambridge University Press)

VII. Aquaculture Ponds

TABLE 7-1: SOIL CLASSIFICATION AFTER BFAR-UNDP/FAO 1981 (Tang, 1982)

CLASS	PERMEABILITY	COMPRESSIBILITY	COMPACTION	SUITABILITY
CLAY	IMPERVIOUS	MEDIUM	FAIR TO GOOD	EXCELLENT
SANDY CLAY	IMPERVIOUS	LOW	GOOD	GOOD
LOAMY	SEMI-PERVIOUS TO IMPERVIOUS	HIGH	FAIR TO VERY POOR	FAIR
SILTY SANDY	SEMI-PERVIOUS TO IMPERVIOUS	MEDIUM TO HIGH	GOOD TO VERY POOR	POOR
PEATY	PERVIOUS	HIGH	GOOD	POOR
PEATY		NEGLIGIBLE		VERY POOR

(Source: A.N. Bose *et al.*, *Coastal Aquaculture Engineering*; 1991)

TABLE 7-2: CLASSIFICATION OF SOIL PARTICLES ACCORDING TO THE INTERNATIONAL SYSTEM AND THE UNITED STATES DEPARTMENT OF AGRICULTURE SYSTEM (after Buckman and Brady, 1960)

NAME OF PARTICLE	DIAMETER LIMITS (mm)	
	INTERNATIONAL SYSTEM	USDA SYSTEM
GRAVEL	>2.00	>2.00
VERY COARSE SAND	2.00 - 0.20	-
COARSE SAND	2.00 - 0.20	1.00 - 0.50
MEDIUM SAND	0.50 - 0.25	-
FINE SAND	0.20 - 0.02	0.25 - 0.10
VERY FINE SAND	0.10 - 0.05	-
SILT	0.02 - 0.002	0.05 - 0.002
CLAY	<0.002	<0.002

(Source: C. E. Boyd, *Water Quality in Ponds for Aquaculture*; 1990)

TABLE 7-3: SETTLEMENT ALLOWANCE FOR DIFFERENT SOIL CONDITION (%) (Tang, 1982)

CONDITION	ALLOWANCE FOR STRUCTURE
POOR MATERIAL / POOR CONSTRUCTION METHODS	15
SOIL EXCEPTIONALLY HIGH IN ORGANIC MATTER	40 OR MORE
COMPACTED BY CONSTRUCTION EQUIPMENT	5 - 10

(Source: A.N. Bose *et al.*, *Coastal Aquaculture Engineering*; 1991)

TABLE 7-4: PERMISSIBLE CANAL VELOCITIES (Fortier and Scobey, 1926)[1] ·

ORIGINAL MATERIAL EXCAVATED FOR CANAL	VELOCITY (ft/s) AFTER AGING; CANALS CARRYING-		
	CLEAR WATER - NO DETRITUS	WATER TRANSPORTING COLLOIDAL SILTS	WATER TRANSPORTING NONCOLLOIDAL SILTS / SANDS / GRAVELS OR ROCK FRAGMENTS
FINE SAND (NONCOLLOIDAL)	1.50	2.50	1.50
SANDY LOAM (NONCOLLOIDAL)	1.75	2.50	2.00
SILT LOAM (NONCOLLOIDAL)	2.00	3.00	2.00
ALLUVIAL SILTS WHEN NONCOLLOIDAL	2.00	3.50	2.00
ORDINARY FIRM LOAM	2.50	3.50	2.25
VOLCANIC ASH	2.50	3.50	2.00
FINE GRAVEL	2.50	5.00	3.75
STIFF CLAY (VERY COLLOIDAL)	3.75	5.00	3.00
GRADED LOAM TO COBBLES (WHEN NONCOLLOIDAL)	3.75	5.00	5.00
ALLUVIAL SILTS WHEN COLLOIDAL	3.75	5.00	3.00
GRADED SILT TO COBBLES WHEN COLLOIDAL	4.00	5.50	5.00
COARSE GRAVEL (NONCOLLOIDAL)	4.00	5.50	5.00
COBBLES AND SHINGLES	5.00	5.50	6.50
SHALES AND HARD PANS	6.00	6.00	5.00

[1] Depth of 3 feet or less

(Source: A.N. Bose *et al.*, *Coastal Aquaculture Engineering*; 1991)

TABLE 7-5: PERMISSIBLE VELOCITIES (m/s) FOR OPEN CHANNELS LINED WITH VEGETATION (Ree, 1949)

COVER	PERCENT SLOPE	SOIL TYPE	
		EROSION RESISTANT	EASILY ERODED
BERMUDA GRASS	0-5	2.50	1.85
	5-10	2.15	1.50
	>10	1.85	1.25
BUFFALO GRASS KENTUCKY BLUEGRASS SMOOTH BROME BLUE GRASS	0-5	2.15	1.50
	5-10	1.85	1.25
	>10	1.50	0.90
CENTIPEDE GRASS BERMUDA GRASS (good turf kept mowed)	0-5	2.75	2.15
	5-10	2.50	1.85
	> 10	2.15	1.50
GRASS MIXTURE	0-5	1.50	1.25
	5-10	1.25	0.90
	>10	Not recommended	
LESPEDEZE SERICEA WEEPING LOVEGRASS ISCHAEMUM (yellow bluestem) KUDZU ALFALFA CRABGRASS	0-5	1.0	0.75
	5-10	Not recommended	
	> 10	Not recommended	
ANNUALS USED AS TEMPORARY COVER (Common Lespedeza and Sudan Grass)	0-5	1.0	0.75
	5-10	Not recommended	
	> 10	Not recommended	

(Source: F.W. Wheaton, *Aquacultural Engineering*; 1977)

TABLE 7-6: END AREAS IN SQUARE FEET OF EMBANKMENT SECTIONS FOR DIFFERENT SIDE SLOPES AND TOP WIDTHS

FILL HEIGHT (feet)	SIDE SLOPES [1] 2.5:1 / 2.5:1 / 2:1 / 3:1	2.5:1 / 3:1 / 2:1 / 3.5:1	3:1 / 3:1 / 2.5:1 / 3.5:1	3.5:1 / 3.5:1 / 3:1 / 4:1	4:1 / 4:1 / 3:1 / 5:1	TOP WIDTH (feet) 8	10	12	14	16
1.0	3	3	3	4	4	8	10	12	14	16
1.2	4	4	4	5	6	10	12	14	17	19
1.4	5	5	6	7	8	11	14	17	20	22
1.6	6	7	8	9	10	13	16	19	22	26
1.8	8	9	10	11	13	14	18	22	25	29
2.0	10	11	12	14	16	16	20	24	28	32
2.2	12	13	15	17	19	18	22	27	31	35
2.4	14	16	17	20	23	19	24	29	34	39
2.6	17	19	20	24	27	21	26	31	36	42
2.8	20	22	23	27	31	22	28	34	39	45
3.0	22	25	27	32	36	24	30	36	42	48
3.2	26	28	31	36	41	26	32	38	45	51
3.4	29	32	35	40	46	27	34	41	47	55
3.6	32	36	39	45	52	29	36	43	50	58
3.8	36	40	43	50	58	30	38	46	53	61
4.0	40	44	48	56	64	32	40	48	56	64
4.2	44	49	53	62	71	34	42	50	59	67
4.4	48	53	58	68	77	35	44	53	61	71
4.6	53	58	63	74	85	37	46	55	64	74
4.8	57	63	69	81	92	38	48	57	67	77
5.0	62	69	75	87	100	40	50	60	70	80
5.2	67	74	81	94	108	42	52	62	73	83
5.4	73	80	87	102	117	43	54	65	75	87
5.6	78	86	94	110	125	45	56	67	78	90
5.8	84	93	101	118	135	46	58	69	81	93
6.0	90	99	108	126	144	48	60	72	84	96
6.2	96	106	115	135	154	50	62	74	87	99
6.4	102	113	123	143	164	51	64	77	89	103
6.6	109	120	131	152	174	53	66	79	92	106
6.8	116	128	139	162	185	54	68	81	95	109
7.0	123	135	147	172	196	56	70	84	98	112
7.2	130	143	156	182	207	58	72	86	101	115
7.4	138	152	165	193	219	59	74	89	103	119
7.6	145	159	174	203	231	61	76	91	106	122
7.8	153	168	183	214	243	62	78	93	109	125
8.0	160	176	192	224	256	64	80	96	112	128
8.2	169	185	202	235	269	66	82	98	115	131
8.4	177	194	212	247	282	67	84	101	117	135
8.6	186	204	222	259	296	69	86	103	120	138
8.8	194	213	232	271	310	70	88	105	123	141
9.0	203	223	243	283	324	72	90	108	126	144
9.2	212	233	254	296	339	74	92	110	129	147
9.4	222	244	266	310	353	75	94	113	131	151
9.6	231	254	277	323	369	77	96	115	134	154
9.8	241	265	289	337	384	78	98	117	137	157
10.0	250	275	300	350	400	80	100	120	140	160
10.2	260	286	313	364	416		102	122	143	163
10.4	271	298	325	379	433		104	125	145	167
10.6	281	309	338	394	449		106	127	148	170
10.8	292	321	350	409	467		108	129	151	173
11.0	302	333	363	424	484		110	132	154	176
11.2	313	344	376	440	502		112	134	157	179
11.4	325	357	390	456	520		114	137	159	183
11.6	336	370	404	472	538		116	139	162	186

TABLE 7-6 (cont):

FILL HEIGHT (feet)	SIDE SLOPES [1] 2.5:1 / 2.5:1 / 2:1 / 3:1	2.5:1 / 3:1 / 2:1 / 3.5:1	3:1 / 3:1 / 2.5:1 / 3.5:1	3.5:1 / 3.5:1 / 3:1 / 4:1	4:1 / 4:1 / 3:1 / 5:1	TOP WIDTH (feet) 8	10	12	14	16
11.8	348	383	418	488	557		118	141	165	189
12.0	360	396	432	504	576		120	144	168	192
12.2	372	409	447	522	595		122	146	171	195
12.4	385	424	462	539	615		124	149	173	199
12.6	397	437	477	557	635		126	151	176	202
12.8	410	451	492	574	655		128	153	179	205
13.0	422	465	507	592	676		130	156	182	208
13.2	436	479	523	610	697		132	158	185	211
13.4	449	494	539	629	718		134	161	187	215
13.6	463	509	555	648	740		136	163	190	218
13.8	476	523	571	667	762		138	166	193	221
14.0	490	539	588	686	784		140	168	196	224
14.2	505	555	605	706	807		142	170	199	227
14.4	519	570	622	726	829		144	173	202	230
14.6	534	586	639	746	853		146	175	204	234
14.8	548	602	657	767	876		148	178	207	237
15.0	563	619	675	788	900		150	180	210	240
15.2	578	635	693	809	924		152	182	213	243
15.4	594	653	711	830	949		154	185	216	246
15.6	609	669	730	852	973		156	187	218	250
15.8	625	687	749	874	999		158	190	221	253
16.0	640	704	768	896	1,024		160	192	224	256
16.2	656	722	787	919	1,050			194	227	259
16.4	673	740	807	942	1,076			197	230	262
16.6	689	758	827	965	1,102			199	232	266
16.8	706	776	847	988	1,129			202	235	269
17.0	723	795	867	1,012	1,156			204	238	272
17.2	740	814	888	1,036	1,183			206	241	275
17.4	757	833	909	1,060	1,211			209	244	278
17.6	774	852	930	1,084	1,239			211	246	282
17.8	792	871	951	1,109	1,267			214	249	285
18.0	810	891	972	1,134	1,296			216	252	288
18.2	828	911	994	1,160	1,325			218	255	291
18.4	846	931	1,016	1,186	1,354			221	258	294
18.6	865	951	1,038	1,212	1,384			223	260	298
18.8	884	972	1,060	1,238	1,414			226	263	301
19.0	903	993	1,083	1,264	1,444			228	266	304
19.2	922	1,014	1,106	1,291	1,475	•		230	269	307
19.4	941	1,035	1,129	1,318	1,505			233	272	310
19.6	960	1,056	1,152	1,345	1,537			235	274	314
19.8	980	1,078	1,176	1,372	1,568			238	277	317
20.0	1,000	1,100	1,200	1,400	1,600			240	280	320
20.2	1,020	1,122	1,224	1,428	1,632			242	283	323
20.4	1,040	1,144	1,248	1,457	1,665			245	286	326
20.6	1,061	1,167	1,273	1,486	1,697			247	288	330
20.8	1,082	1,190	1,298	1,515	1,731			250	291	333
21.0	1,103	1,213	1,323	1,544	1,764			252	294	336
21.2	1,124	1,236	1,348	1,574	1,798			254	297	339
21.4	1,145	1,254	1,374	1,604	1,832			257	300	342
21.6	1,166	1,283	1,400	1,634	1,866			259	302	346
21.8	1,188	1,307	1,426	1,664	1,901			262	305	349
22.0	1,210	1,331	1,452	1,694	1,936			264	308	352
22.2	1,232	1,356	1,479	1,725	1,971			266	311	355
22.4	1,254	1,380	1,506	1,756	2,007			269	314	358
22.6	1,277	1,405	1,533	1,788	2,043			271	316	362
22.8	1,300	1,430	1,560	1,820	2,079			274	319	365
23.0	1,323	1,455	1,587	1,852	2,116			276	322	368

(continued on page 151)

(continued from page 150)

To find the end area for any fill height, add square feet given under staked side slopes to that under the top width for the total section.

[1] Any combination of slopes that adds to 5, 6, or 7 may be used. A combination of 3.5:1 front and 2.5:1 back gives the same results as 3:1 front and back.

EXAMPLE:

To calculate the end area for a 6.4-foot fill with 3:1 front and back slopes and a 14-foot top width. From the table, 123 plus 89 or 212 square feet for the section. Any combination of slopes that adds to 5, 6, or 7 may be used. A combination of 3.5:1 front and 2.5:1 back gives the same results as 3:1 front and back.

(Source: USDA, *Ponds- Planning, Design, Construction;* 1971)

TABLE 7-7: VOLUME OF EARTH NEEDED FOR AN EARTHFILL DAM USING "SUM-OF-END-AREAS" METHOD

This table serves as an example for calculating the volume of earthfill required for an embankment based on the "Sum-of-End-Area" method.

STATION (ft.)	GROUND ELEVATION (ft.)	FILL HEIGHT [1] (ft.)	END AREA [2] (ft.^{2})	SUM OF END AREAS (ft.^{2})	DISTANCE	DOUBLE VOLUME (ft.^{3})
0 + 50	35.0	0	0			
				44	18	792
+ 68	32.7	2.3	44-			
				401	32	12,832
1 + 00	25.9	9.1	357-			
				1,066	37	3,9442
+ 37	21.5	13.5	709-			
				1,564	16	25,024
+ 53	20.0	15.0	855-			
				1,730	22	38,060
+ 75	19.8	15.2	875-			
				1,781	25	44,525
2 + 00	19.5	15.5	906-			
				1,730	19	32,870
+ 19	20.3	14.7	824-			
				1,648	13	21,424
+ 32	20.3	14.7	824-			
				1,805	4	7,220
+ 36	18.8	16.2	981-			
				2,030	4	8,120
+ 40	18.2	16.8	1,049-			
				2,064	3	6,192
+ 43	18.5	16.5	1,015-			
				1,911	3	5,733
+ 46	19.6	15.4	896-			
				1,771	13	23,023
+ 59	19.8	15.2	875-			
				1,650	41	67,650
3 + 00	20.8	14.2	775-			
				1,023	35	35,805
+ 35	27.7	7.3	248-			
				324	25	8,100
+ 60	31.6	3.4	76-			
				76	36	2,736
3 + 96	35.0	.0	0-			

(continued on page 152)

151

(cointinued from page 151

(1) Elevation of top of dam without allowance for settlement.

(2) End areas based on 12-foot top width and 3:1 slopes on both sides.

EXAMPLE:

To calculate the volume of earthfill required for the embankment described in the table above:

Set stakes along the centerline of the embankment at intervals of 100 feet or less. Set the fill and slope stakes upstream and downstream from the centerline stakes to mark the points of intersection of the side slopes with the ground surface and to mark the outer limits of construction. With the fill settled fill heights, side slopes and top width established, find the net areas at each station point along the centerline of the embankment from the table. The double volume (in cubic feet) of fill between two points on the centerline of the embankment is equal to the sum of the end areas at these points multiplied by the distance between the points. Divide double volume in cubic feet by 54 to obtain volume in cubic yards.

EXAMPLE:

$$\frac{379,548}{54} = 7,029 \ cubic \ yards$$

Allowance for settlement (10%) = 703 cubic yards

Total volume = 7,732 cubic yards

(Source: USDA, *Ponds- Planning, Design, Construction*; 1971)

TABLE 7-8: RECOMMENDED SIDE SLOPES AND TOP WIDTH OF POND DIKES.

TYPE OF SOIL	INSIDE SLOPE	OUTSIDE SLOPE	WATER DEPTH IN POND (m)	TOP WIDTH OF DIKE (m)	FREEBOARD (m)
SANDY LOAM	1:2 - 1:3	1:1.5 -1:2	0.50	0.50	0.40
SANDY CLAY	1:1.5	1:1.5	0.50 - 0.80	0.50 - 1.00	0.40 - 0.50
FIRM CLAY	1:1	1:1	0.80 - 1.20	1.50	0.50
WITH BRICK LINING INSIDE	1:1 - 1:1.5	1:1.5 - 1:2	1.20 - 2.00	2.00 - 2.50	0.50
WITH CONCRETE LINING INSIDE	0.75 - 1:1	1.5 - 1:2	2.00 - 3.00	2.50 - 4.00	0.50 - 0.60

(Source: T.V.R. Pillay, *Aquaculture Principles and Practices*; 1990)

TABLE 7-9: MINIMUM CROWN WIDTH, IN FEET (m) AS RELATED TO HEIGHT OF POND EMBANKMENTS.

HEIGHT OF DAM IN FEET (meters)	MINIMUM CROWN WIDTH IN FEET (meters)
Under 10 (3.3)	8 (2.4)
10-15 (3.3 - 4.6)	10 (3.3)
15-20 (4.6 - 6.6)	12 (3.7)
20-25 (6.6 - 7.6)	14 (4.3)

(Source: Courtesy of Dr. William McLarney, *The Freshwater Aquaculture Book*; copyright © 1984 - with permission of Hartley & Marks, Inc.)

TABLE 7-10: SIZE OF OUTLET PIPES FOR VARIOUS SIZES OF PONDS

SIZE OF POND	DIAMETER OF OUTLET PIPE
1/20 acre (0.02 ha) or less	2 inches (5.1 cm)
1/20 - 1/5 acre (0.02-0.08 ha)	3 - 4 inches (7.6-10.2 cm)
1/5 - 3/4 acre (0.08-0.3 ha)	6 - 9 inches (15.2-22.9 cm)
1/4 - 1 acre (0.3-0.4 ha)	12 inches (30.5 cm)
More than 1 acre (0.4 ha)	Sluice or monk

(Source: Courtesy of Dr. William McLarney, *The Freshwater Aquaculture Book*; copyright © 1984 - with permission of Hartley & Marks, Inc.)

TABLE 7-11: CHARACTERISTIC DIMENSIONS OF OPTIMUM TRAPEZOIDAL CHANNEL FOR GIVEN CROSS-SECTIONAL AREA AND SIDE SLOPE

SIDE SLOPE	$\frac{h}{\sqrt{A}}$	$\frac{b}{\sqrt{a}}$	$\frac{B}{\sqrt{A}}$	$\frac{P_w}{\sqrt{A}}$	$\frac{R}{\sqrt{A}}$
0.5:1	0.759	0.938	1.698	2.640	0.379
1:1	0.739	0.612	2.092	2.705	0.370
1.5:1	0.689	0.417	2.483	2.905	0.344
2:1	0.636	0.300	2.844	3.145	0.318
2.5:1	0.589	0.227	3.169	3.395	0.295
3:1	0.549	0.174	3.502	3.645	0.275

WHERE:

h = Water Depth (m) b = Bottom Width (m) B = Surface Width (m)
P_w = Wetted Perimeter (m) R = Hydraulic Radius (m) A = Cross-sectional Area (m^2)

(Source: A.N. Bose *et al.*, *Coastal Aquaculture Engineering*; copyright © 1991 - with permission of Cambridge University Press)

TABLE 7-12: RELATION BETWEEN HEAD AND DISCHARGE FOR CIPPOLETTI AND RECTANGULAR WEIRS[1]

HEAD (in.)	DISCHARGE (gpm)	HEAD (in.)	DISCHARGE (gpm)	HEAD (in.)	DISCHARGE (gpm)
0.250	5.00	4.25	317	8.25	860
0.500	14.0	4.50	346	8.50	900
0.750	23.0	4.75	375	8.75	939
1.00	36.0	5.00	405	9.00	978
1.25	50.0	5.25	436	9.25	1020
1.50	66.0	5.50	468	9.50	1060
1.75	84.0	5.75	500	9.75	1100
2.00	102	6.00	533	10.0	1150
2.25	122	6.25	567	10.3	1190
2.50	143	6.50	601	10.5	1230
2.75	165	6.75	636	10.8	1280
3.00	188	7.00	672	11.0	1320
3.25	212	7.25	708	11.3	1370
3.50	237	7.50	745	11.5	1410
3.75	263	7.75	783	11.8	1460
4.00	290	8.00	820	12.0	1510

[1] Discharge values assume a 1-foot-long weir crest. For shorter or longer crests, multiply these values by the actual length in feet

(Source: U.S. Fish and Wildlife Service, *Fish Hatchery Management*; 1982)

TABLE 7-13: DISCHARGE THROUGH CIPOLLETTI TRAPEZOIDAL WEIR WITH END CONTRACTIONS (liters/s)

	WEIR OPENING WIDTH (cm)											
HEAD (cm)	**5.0**	**10.0**	**15.0**	**20.0**	**25.0**	**30.0**	**60.0**	**90.0**	**120.0**	**180.0**	**240.0**	**300.0**
2	.19	.39	.58	.77	.97	1.16	2.32	3.49	4.65	6.97	9.30	11.62
4	.55	1.10	1.64	2.19	2.74	3.29	6.58	9.86	13.15	19.73	26.30	32.88
6	1.01	2.01	3.02	4.03	5.03	6.04	12.08	18.12	24.16	36.24	48.32	60.40
8	1.55	3.10	4.65	6.20	7.75	9.30	18.60	27.90	37.20	55.80	74.40	93.00
10	2.17	4.33	6.50	8.66	10.83	13.00	25.99	38.99	51.99	77.98	103.98	129.97
12	2.85	5.6i	8.54	11.39	14.24	17.08	34.17	51.25	68.34	102.51	136.68	170.85
14	3.59	7.18	10.76	14.35	17.94	21.53	43.06	64.59	86.12	129.18	172.24	215.29
16	4.38	8.77	13.15	17.54	21.92	26.30	52.61	78.91	105.22	157.82	210.43	263.04
18	5.23	10.46	15.69	20.92	26.16	31.39	62.77	94.16	125.55	188.32	251.10	313.87
20	6.13	12.25	18.38	24.51	30.63	36.76	73.52	110.28	147.04	220.57	294.09	367.61
22	7.07	14.14	21.21	28.27	35.34	42.41	84.82	127.23	169.64	254.46	339.29	424.11
24	8.05	16.11	24.16	32.22	40.27	48.32	96.65	144.97	193.29	289.94	386.59	483.24
26	9.08	18.16	27.24	36.33	45.41	54.49	108.98	163.46	217.95	326.93	435.90	544.88
28	10.15	20.30	30.45	40.60	50.75	60.89	121.79	182.68	243.58	365.37	487.16	608.95
30	11.26	22.51	33.77	45.02	56.28	67.53	135.07	202.60	270.14	405.21	540.27	675.34
32	12.40	24.80	37.20	49.60	62.00	74.40	148.80	223.20	297.60	446.39	595.19	743.99
34	13.58	27.16	40.74	54.32	67.90	81.48	162.96	244.45	325.93	488.89	651.85	814.82
36	14.80	29.59	44.39	59.18	73.98	88.78	177.55	266.33	355.10	532.66	710.21	887.76
38	16.05	32.09	48.14	64.18	80.23	96.28	192.55	288.83	385.10	577.65	770.21	962.76
40	17.33	34.66	51.99	69.32	86.65	103.98	207.95	311.93	415.90	623.85	831.81	1039.76
42	18.65	37.29	55.94	74.58	93.23	111.87	223.74	335.61	447.48	671.22	894.96	1118.71
44	19.99	39.99	59.98	79.97	99.96	119.96	239.91	359.87	479.82	719.73	959.65	1199.56
46	21.37	42.74	64.11	85.48	106.86	128.23	256.45	384.68	512.91	769.36	1025.81	1282.27
48	22.78	45.56	68.34	91.12	113.90	136.68	273.36	410.04	546.72	820.08	1093.44	1366.80
50	24.22	48.44	72.66	96.87	121.09	145.31	290.62	435.93	581.24	871.86	1162.48	1453.10
52	25.69	51.37	77.06	102.74	128.43	154.12	308.23	462.35	616.46	924.69	1232.93	1541.16
54	27.18	54.36	81.55	108.73	135.91	163.09	326.18	489.28	652.37	978.55	1304.74	1630.92
56	28.71	57.41	86.12	114.82	143.53	172.24	344.47	516.71	688.94	1033.42	1377.89	1722.36
58	29.58	30.26	60.51	90.77	121.03	151.29	181.54	363.09	544.63	726.18	1089.27	1452.36
60	31.84	63.67	95.51	127.34	159.18	191.02	382.03	573.05	764.06	1146.09	1528.12	1910.16
62	33.44	66.88	100.32	133.76	167.20	200.65	401.29	601.94	802.58	1203.87	1605.16	2006.45
64	35.07	70.14	105.22	140.29	175.36	210.43	420.86	631.30	841.73	1262.59	1683.46	2104.32
66	36 73	73.46	110.19	146.92	183.64	220.37	440.75	661.12	981.49	1322.24	1762.98	2203.73
68	38 41	76.82	115.23	153.64	192.05	230.47	460.93	691.40	921.86	1382.79	1843.72	2304.65
70	40.12	80.24	120.35	160.47	200.59	240.71	481.41	722.12	962.83	1444.24	1925.66	2407.07
72	41.85	83.70	125.55	167.40	239.25	251.10	502.19	753.29	1004.39	1506.58	2008.77	2510.96
74	43.61	87.21	130.82	174.42	218.03	261.63	523.26	784.89	1046.52	1569.79	2093.05	2616.31
76	45.38	90.77	136.15	181.54	226.92	272.31	544.62	816.93	1089.24	1633.85	2178.47	2723.09
78	47.19	94.38	141.56	188.75	235.94	283.13	566.26	849.39	1132.51	1698.77	2265.03	2831.29
80	49.01	98.03	147.04	196.06	245.07	294.09	588.18	882.26	1176.35	1764.53	2352.70	2940.88
82	50.86	101.73	152.59	203.46	254.32	305.18	610.37	915.55	1220.74	1831.11	2441.48	3051.85
84	52 74	105.47	158.21	210.95	263.68	316.42	632.84	949.25	1265.67	1898.51	2531.34	3164.18
86	54.63	109.26	163.89	218.52	273.15	327.79	655.57	983.36	1311.14	1966.71	2622.28	3277.85
88	56.55	113.10	169.64	226.19	282.74	339.29	678.57	1017.86	1357.14	2035.72	2714.29	3392.86
90	58.49	116.97	175.46	233.95	292.43	350.92	701.84	1052.75	1403.67	2105.51	2807.34	3509.18
92	60.45	120.89	181.34	241.79	302.23	362.68	725.36	1088.04	1450.72	2176.08	2901.44	3626.80
94	62.43	124.86	187.29	249.71	312.14	374.57	749.14	1123.71	1498.28	2247.42	2996.56	3745.70
96	64.43	128.86	193.29	257.73	322.16	386.59	773.18	1159.76	1546.35	2319.53	3092.71	3865.88
98	66.46	132.91	199.37	265.82	332.28	398.73	797.46	1196.20	1594.93	2392.39	3189.85	3987.32
100	68.50	137.00	205.50	274.00	342.50	411.00	822.00	1233.00	1644.00	2466.00	3288.00	4110.00

DISCHARGE FROM A CIPOLLETTI'S WEIR IS CALCULATED AS:

$$Q = 1.37\ L_w (h)^{3/2}$$

WHERE:

Q = discharge (meters^3/second)
L_w = crest length (meters)
h = head (meters)

(Source: F.W. Wheaton, *Aquacultural Engineering*; copyright © 1985 - reprinted with permission of John Wiley & Sons, Inc.)

TABLE 7-14: DISCHARGE THROUGH A 90° TRIANGULAR WEIR (liters/s)

HEAD (cm)	DISCHARGE	HEAD (cm)	DISCHARGE	HEAD (cm)	DISCHARGE	HEAD (cm)	DISCHARGE
0.5	.00	15.0	14.03	31.5	76.30	47.0	207.47
1.0	.01	16.5	15.15	32.0	79.36	47.5	213.04
1.5	.04	17.0	16.32	32.5	82.50	48.0	218.69
2.0	.08	17.5	17.55	33.0	85.70	48.5	224.43
2.5	.14	18.0	18.83	33.5	88.99	49.0	230.26
3.0	.21	18.5	20.17	34.0	92.35	49.5	236.17
3.5	.31	19.0	21.56	34.5	95.78	50.0	242.18
4.0	.44	19.5	23.00	35.0	99.29	50.5	248.28
4.5	.59	20.0	24.51	35.5	102.87	51.0	254.48
5.0	.77	20.5	26.07	36.0	106.53	51.5	260.76
5.5	.97	21.0	27.69	36.5	110.27	52.0	267.13
6.0	1.21	21.5	29.36	37.0	114.08	52.5	273.60
6.5	1.48	22.0	31.10	37.5	117.98	53.0	280.16
7.0	1.78	22.5	32.90	38.0	121.95	53.5	286.82
7.5	2.11	23.0	34.76	38.5	126.00	54.0	293.57
8.0	2.48	23.5	36.68	39.0	130.13	54.5	300.41
8.5	2.89	24.0	38.66	39.5	134.34	55.0	307.35
9.0	3.33	24.5	40.70	40.0	138.63	55.5	314.38
9.5	3.81	25.0	42.81	40.5	143.01	56.0	321.51
10.0	4.33	25.5	44.99	41.0	147.46	56.5	328.73
10.5	4.89	26.0	47.22	41.5	152.00	57.0	336.05
11.0	5.50	26.5	49.53	42.0	156.62	57.5	343.47
11.5	6.14	27.0	51.90	42.5	161.32	58.0	350.99
12.0	6.83	27.5	54.33	43.0	166.11	58.5	358.60
12.5	7.57	28.0	56.83	43.5	170.98	59.0	366.31
13.0	8.35	28.5	59.41	44.0	175.93	59.5	374.12
13.5	9.17	29.0	62.05	44.5	180.98	60.0	382.03
14 0	10.05	29.5	64.76	45.0	186.10	60.5	390.04
14.5	10.97	30.0	67.53	45.5	191.32	61.0	398.1S
15.0	11.94	30.5	70.38	46.0	196.61	61.5	406.36
15.5	12.96	31.0	73.30	46.5	202.00	62.0	414.67
62.5	423.08	72.0	602.63	81.5	821.51	91.0	1,082.24
63.0	431.59	72.5	613.15	82.0	834.17	91.5	1,097.17
63.5	440.20	73.0	623.77	82.5	846.95	92.0	1,112.22
64.0	448.92	73.5	634.51	83.0	859.84	92.5	1,127.39
64.5	457.74	74.0	645.36	83.5	872.84	93.0	1,142.69
65.0	466.66	74.5	656.31	84.0	885.97	93.5	1,158.11
65.5	475.69	75.0	667.38	84.5	899.21	54.0	1,173.65
66.0	484.82	75.5	678.56	85.0	912.57	94.5	1,189.32
66.5	494.05	76.0	689.85	85.5	926.05	95.0	1,205.12
67.0	503.39	76 5	701.25	86.0	939.65	95.5	1,221.04
67.5	512.84	77.0	712.77	86.5	953.37	96.0	1,237.08
68.0	522.39	77.5	724.39	87.0	967.21	96.5	1,253.25
68.5	532.04	78.0	736 13	87.5	981.16	97.0	1,269.55
69.0	541.81	78.5	747.99	88.0	995.24	97.5	1,285.97
69.5	551.67	79.0	759.96	88.5	1,009.44	98.0	1,302.52
70.0	561.65	79 5	772.04	89.0	1,023.75	98.5	1,319.20
70.5	571.73	80.0	784.23	89.5	1,038.19	99.0	1,336.01
71.0	581.92	80.5	796.54	90.0	1,052.75	99.5	1,352.94
71.5	592.22	81.0	808.97	90.5	1,067.44	100.0	1,370.00

DISCHARGE FROM A 90° TRIANGULAR WEIR IS CALCULATED AS:

$$Q = 1.37h^{5/2}$$

WHERE:

Q = discharge (m^3/s)
h = head (m)

(Source: F.W. Wheaton, *Aquacultural Engineering*; copyright © 1985 - reprinted with permission of John Wiley & Sons, Inc.)

TABLE 7-15: DISCHARGE THROUGH A PARSHALL FLUME (liters/s) UNDER FREE FLOW [1]

HEAD (cm)	FLUME OPENING WIDTH (cm)											
	5.0	10.0	15.0	20.0	25.0	30.0	60.0	90.0	120.0	180.0	240.0	300.0
2	.36	.66	.95	1.23	1.50	1.77	3.28	4.70	6.07	8.71	11.23	14.06
4	.97	1.85	2.68	3.49	4.29	5.07	9.60	13.92	18.12	26.28	34.20	42.63
6	1.75	3.36	4.91	6.43	7.93	9.40	17.98	26.26	34.35	50.15	65.58	81.55
8	2.66	5.14	7.55	9.92	12.26	14.57	28.08	41.20	54.08	79.33	104.09	129.22
10	3.68	7.15	10.54	13.88	17.18	20.46	39.67	58.43	76.89	113.21	148.95	184.67
12	4.80	9.37	13.84	18.27	22.65	27.00	52.62	77.73	102.51	151.39	199.61	247.22
14	6.00	11.76	17.43	23.04	28.60	34.13	66.81	98.94	130.72	193.55	255.68	316.37
16	7.29	14.33	21.28	28.17	35.01	41.82	82.16	121.95	161.37	239.46	316.83	391.72
18	8.65	17.06	25.37	33.63	41.85	50.03	98.61	146.64	194.31	288.91	382.79	472.96
20	10.08	19.93	29.70	39.41	49.09	58.72	116.09	172.93	229.44	341.74	453.36	559.81
22	11.57	22.95	34.25	45.50	56.71	67.89	134.56	200.76	266.66	397.81	528.35	652.03
24	13.13	26.10	39.00	51.86	64.69	77.50	153.97	230.06	305.88	456.99	607.58	749.42
26	14.75	29.38	43.96	58.50	73.03	87.53	174.30	260.77	347.04	519.18	690.92	851.82
28	16.43	32.78	49.10	65.41	81.70	97.98	195.50	292.84	390.07	584.27	778.24	959.05
30	18.16	36.30	54.44	72.57	90.70	108.82	217.55	326.24	434.91	652.19	869.42	1070.98
32	19.94	39.93	59.95	79.97	100.01	120.05	240.43	360.93	481.51	722.85	964.37	1187.49
34	21.78	43.68	65.63	87.61	109.62	131.65	264.10	396.86	529.82	796.19	1062.98	1308.44
36	23.66	47.53	71.48	95.49	119.53	143.61	288 55	434.01	579.81	872.13	1165.16	1433.75
38	25.59	51.49	77.50	103.58	129.73	155.92	313.75	472.34	631.42	950.63	1270.85	1563.30
40	27.57	55.54	83.67	111.90	140.21	168.58	339.70	511.83	684.63	1031.61	1379.97	1697.01
42	29.60	59.70	90.00	120.43	150.96	181.57	366.37	552.46	739.39	1115.04	1492.44	1834.80
44	31.67	63.95	96.47	129.16	161.97	194.88	393.75	594.19	795.68	1200.86	1608.20	1976.57
46	33.78	68.29	103.10	138.10	173.25	208.52	421.81	637.01	853.47	1289.04	1727.20	2122.27
48	35.93	72.73	109.87	147.24	184.78	222.47	450.56	680.90	912.72	1379.51	1849.37	2271.82
50	38.12	77.25	116.78	156.57	196.57	236.72	479.97	725.83	973.42	1472.26	1974.67	2425.16
52	40.36	81.86	123.83	166.10	208.59	251.28	510.04	771.79	1035.54	1567.23	2103.05	2582.23
54	42.63	86.56	131.01	175.81	220.86	266.13	540.74	818.76	1099.05	1664.41	2234.45	2742.96
56	44.94	91.34	138.33	185.70	233.36	281.27	572.08	866.73	1163.93	1763.74	2368.83	2907.30
58	47.29	96.21	145.77	195.77	246.10	296.69	604.04	915.67	1230.17	1865.20	2506.15	3075.20
60	49.68	101.15	153.35	206.02	259.06	312.39	636.60	965.58	1297.73	1968.76	2646.38	3246.62
62	52.10	106.18	161.05	216.45	272.25	328.37	669.77	1016.44	1366.61	2074.39	2789.46	3421.49
64	54.56	111.28	168.87	227.05	285.66	344.62	703.54	1068.24	1436.79	2182.07	2935.37	3599.79
66	57.05	116.46	176.82	237.81	299.28	361.14	737.88	1120.95	1508.24	2291.76	3084.07	3781.46
68	59.58	121.72	184.88	248.74	313.12	377.92	772.81	1174.58	1580.95	2403.45	3235.52	3966.46
70	62.14	127.05	193.07	259.84	327.17	394.96	808.30	1229.11	1654.91	2517.10	3389.70	4154.76
72	64.74	132.45	201.37	271.09	341.43	412.26	844.35	1284.53	1730.10	2632.70	3546.57	4346.31
74	67.37	137.93	209.78	282.51	355.89	429.80	880.96	1340.82	1806.50	2750.23	3706.10	4541.08
76	70.03	143.47	218.31	294.08	370.56	447.60	918.12	1397.99	1884.11	2869.65	3868.26	4739.04
78	72.72	149.09	226.95	305.81	385.42	465.64	955.81	1456.00	1962.90	2990.96	4033.04	4940.15
80	75.44	154.78	235.70	317.69	400.48	483.93	994.04	1514.87	2042.87	3114.12	4200.39	5144.38
82	78.20	160.53	244.56	329.72	415.74	502.45	1032.80	1574.57	2123.99	3239.13	4370.30	5351.69
84	80.98	166.35	253.53	341.90	431.19	521.22	1072.08	1635.11	2206.27	3365.97	4542.73	5562.06
86	83.79	172.24	262.60	354.23	446.83	540.21	1111.88	1696.46	2289.69	3494.60	4717.68	5775.46
88	86.64	178.20	271.78	366.70	462.66	559.44	1152.19	1758.63	2374.24	3625.04	4895.10	5991.85
90	89.51	184.22	281.06	379.32	478.67	578.90	1193.00	1821.60	2459.90	3757.24	5074.99	6211.22
92	92.42	190.30	290.44	392.08	494.87	598.58	1234.32	1885.36	2546.67	3891.20	5257.31	6433.53
94	95.35	196.45	299.93	404.99	511.25	618.49	1276.13	1949.92	2634.54	4026.90	5442.05	6658.76
96	98.31	202.66	309.51	418.03	527.81	638.62	1318.44	2015.26	2723.49	4164.32	5629.19	6886.88
98	101.30	208.94	319.19	431.21	544.55	658.97	1361.23	2081.37	2813.51	4303.46	5818.71	7117.88
100	104.31	215.27	328.98	444.52	561.47	679.54	1404.50	2148.24	2904.60	4444.30	6010.58	7351.72

[1] Head measurement is the upstream head.

(Source: F.W. Wheaton, *Aquacultural Engineering*; copyright © 1985 - reprinted with permission of John Wiley & Sons, Inc.)

TABLE 7-16: DISCHARGE FROM A RECTANGULAR WEIR WITH FULL END CONTRACTIONS (liters/s)

HEAD (cm)	WEIR OPENING WIDTH (cm)											
	5.0	10.0	15.0	20.0	25.0	30.0	60.0	90.0	120.0	180.0	240.0	300.0
2	.24	.50	.76	1.02	1.28	1.54	3.10	4.66	6.22	9.35	12.47	15.59
4	.62	1.35	2.09	2.83	3.56	4.30	8.71	13.13	17.55	26.38	35.21	44.04
6	1.03	2.38	3.73	5.08	6.44	7.79	15.90	24.01	32.13	48.35	64.58	80.80
8	1.42	3.50	5.58	7.66	9.74	11.82	24.31	36.80	49.30	74.28	99.26	124.24
10	1.75	4.65	7.56	10.47	13.38	16.29	33.75	51.20	68.66	103.57	138.48	173.39
12	1.99	5.81	9.64	13.46	17.29	21.11	44.06	67.00	89.95	135.84	181.73	227.63
14	2.12	6.94	11.76	16.58	21.40	26.22	55.13	84.05	112.96	170.79	228.63	286.46
16	2.12	8.01	13.90	19.78	25.67	31.56	66.89	102.22	137.54	208.20	278.86	349.51
18	1.97	8.99	16.02	23.04	30.07	37.10	79.25	121.41	163.56	247.87	332.18	416.49
20	1.65	9.87	18.10	26.33	34.56	42.79	92.16	141.53	190.91	289.65	388.40	487.14
22	1.14	10.63	20.13	29.62	39.11	48.61	105.57	162.53	219.49	333.41	447.33	561.25
24	.43	11.25	22.07	32.88	43.70	54.52	119.42	184.32	249.22	379.03	508.83	638.63
26		11.71	23.91	36.10	48.30	60.50	133.68	206.86	280.04	426.40	572.76	719.13
28		12.00	25.63	39.26	52.89	66.52	148.30	230.09	311.88	475.45	639.02	802.59
30		12.09	27.21	42.33	57.45	72.56	163.27	253.97	344.67	526.08	707.48	888.89
32		11.99	28.64	45.30	61.95	78.61	178.53	278.45	378.37	578.22	778.06	977.91
34		11.67	29.91	48.15	66.39	84.63	194.07	303.50	412.94	631.81	850.68	1069.55
36		11.13	31.00	50.87	70.74	90.62	209.85	329.08	448.31	686.78	925.24	1163.70
38		10.34	31.90	53.45	75.00	96.55	225.85	355.16	484.46	743.07	1001.68	1260.29
40		9.31	32.58	55.86	79.13	102.41	242.05	381.70	521.35	800.64	1079.93	1359.22
42		8.01	33.05	58.10	83.14	108.18	258.43	408.68	558.93	859.43	1159.93	1460.42
44		6.44	33.30	60.15	87.00	113.85	274.96	436.07	597.17	919.39	1241.61	1563.83
46		4.59	33.30	62.00	90.70	119.40	291.62	463.84	636.05	980.49	1324.92	1669.36
48		2.45	33.04	63.64	94.23	124.83	308.40	491.97	675.54	1042.68	1409.81	1776.95
50			32.53	65.05	97.58	130.11	325.27	520.43	715.59	1105.91	1496.24	1886.56
52			31.74	66.24	100.73	135.23	342.22	549.21	756.19	1170.17	1584.14	1998.12
54			30.67	67.17	103.68	140.19	359.23	578.27	797.32	1235.40	1673.49	2111.58
56			29.30	67.86	106.41	144.96	376.29	607.61	838.94	1301.58	1764.23	2226.88
58			27.63	68.27	108.91	149.55	393.37	637.20	881.03	1368.68	1856.33	2343.99
60			25.65	68.41	111.17	153.93	410.47	667.02	923.57	1436.66	1949.75	2462.85
62			23.35	68.27	113.18	158.09	427.57	697.06	966.54	1505.50	2044.46	2583.42
64			20.73	67.83	114.93	162.04	444.66	727.29	1009.91	1575.16	2140.41	2705.65
66			17.76	67.09	116.42	165.75	461.72	757.70	1053.67	1645.62	2237.57	2829.52
68			14.44	66.03	117.62	169.21	478.74	788.27	1097.80	1716.86	2335.92	2954.98
70			10.78	64.66	118.54	172.42	495.70	818.99	1142.28	1788.85	2435.42	3081.99
72			6.74	62.95	119.16	175.36	512.60	849.84	1187.08	1861.56	2536.04	3210.52
74			2.34	60.91	119.47	178.04	529.42	880.81	1232.20	1934.98	2637.75	3340.53
76				58.52	119.47	180.43	546.16	911.88	1277.61	2009.07	2740.53	3471.99
78				55.77	119.15	182.52	562.79	943.05	1323.31	2083.83	2844.35	3604.87
80				52.66	118.49	184.32	579.30	974.28	1369.26	2159.22	2949.18	3739.13
82				49.19	117.50	185.81	595.70	1005.58	1415.46	2235.23	3054.99	3874.76
84				45.33	116.16	186.99	611.96	1036.93	1461.90	2311.84	3161.77	4011.71
86				41.09	114.46	187.83	628.07	1068.31	1508.55	2389.02	3269.49	4149.97
88				36.45	112.40	188.35	644.03	1099.72	1555.40	2466.77	3378.13	4289.50
90				31.42	109.97	188.52	659.83	1131.13	1602.44	2545.05	3487.66	4430.27
92				25.98	107.16	188.35	675.45	1162.55	1649.66	2623.86	3598.07	4572.27
94				20.12	103.97	187.81	690.89	1193.96	1697.03	2703.18	3709.32	4715.47
96				13.85	100.38	186.92	706.13	1225.34	1744.56	2782.98	3821.41	4859.84
98				7.14	96.39	185.65	721.17	1256.69	1792.22	2863.26	3934.31	5005.36
100					92.00	184.00	736.00	1288.00	1840.00	2944.00	4048.00	5152.00

DISCHARGE FROM A RECTANGULAR WEIR IS CALCULATED AS:

$$Q = 1.84\ L_w(h)^{3/2}$$

WHERE:

Q = Discharge (meters^3/second)
L_w = Crest length (meters)
h = Head (meters)

(Source: F.W. Wheaton, *Aquacultural Engineering*; copyright © 1985 - reprinted with permission of John Wiley & Sons, Inc.)

TABLE 7-17: RATES OF FLOW THROUGH DITCH-TO-FURROW PIPES FOR VARIOUS HEADS

RATE OF FLOW THROUGH SMALL PIPE

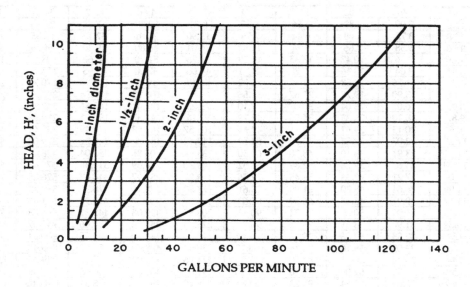

RATE OF FLOW THROUGH LARGE PIPE

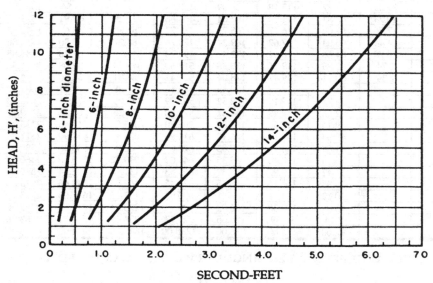

(Source: U.S. Fish and Wildlife Service, *Water Measurement Manual*; 1981)

TABLE 7-18: RATES OF FLOW THROUGH DITCH-TO-FURROW SIPHONS FOR VARIOUS HEADS.

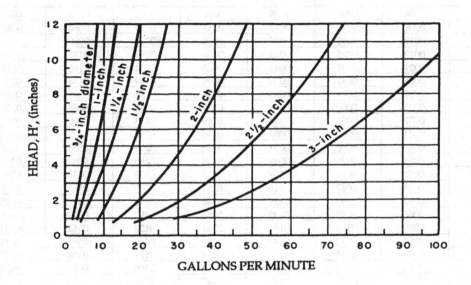

RATE OF FLOW THROUGH SMALL SIPHONS

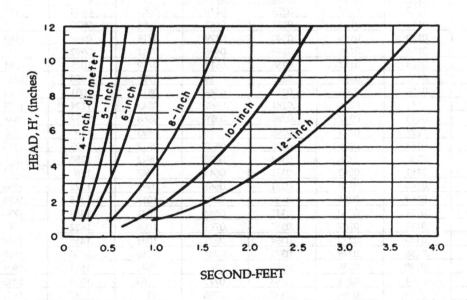

RATE OF FLOW THROUGH LARGE SIPHONS

(Source: U.S. Fish and Wildlife Service, *Water Measurement Manual*; 1981)

159

TABLE 7-19: ACRE-FEET EQUIVALENT TO A GIVEN NUMBER OF SECOND-FEET FLOWING FOR A GIVEN LENGTH OF TIME

SECOND-FEET	MINUTES			HOURS					
	15	30	45	1	2	3	4	5	6
0.01	.0002	.0004	.0006	.0008	.0017	.0025	.0033	.0041	.0050
0.02	.0004	.0008	.0012	.0017	.0033	.0050	.0066	.0083	.0099
0.03	.0006	.0012	.0019	.0025	.0050	.0074	.0099	.0124	.0149
0.04	.0008	.0017	.0025	.0033	.0066	.0099	.0132	.0165	.0198
0.05	.0010	.0021	.0031	.0041	.0083	.0124	.0165	.0207	.0248
0.06	.0012	.0025	.0037	.0050	.0099	.0149	.0198	.0248	.0298
0.07	.0015	.0029	.0043	.0058	.0116	.0174	.0231	.0289	.0347
0.08	.0017	.0033	.0050	.0066	.0132	.0198	.0265	.0331	.0397
0.09	.0019	.0037	.0056	.0074	.0149	.0223	.0298	.0372	.0446
0.10	.0021	.0041	.0062	.0083	.0165	.0248	.0331	.0413	.0496
0.11	.0023	.0046	.0068	.0091	.0182	.0273	.0364	.0455	.0546
0.12	.0025	.0051	.0074	.0099	.0198	.0298	.0397	.0496	.0595
0.13	.0027	.0054	.0081	.0107	.0215	.0322	.0430	.0537	.0645
0.14	.0029	.0058	.0087	.0116	.0231	.0347	.0463	.0579	.0694
0.15	.0031	.0062	.0093	.0124	.0248	.0372	.0496	.0620	.0744
0.16	.0033	.0066	.0099	.0132	.0265	.0397	.0529	.0661	.0793
0.17	.0035	.0070	.0105	.0141	.0281	.0422	.0562	.0703	.0843
0.18	.0037	.0074	.0112	.0149	.0298	.0446	.0595	.0744	.0893
0.19	.0039	.0079	.0118	.0157	.0314	.0471	.0628	.0785	.0942
0.20	.0041	.0083	.0124	.0165	.0331	.0496	.0661	.0826	.0992
0.21	.0043	.0087	.0130	.0174	.0347	.0521	.0694	.0868	.1041
0.22	.0046	.0091	.0136	.0182	.0364	.0546	.0727	.0909	.1091
0.23	.0048	.0095	.0143	.0190	.0380	.0570	.0760	.0950	.1141
0.24	.0050	.0099	.0149	.0198	.0397	.0595	.0793	.0992	.1190
0.25	.0052	.0103	.0155	.0207	.0413	.0620	.082d	.1033	.1240
0.26	.0054	.0107	.0161	.0215	.0430	.0645	.0860	.1074	.1289
0.27	.0056	.0112	.0167	.0223	.0446	.0669	.0893	.1116	.1339
0.28	.0058	.0116	.0174	.0231	.0463	.0694	.0926	.1157	.1388
0.29	.0060	.0120	.0180	.0240	.0479	.0719	.0959	.1198	.1438
0.30	.0062	.0124	.0186	.0248	.0496	.0744	.0992	.1240	.1188
0.31	.0064	.0128	.0192	.0256	.0512	.0769	.1025	.1281	.1537
0.32	.0066	.0132	.0198	.0265	.0529	.0793	.1058	.1322	.1587
0.33	.0068	.0136	.0205	.0273	.0546	.0818	.1091	.1364	.1636
0.34	.0070	.0141	.0211	.0281	.0562	.0843	.1124	.1405	.1686
0.35	.0072	.0145	.0217	.0289	.0579	.0868	.1157	.1446	.1736
0.36	.0074	.0149	.0223	.0298	.0595	.0893	.1190	.1488	.1785
0.37	.0076	.0153	.0229	.0306	.0612	.0917	.1223	.1529	.1835
0.38	.0079	.0157	.0236	.0314	.0628	.0942	.1256	.1570	.1884
0.39	.0081	.0161	.0242	.0322	.0645	.0967	.1289	.1612	.1934
0.40	.0083	.0165	.0248	.0331	.0661	.0992	.1322	.1653	.1984
0.41	.0085	.0169	.0254	.0339	.0678	.1017	.1355	.1694	.2033
0.42	.0087	.0174	.0260	.0347	.0694	.1041	.1388	.1736	.2083
0.43	.0089	.0178	.0267	.0355	.0711	.1066	.1422	.1777	.2132
0.44	.0091	.0186	.0273	.0364	.0727	.1091	.1455	.1818	.2182
0.45	.0093	.0186	.0279	.0372	.0744	.1116	.1488	.1860	.2231
0.46	.0095	.0190	.0285	.0380	.0760	.1141	.1521	.1901	.2281
0.47	.0097	.0194	.0291	.0388	.0777	.1165	.1554	.1942	.2331
0.48	.0099	.0198	.0298	.0397	.0793	.1190	.1587	.1984	.2380
0.49	.0101	.0203	.0304	.0405	.0810	.1215	.1620	.2025	.2430
0.50	.0103	.0207	.0310	.0413	.0826	.1240	.1653	.2066	.2479

TABLE 7-19 (cont.)

SECOND-FEET	MINUTES 15	30	45	HOURS 1	2	3	4	5	6
0.51	.0105	.0211	.0316	.0422	.0843	.1265	.1686	.2107	.2529
0.52	.0107	.0215	.0322	.0430	.0860	.1289	.1719	.2149	.2579
0.53	.0110	.0219	.0329	.0438	.0876	.1314	.1752	.2190	.2628
0.54	.0112	.0223	.0335	.0446	.0893	.1339	.1785	.2231	.2678
0.55	.0114	.0227	.0341	.0455	.0~09	.1364	.1818	.2273	.2727
0.56	.0116	.0231	.0347	.0463	.0926	.1388	.1851	.2314	.2777
0.57	.0118	.0236	.0353	.0471	.0942	.1413	.1884	.2355	.2826
0.58	.0120	.0240	.0360	.0479	.0959	.1438	.1917	.2397	.2876
0.59	.0122	.0244	.0366	.0488	.0975	.1463	.1950	.2438	.2926
0.60	.0124	.0248	.0372	.0496	.0992	.1488	.1984	.2479	.2975
0.61	.0126	.0252	.0378	.0504	.1008	.1512	.2017	.2521	.3025
0.62	.0128	.0256	.0384	.0512	.1025	.1537	.2050	.2562	.3074
0.63	.0130	.0260	.0391	.0521	.1041	.1562	.2083	.2603	.3124
0.64	.0132	.0265	.0397	.0529	.1058	.1587	.2116	.2645	.3174
0.65	.0134	.0269	.0403	.0537	.1074	.1612	.2149	.2686	.3223
0.66	.0136	.0273	.0409	.0546	.1091	.1636	.2182	.2727	.3273
0.67	.0138	.0277	.0415	.0554	.1107	.1661	.2215	.2769	.3322
0.68	.0141	.0281	.0422	.0582	.1124	.1686	.2248	.2810	.3372
0.69	.0143	.0285	.0428	.0570	.1141	.1711	.2281	.2851	.3422
0.70	.0145	.0289	.0434	.0579	.1157	.1736	.2314	.2893	.3471
0.71	.0147	.0293	.0440	.0587	.1174	.1760	.2348	.2934	.3521
0.72	.0149	.0298	.0446	.0595	.1190	.1785	.2380	.2975	.3570
0.73	.0151	.0302	.0453	.0603	.1207	.1810	.2413	.3017	.3620
0.74	.0153	.0306	.0459	.0612	.1223	.1835	.2446	.3058	.3669
0.75	.0155	.0310	.0465	.0620	.1240	.1860	.2479	.3099	.3719
0.76	.0157	.0314	.0471	.0628	.1256	.1884	.2512	.3141	.3769
0.77	.0159	.0318	.0477	.0636	.1273	.1909	.2546	,3182	.3818
0.78	.0161	.0322	.0484	.0645	.1289	.1934	.2579	.3223	.3868
0.79	.0163	.0326	.0490	.0653	.1306	.1859	.2612	.3265	.3917
0.80	.0165	.0031	.0496	.0661	.1322	.1984	.2645	.3306	.3697
0.81	.0167	.0335	.0502	.0669	.1339	.2908	.2678	.3347	.4017
0.82	.0169	.0339	.0508	.0678	.1355	.2033	.2711	.3988	.4066
0.83	.0172	.0343	.0515	.0686	.1372	.2058	.2744	.3430	.4116
0.84	.0174	.0347	.0521	.0694	.1388	.2083	.2777	.3471	.4165
0.85	.0176	.0351	.0527	.0703	.1405	.2107	.2810	.3512	.4215
0.86	.0178	.0355	.0533	.0711	.1422	.2132	.2843	.3554	.4265
0.87	.0180	.0360	.0539	.0719	.1438	.2157	.2876	.3595	.4314
0.88	.0182	.0364	.0546	.0727	.1455	.2182	.2609	.3636	.4364
0.89	.0184	.0368	.0552	.0736	.1471	.2207	.2942	.3678	.4413
0.90	.0186	.0372	.0558	.0744	.1488	.2231	.2875	.3719	.4463
0.91	.0188	.0376	.0564	.0752	.1504	.2256	.3008	.3760	.4512
0.92	.0190	.0360	.0570	.0760	.1521	.2281	.3041	.3802	.4562
0.93	.0192	.0384	.0576	.0769	.1537	.2306	.3074	.3843	.4612
0.94	.0194	.0388	.0583	.0777	.1554	.2331	.3107	.3884	.4661
0.95	.0196	.0393	.0589	.0785	.1570	.2355	.3141	.3926	.4711
0.96	.0198	.0397	.0595	.0793	.1587	.2380	.3174	.3967	.4760
0.97	.0200	.0401	.0601	.0802	.1603	.2405	.3207	.4008	.4810
0.88	.0203	.0405	.0607	.0810	.1620	.2430	.3240	.9050	.4860
0.99	.0205	.0409	.0614	.0818	.1636	.2455	.3273	.4091	.4909
1.00	.0207	.0413	.0620	.0826	.1653	.2479	.3306	.4132	.4959

TABLE 7-19 (cont.)

SECOND-FEET					HOURS				
	7	8	9	10	11	12	13	14	15
0.01	.006	.007	.007	.008	.009	.010	.011	.012	.012
0.02	.012	.013	.015	.017	.018	.020	.021	.023	.025
0.03	.017	.020	.022	.025	.027	.030	.032	.035	.037
0.04	.023	.026	.030	.033	.036	.040	.043	.046	.050
0.05	.029	.033	.037	.041	.045	.0$0	.054	.058	.062
0.06	.035	.040	.045	.050	.055	.060	.064	.069	.074
0.07	.040	.046	.052	.058	.064	.069	.075	.081	.087
0.08	.046	.053	.060	.036	.073	.079	.086	.093	.099
0.09	.052	.060	.067	.074	.082	.089	.097	.104	.112
0.10	.058	.066	.074	.083	.091	.099	.107	.116	.124
0.11	.064	.073	.082	.091	.100	.109	.118	.127	.136
0.12	.069	.079	.089	.099	.109	.119	.129	.139	.149
0.13	.075	.086	.097	.107	.118	.129	.140	.150	.161
0.14	.081	.093	.104	.116	.127	.139	.150	.162	.174
0.15	.087	.099	.112	.124	.136	.149	.161	.174	.186
0.16	.093	.106	.119	.132	.145	.159	.172	.185	.198
0.17	.098	.112	.126	.140	.155	.169	.183	.197	.211
0.18	.104	.119	.134	.149	.164	.179	.193	.208	.223
0.19	.110	.126	.141	.157	.173	.188	.204	.220	.236
0.20	.116	.132	.149	.165	.182	.198	.215	.231	.248
0.21	.121	.139	.156	.174	.191	.208	.226	.243	.260
0.22	.127	.145	.164	.182	.200	.218	.236	.255	.273
0.23	.133	.152	.171	.190	.209	.228	.247	.266	.285
0.24	.139	.159	.179	.198	.218	.238	.258	.278	.298
0.25	.145	.165	.186	.207	.227	.248	.269	.289	.310
0.26	.150	.172	.193	.215	.236	.258	.279	.301	.322
0.27	.156	.179	.201	.223	.245	.268	.290	.312	.335
0.28	.162	.185	.208	.231	.255	.278	.301	.324	.347
0.29	.168	.192	.216	.240	.264	.288	.312	.336	.360
0.30	.174	.198	.223	.248	.273	.298	.322	.347	.372
0.31	.179	.205	.231	.256	.282	.307	.333	.359	.384
0.32	.185	.212	.238	.264	.291	.317	.344	.370	.397
0.33	.191	.218	.245	.273	.300	.327	.355	.382	.409
0.34	.197	.225	.253	.281	.309	.337	.365	.393	.421
0.35	.202	.231	.260	.289	.318	.347	.376	.405	.434
0.36	.208	.238	.268	.298	.327	.357	.387	.417	.446
0.37	.214	.245	.275	.306	.336	.367	.398	.428	.459
0.38	.220	.251	.283	.314	.345	.377	.408	.440	.471
0.39	.226	.258	.290	.322	.355	.387	.419	.451	.483
0.40	.231	.264	.298	.331	.364	.397	.430	.463	.496
0.41	.237	.271	.305	.339	.373	.407	.440	.474	.508
0.42	.243	.278	.312	.347	.382	.417	.451	.486	.521
0.43	.249	.284	.320	.355	.391	.426	.462	.498	.533
0.44	.255	.291	.327	.364	.400	.436	.473	.509	.545
0.45	.260	.298	.335	.372	.409	.446	.483	.521	.558
0.46	.266	.304	.342	.380	.418	.456	.494	.532	.570
0.47	.272	.311	.350	.388	.427	.466	.505	.544	.583
0.48	.278	.317	.357	.397	.436	.476	.516	.555	.595
0.49	.283	.324	.364	.405	.445	.486	.526	.567	.6O7
0.50	.289	.331	.372	.413	.455	.496	.537	.579	.620

TABLE 7-19 (cont.)

SECOND-FEET				HOURS					
	7	8	9	10	11	12	13	14	15
0.51	.296	.337	.379	.421	.404	.506	.548	.590	.632
0.52	.301	.344	.387	.430	.473	.516	.559	.602	.645
0.53	.307	.350	.394	.438	.482	.526	.569	.613	.657
0.54	.312	.357	.402	.446	.491	.536	.580	.625	.669
0.55	.318	.364	.409	.455	.500	.545	.591	.636	.682
0.56	.324	.370	.417	.463	.509	.555	.602	.648	.694
0.57	.330	.377	.424	.471	.518	.565	.612	.660	.707
0.58	.336	.383	.431	.479	.527	.575	.623	.671	.719
0.59	.342	.390	.439	.488	.536	.585	.634	.683	.731
0.60	.347	.397	.446	.496	.545	.595	.645	.694	.744
0.61	.353	.403	.454	.504	.555	.605	.655	.706	.756
0.62	.359	.410	.461	.512	.564	.615	.666	.717	.769
0.63	.364	.417	.469	.521	.573	.625	.677	.729	.781
0.64	.370	.423	.476	.529	.582	.635	.688	.740	.793
0.65	.376	.430	.483	.537	.59]	.645	.698	.752	.806
0.66	.382	.436	.490	.545	.600	.655	.709	.764	.818
0.67	.388	.443	.498	.554	.609	.664	.720	.775	.831
0.68	.393	.450	.506	.562	.618	.674	.731	.787	.843
0.69	.399	.456	.513	.570	.627	.684	.741	.897	.855
0.70	.405	.463	.521	.579	.636	.694	.752	.810	.868
0.71	.410	.469	.528	.587	.645	.704	.763	.821	.880
0.72	.417	.476	.563	.595	.655	.714	.774	.833	.893
0.73	.422	.483	.543	.603	.664	.724	.784	.845	.905
0.74	.428	.489	.550	.612	.673	.734	.795	.856	.917
0.75	.434	.496	.558	.620	.682	.744	.806	.868	.930
0.76	.440	.502	.565	.628	.691	.754	.817	.879	.942
0.77	.445	.509	.573	.636	.700	.704	.827	.891	.955
0.78	.451	.516	.580	.645	.709	.774	.838	.902	.967
0.79	.457	.522	.588	.653	.718	.783	.849	.914	.979
0.80	.463	.529	.595	.661	.727	.793	.860	.929	.992
0.81	.469	.535	.602	.669	.736	.803	.870	.937	1.009
0.82	.474	.542	.610	.678	.745	.813	.881	.949	1.017
0.83	.480	.549	.617	.686	.755	.823	.892	.960	1.029
0.84	.486	.555	.625	.699	.764	.833	.902	.972	1.041
0.85	.492	.562	.632	.702	.773	.843	.913	.983	1.054
0.86	.498	.569	.640	.711	.782	.853	.924	.995	1.066
0.87	.503	.575	.647	.719	.791	.863	.935	1.007	1.079
0.88	.509	.582	.655	.727	.800	.873	.945	1.018	1.091
0.89	.515	.588	.662	.736	.809	.883	.956	1.030	1.103
0.90	.521	.595	.669	.749	.818	.893	.967	1.041	1.116
0.91	.526	.602	.677	.752	.827	.902	.978	1.053	1.128
0.92	.532	.608	.684	.760	.836	.912	.988	1.064	1.140
0.93	.538	.615	.692	.769	.845	.922	.999	1.076	1.153
0.94	.544	.621	.699	.777	.855	.932	1.010	1.088	1.165
0.95	.550	.628	.707	.785	.864	.942	1.021	1.099	1.178
0.96	.555	.635	.714	.793	.873	.952	1.031	1.111	1.190
0.97	.561	.641	.721	.802	.882	.962	1.042	1.122	1.202
0.98	.547	.648	.729	.810	.891	.972	1.053	1.134	1.215
0.99	.573	.655	.736	.818	.900	.982	1.064	1.145	1.227
1.00	.579	.661	.744	.826	.909	.992	1.074	1.157	1.240

163

TABLE 7-19 (cont.):

SECOND-FEET	HOURS								
	16	17	18	19	20	21	22	23	24
0.01	.013	.014	.015	.016	.017	.017	.018	.019	.020
0.02	.026	.028	.030	.031	.033	.035	.036	.038	.040
0.03	.040	.042	.045	.047	.050	.052	.055	.057	.060
0.04	.053	.056	.060	.063	.066	.069	.073	.076	.079
0.05	.066	.070	.074	.079	.083	.087	.091	.095	.099
0.06	.079	.084	.089	.094	.099	.104	.109	.114	.119
0.07	.093	.098	.104	.110	.116	.121	.127	.133	.139
0.08	.106	.112	.119	.126	.132	.139	.145	.152	.159
0.09	.119	.126	.134	.141	.149	.156	.164	.171	.179
0.10	.132	.140	.149	.157	.165	.174	.182	.190	.198
0.11	.145	.155	.164	.173	.182	.191	.200	.209	.218
0.12	.159	.169	.179	.188	.198	.208	.218	.228	.238
0.13	.172	.183	.193	.204	.215	.226	.236	.247	.258
0.14	.185	.197	.208	.220	.231	.243	.255	.266	.278
0.15	.198	.211	.223	.236	.248	.200	.273	.285	.298
0.16	.212	.225	.239	.251	.264	.278	.291	.304	.317
0.17	.225	.239	.253	.267	.281	.295	.309	.323	.337
0.18	.238	.253	.268	.283	.298	.312	.327	.342	.357
0.19	.251	.267	.283	.298	.314	.330	.345	.361	.377
0.20	.264	.281	.298	.314	.331	.347	.364	.380	.397
0.21	.278	.295	.312	.330	.347	.364	.382	.399	.417
0.22	.291	.309	.327	.345	.364	.382	.400	.418	.433
0.23	.304	.323	.342	.361	.380	.399	.418	.437	.456
0.24	.317	.337	.357	.377	.397	.417	.436	.456	.476
0.25	.331	.351	.372	.393	.413	.434	.455	.475	.498
0.26	.344	.365	.387	.408	.430	.451	.473	.494	.516
0.27	.357	.379	.402	.424	.446	.469	.491	.513	.536
0.28	.370	.393	.417	.440	.463	.486	.509	.532	.555
0.29	.383	.407	.431	.455	.479	.503	.527	.551	.575
0.30	.397	.421	.446	.471	.496	.521	.545	.570	.595
0.31	.410	.436	.461	.487	.512	.538	.564	.590	.615
0.32	.423	.450	.476	.502	.529	.555	.582	.608	.635
0.33	.436	.464	.491	.518	.545	.573	.600	.627	.655
0.34	.450	.478	.506	.534	.562	.590	.618	.646	.674
0.35	.463	.492	.521	.550	.579	.607	.636	.665	.694
0.36	.476	.506	.536	.565	.595	.625	.655	.684	.714
0.37	.489	.520	.550	.581	.612	.642	.673	.703	.734
0.38	.502	.534	.565	.597	.628	.660	.691	.722	.754
0.39	.516	.548	.580	.612	.645	.677	.709	.741	.774
0.40	.529	.562	.595	.628	.661	.694	.727	.760	.793
0.41	.542	.576	.610	.644	.678	.712	.745	.779	.813
0.42	.555	.590	.625	.660	.694	.729	.764	.798	.833
0.43	.569	.6O4	.640	.675	.711	.746	.782	.817	.853
0.44	.582	.618	.655	.691	.727	.764	.800	.836	.873
0.45	.595	.632	.669	.707	.744	.781	.818	.855	.893
0.46	.608	.646	.684	.722	.760	.798	.836	.874	.912
0.47	.621	.660	.699	.738	.777	.816	.855	.893	.932
0.48	.635	.674	.714	.754	.793	.833	.873	.912	.952
0.49	.648	.688	.729	.769	.810	.850	.891	.931	.972
0.50	.661	.702	.744	.785	.826	.868	.909	.950	.992

164

TABLE 7-19 (cont.):

SECOND-FEET	HOURS								
	16	17	18	19	20	21	22	23	24
0.51	.67	.72	.76	.80	.84	.89	.93	.97	1.01
0.52	.69	.73	.77	.82	.86	.90	.95	.99	1.03
0.53	.70	.74	.79	.83	.88	.92	.96	1.01	1.05
0.54	.71	.76	.80	.85	.89	.94	.98	1.03	1.07
0.55	.73	.77	.82	.86	.91	.95	1.00	1.05	1.09
0.56	.74	.79	.83	.88	.93	.97	1.02	1.06	1.11
0.57	.75	.80	.85	.90	.94	.99	1.04	1.08	1.13
0.58	.77	.81	.86	.91	.96	1.01	1.05	1.10	1.15
0.59	.78	.83	.88	.93	.98	1.02	1.07	1.12	1.17
0.60	.79	.84	.89	.94	.99	1.04	1.09	1.14	1.19
0.61	.81	.86	.91	.96	1.01	1.06	1.11	1.16	1.21
0.62	.82	.87	.92	.97	1.02	1.08	1.13	1.18	1.23
0.63	.83	.89	.94	.99	1.04	1.09	1.15	1.20	1.25
0.64	.85	.90	.95	1.00	1.06	1.11	1.16	1.22	1.27
0.65	.86	.91	.97	1.02	1.07	1.13	1.18	1.24	1.29
0.66	.87	.93	.98	1.04	1.09	1.15	1.20	1.25	1.31
0.67	.89	.94	1.00	1.05	1.11	1.16	1.22	1.27	1.33
0.68	.90	.96	1 01	1.07	1.12	1.18	1.24	1.29	1.35
0.69	.91	.97	1.03	1.08	1.14	1.20	1.25	1.31	1.37
0.70	.93	.98	1.04	1.10	1.16	1.21	1.27	1.33	1.39
0.71	.94	1.00	1.06	1.11	1.17	1.23	1.29	1.35	1.41
0.72	.95	1.01	1.07	1.13	1.19	1.25	1.31	1.37	1.43
0.73	.97	1.03	1.09	1.14	1.21	1.27	1.33	1.39	1.45
0.74	.98	1.04	1.10	1.10	1.22	1.28	1.35	1.41	1.47
0.75	.99	1.05	1.12	1.18	1.24	1.30	1.36	1.43	1.49
0.76	1.00	1.07	1.13	1.19	1.26	1.31	1.38	1.44	1.51
0.77	1.02	1.08	1.15	1.21	1.27	1.34	1.40	1.46	1.53
0.78	1.03	1.10	1.16	1.22	1.29	1.35	1.42	1.48	1.55
0.79	1.04	1.11	1.18	1.24	1.31	1.37	1.44	1.50	1.57
0.80	1.06	1.12	1.19	1.26	1.32	1.39	1.45	1.52	1.59
0.81	1.07	1.14	1.20	1.27	1.34	1.41	1.47	1.54	1.61
0.82	1.08	1.15	1.22	1.29	1.36	1.43	1.49	1.56	1.63
0.83	1.10	1.17	1.23	1.30	1.37	1.44	1.51	1.58	1.95
0.84	1.11	1.18	1.25	1.32	1.39	1.46	1.53	1.60	1.67
0.85	1.12	1.19	1.26	1.33	1.40	1.48	1.55	1.62	1.69
0.86	1.14	1.21	1.28	1.35	1.42	1.49	1.56	1.63	1.71
0.87	1.15	1.22	1.29	1.37	1.44	1.51	1.58	1.65	1.73
0.88	1.16	1.24	1.31	1.38	1.45	1.53	1.60	1.67	1.75
0.89	1.17	1.25	1.32	1.40	1.47	1.54	1.62	1.69	1.77
0.90	1.19	1.26	1.34	1.41	1.49	1.56	1.64	1.71	1.79
0.91	1.20	1.28	1.35	1.43	1.50	1.58	1.65	1.73	1.80
0.92	1.22	1.29	1.37	1.44	1.52	1.60	1.67	1.75	1.82
0.93	1.23	1.31	1.38	1.46	1.54	1.61	1.69	1.77	1.84
0.94	1.24	1.32	1.40	1.49	1.55	1.63	1.71	1.79	1.86
0.95	1.26	1.33	1.41	1.50	1.57	1.65	1.73	1.81	1.88
0.96	1.27	1.35	1.43	1.51	1.59	1.67	1.75	1.82	1.90
0.97	1.28	1.36	1.44	1.52	1.60	1.68	1.76	1.84	1.92
0.98	1.30	1.38	1.49	1.54	1.62	1.70	1.78	1.86	1.94
0.99	1.31	1.40	1.47	1.55	1.64	1.72	1.80	1.88	1.96
1.00	1.32	1.41	1.49	1.57	1.65	1.74	9.82	1.90	1.98

TABLE 7-19 (cont.):

SECOND-FEET	DAYS OF 24 HOURS								
	2	3	4	5	6	7	8	9	10
0.01	.04	.06	.08	.10	.12	.14	.16	.18	.20
0.02	.08	.12	.16	.20	.24	.28	.32	.38	.40
0.03	.12	.18	.24	.30	.36	.42	.48	.54	.60
0.04	.16	.24	.32	.40	.48	.56	.63	.71	.79
0.05	.20	.30	.40	.50	.60	.69	.79	.89	.99
0.06	.24	.36	.48	.60	.71	.83	.95	1.07	1.19
0.07	.28	.42	.56	.69	.83	.97	1.11	1.25	1.39
0.08	.32	.48	.63	.79	.95	1.11	1.27	1.43	1.59
0.09	.36	.54	.71	.89	1.07	1.25	1.43	1.61	1.79
0.10	.40	.60	.79	.99	1.19	1.39	1.59	1.79	1.98
0.11	.44	.65	.87	1.09	1.31	1.53	1.75	1.96	2.18
0.12	.48	.71	.95	1.19	1.43	1.67	1.90	2.14	2.38
0.13	.52	.77	1.03	1.29	1.55	1.80	2.02	2.32	2.58
0.14	.56	.83	1.11	1.39	1.67	1.94	2.22	2.50	2.78
0.15	.60	.89	1.19	1.49	1.79	2.08	2.38	2.68	2.98
0.16	.63	.95	1.27	1.59	1.90	2.22	2.54	2.86	3.17
0.17	.67	1.01	1.35	1.69	2.02	2.36	2.70	3.03	3.37
0.18	.71	1.07	1.43	1.79	2.14	2.50	2.86	3.21	3.57
0.19	.75	1.13	1.51	1.88	2.26	2.64	3.01	3.39	3.77
0.20	.79	1.19	1.59	1.98	2.38	2.78	3.17	3.57	3.97
0.21	.83	1.25	1.67	2.08	2.50	2.92	3.33	3.75	4.17
0.22	.87	1.31	1.75	2.18	2.62	3.05	3.50	3.93	4.36
0.23	.91	1.37	1.82	2.28	2.74	3.19	3.65	4.11	4.56
0.24	.95	1.43	1.90	2.38	2.86	3.33	3.81	4.28	4.76
0.25	.99	1.49	1.98	2.48	2.98	3.47	3.97	4.46	4.96
0.26	1.03	1.55	2.06	2.58	3.09	3.61	4.13	4.64	5.16
0.27	1.07	1.61	2.14	2.68	3.21	3.75	4.28	4.82	5.36
0.28	1.11	1.67	2.22	2.78	3.33	3.89	4.44	5.00	5.55
0.29	1.15	1.73	2.30	2.88	3.45	4.03	4.60	5.18	5.75
0.30	1.19	1.79	2.38	2.98	3.56	4.17	4.76	5.35	5.95
0.31	1.23	1.84	2.46	3.07	3.69	4.30	4.92	4.53	6.15
0.32	1.27	1.90	2.54	3.17	3.81	4.44	5.08	5.71	6.35
0.33	1.31	1.96	2.62	3.27	3.93	4.58	5.24	5.89	6.55
0.34	1.35	2.02	2.70	3.37	4.05	4.72	5.40	6.07	6.74
0.35	1.39	2.08	2.78	3.47	4.17	4.86	5.55	6.25	6.94
0.36	1.43	2.14	2.86	3.57	4.28	5.00	5.71	6.43	7.14
0.37	1.47	2.20	2.94	3.67	4.40	5.14	5.87	6.60	7.34
0.38	1.51	2.26	3.01	3.77	4.52	5.28	6.03	6.78	7.54
0.39	1.55	2.32	3.09	3.87	4.64	5.41	6.19	6.96	7.74
0.40	1.59	2.38	3.17	3.97	4.76	5.55	6.35	7.14	7.93
0.41	1.63	2.44	3.25	4.07	4.88	5.69	6.51	7.32	8.13
0.42	1.67	2.50	3.33	4.17	5.00	5.83	6.66	7.50	8.33
0.43	1.71	2.56	3.41	4.26	5.12	5.97	6.82	7.68	8.53
0.44	1.75	2.62	3.49	4.36	5.24	6.11	6.98	7.85	8.73
0.45	1.79	2.68	3.57	4.46	5.36	6.25	7.14	8.03	8.93
0.46	1.82	2.74	3.65	4.56	5.47	6.39	7.30	8.21	9.12
0.47	1.86	2.80	3.73	4.66	5.59	6.53	7.46	8.39	9.32
0.48	1.90	2.86	3.81	4.76	5.71	6.66	7.62	8.57	9.52
0.49	1.94	2.92	3.89	4.86	5.83	6.80	7.78	8.75	9.72
0.50	1.98	2.98	3.97	4.96	5.95	6.94	7.93	8.93	9.92

TABLE 7-19 (cont.):

SECOND-FEET	DAYS OF 24 HOURS								
	2	3	4	5	6	7	8	9	10
0.51	2.02	3.03	4.05	5.06	6.07	7.08	8.09	9.10	10.12
0.52	2.06	3.09	4.13	5.16	6.19	7.22	8.25	9.28	10.31
0.53	2.10	3.15	4.20	5.26	6.31	7.36	8.41	9.46	10.52
0.54	2.14	3.21	4.28	5.36	6.43	7.50	8.57	9.64	10.71
0.55	2.18	3.27	4.36	5.45	6.55	7.64	8.73	9.82	10.91
0.56	2.22	3.33	4.44	5.55	6.66	7.78	8.89	10.00	11.11
0.57	2.26	3.39	4.52	5.65	6.78	7.91	9.04	10.18	11.31
0.58	2.30	3.45	4.60	5.75	6.90	8.05	9.20	10.35	11.50
0.59	2.34	3.51	4.68	5.85	7.02	8.19	9.36	10.53	11.70
0.60	2.38	3.57	4.76	5.95	7.14	8.33	9.52	10.71	11.90
0.61	2.42	3.63	4.84	6.05	7.26	8.47	9.68	10.89	12.10
0.62	2.46	3.69	4.92	6.15	7.38	8.61	9.84	11.07	12.30
0.63	2.50	3.75	5.00	6.25	7.50	8.75	10.00	11.25	12.50
0.64	2.54	3.81	5.08	6.35	7.62	8.89	10.16	11.42	12.69
0.65	2.58	3.87	5.16	6.45	7.74	9.02	10.31	11.60	12.89
0.66	2.62	3.93	5.24	6.55	7.85	9.16	10.47	11.78	13.09
0.67	2.66	3.99	5.32	6.64	7.97	9.30	10.63	11.96	13.29
0.68	2.70	4.05	5.40	6.74	8.09	9.44	10.79	12.14	13.49
0.69	2.74	4.11	5.47	6.84	8.21	9.58	10.95	12.32	13.69
0.70	2.78	4.17	5.55	6.94	8.33	9.71	11.11	12.50	13.88
0.71	2.82	4.22	5.63	7.04	8.45	9.89	11.27	12.67	14.08
0.72	2.86	4.28	5.71	7.14	8.57	10.00	11.42	12.85	14.28
0.73	2.90	4.34	5.79	7.24	8.69	10.14	11.58	13.03	14.48
0.74	2.94	4.40	5.87	7.34	8.81	10.27	11.74	13.21	14.68
0.75	2.98	4.46	5.95	7.44	8.93	10.41	11.90	13.39	14.88
0.76	3.01	4.52	6.03	7.59	9.04	10.55	12.06	13.57	15.07
0.77	3.05	4.58	6.11	7.64	9.16	10.69	12.22	13.75	15.27
0.78	3.09	4.64	6.19	7.74	9.28	10.83	12.38	13.92	15.47
0.79	3.13	4.70	6.27	7.83	9.40	10.97	12.54	14.10	15.67
0.80	3.17	4.76	6.35	7.93	9.52	11.11	12.70	14.28	15.87
0.81	3.21	4.82	6.43	8.03	9.64	11.25	12.85	14.46	16.07
0.82	3.25	4.88	6.51	8.13	9.76	11.39	13.01	14.64	16.26
0.83	3.29	4.94	6.59	8.23	9.88	11.52	13.17	14.82	16.46
0.84	3.33	5.00	6.66	8.33	10.00	11.66	13.33	15.00	16.66
0.85	3.37	5.06	6.74	8.43	10.12	11.80	13.49	15.17	16.86
0.86	3.41	5.12	6.82	8.53	10.23	11.94	13.65	15.35	17.06
0.87	3.45	5.18	6.90	8.63	10.35	12.08	13.80	15.53	17.26
0.88	3.50	5.24	9.99	8.73	10.47	12.22	13.96	15.71	17.45
0.89	3.53	5.30	7.06	8.83	10.59	12.36	14.12	15.89	17.65
0.90	3.57	5.36	7.14	8.93	10.71	12.50	14.28	16.07	17.85
0.91	3.61	5.41	7.92	9.02	10.83	12.63	14.44	16.24	18.05
0.92	3.65	5.47	7.30	9.12	10.95	12.77	14.60	16.42	18.25
0.93	3.69	5.53	7.38	9.22	11.07	12.91	14.76	16.60	18.45
0.94	3.73	5.59	7.46	9.32	11.19	13.05	14.92	16.86	18.64
0.95	3.77	5.65	7.54	9.42	11.31	13.19	15.07	16.96	18.84
0.96	3.81	5.71	7.62	9.52	11.42	13.33	15.23	17.14	19.04
0.97	3.85	5.77	7.70	9.62	11.54	13.47	15.39	17.32	19.24
0.98	3.89	5.83	7.78	9.72	11.66	13.61	15.55	17.49	19.44
0.99	3.93	5.89	7.85	9.82	11.78	13.74	15.71	17.67	19.64
1.00	3.97	5.95	7.93	9.92	11.90	13.88	15.87	17.85	19.83

(Source: U.S. Dept. of Interior *Water Measurement Manual*; 1981)

TABLE 7-20: ESTIMATED LIME REQUIREMENT (kg CACO₃/ha) NEEDED TO INCREASE THE TOTAL HARDNESS AND ALKALINITY OF POND WATER TO 20 mg/l OR GREATER (Boyd, 1982)

MUD PH IN WATER	CALCIUM CARBONATE REQUIRED ACCORDING TO MUD PH IN BUFFERED SOLUTION									
	7.9	7.8	7.7	7.6	7.5	7.4	7.3	7.2	7.1	7.0
5.7	91	182	272	363	454	544	635	726	817	908
5.6	126	252	378	504	630	756	882	1008	1,34	1260
5.5	202	404	604	806	1008	1210	1411	1612	1814	2016
5.4	290	580	869	1160	1449	1738	2029	2318	2608	2898
5.3	340	680	1021	1360	1701	2041	2381	2722	3062	3402
5.2	391	782	1172	1562	1548	2344	2734	3124	3515	3906
5.1	441	882	1323	1765	2205	2646	3087	3528	3969	4410
5.0	504	1008	1512	2016	2520	3024	3528	4032	4536	5040
4.9	656	1310	1966	2620	3276	3932	4586	5242	5980	6552
4.8	672	1344	2016	2688	3360	4032	4704	5390	6048	6720
4.7	706	1412	2116	2822	3528	4234	4940	5644	6350	7056

Lime required (as CaCO₃) is estimated from the pH of the pond muds before and after the addition of a buffer solution. The mud sample for lime requirement measurement should be dried at room temperature by spreading in a thin layer on a plastic sheet. The dried mud sample is then ground using a pestle and mortar and passed through a 20-mesh sieve (0.85mm openings) for pH analysis. The buffer solution is prepared by dissolving 20g of p-nitrophenol, 15g of boric acid, 74g of potassium chloride, and 10.5g of potassium hydroxide in distilled water and diluting to one liter in a volumetric flask. Place 20.0 g of the dried and ground mud sample into a 100 ml beaker, adding 20 ml of distilled water, and stir intermittently.

(Source: C.E. Boyd, *Water Quality in Ponds for Aquaculture*; 1990)

TABLE 7-21: LIME REQUIREMENTS OF BOTTOM MUDS BASED ON pH AND TEXTURE OF MUD (after Schaeperclaus, 1933)

MUD PH	LIME REQUIREMENT (kg/ha as CACO₃)		
	HEAVY LOAMS OR CLAYS	SANDY LOAM	SAND
4.0	14320	7160	4475
4.0 - 4.5	10,740	5370	4475
4.6 - 5.0	8950	4475	3580
5.5 - 5.1	5370	3580	1790
5.6 - 6.0	3580	1790	895
6.1 - 6.5	1790	1790	0
6.5	0	0	0

(Source: C. E. Boyd, *Water Quality in Ponds for Aquaculture*; 1990)

TABLE 7-22: CHEMICAL COMPOUNDS COMMONLY USED TO DISINFECT POND BOTTOMS AND ELIMINATE PREDATORS.
Dosages refer to commercial product, unless otherwise indicated.

COMPOUND	DISINFECTANT (D) OR PISCICIDE (P)	RECOMMENDED DOSAGE
BENZALKONIUM CHLORIDE	D	0.5-1.0 mg/l
FORMALIN	D	5-10 mg/l
POTASSIUM PERMANGANATE	D	24 mg/l
HYAMIN	D	0.5-1.0 mg/l
ORGANIC SILVER	D	1-10 mg/l
ORGANIC IODINE	D	1-5mg/l
CALCIUM CARBIDE	D	150-250 kg/ha
MALACHITE GREEN	D,P	NO DATA
SODIUM HYPOCHLONTE (5.25%)	D,P	100-300 mg/l
CALCIUM HYPOCHLORITE (HTH (65%)	D,P	10-300 mg/l
CALCIUM OXIDE	D,P	1,000-1,500 kg/ha
CALCIUM HYDROXIDE	D,P	1,000-2,000 kg/ha
ROTENONE (5%)	P	14 mg/l
"CHEM FISH" (Tifa)	P	2-8 mg/l
TEA-SEED CAKE (7% saponin)	P	10-25 mg/l
POTASSIUM CYANIDE	P	1 mg/l
LIME + AMMONIUM SULFATE (5-10:1)	P	1,100-1,200 kg/ha[1]
TOBACCO DUST	P	200-400 kg/ha
ANHYDROUS AMMONIA	P	30 mg/l
ROSIN AMINE D-ACETATE (RADA)	P	24 mg/l
SODIUM PENTACHLOROPHENATE	P	0.5-1.0 mg/l
GUTHION	P	NO DATA
SEVIN	P	NO DATA
MALATHION	P	25-50 mg/l
THIODAN	P	NO DATA
ENDRIN	P	NO DATA

(1) 1,000 kg/ha CaO or Ca(OH)$_2$ + 100-200 kg/ha ammonium sulfate

(Source: H.C. Clifford, In: J. Wyban (ed.) *Proceedings of the World Aquaculture Society Special Session on Shrimp Farming;* 1992)

TABLE 7-23: AMOUNT OF CUBE, DERRIS, OR EMULSIFIABLE ROTENONE REQUIRED TO KILL FISH IN PONDS OF VARIOUS SIZES AND DEPTHS

SURFACE AREA m^2	Acres	AVERAGE DEPTH OF WATER					
		15 cm.	6 in.	30 cm.	12 in.	45 cm	18 in.
		cl.[1]	oz.	cl.	oz.	cl.	oz.
400	1/10	8.9	3	14.8	5	23.7	8
800	1/5	14.8	5	29.6	10	44.4	15
1000	1/4	17.8	6	35.5	12	53.3	18
1200	3/10	23.7	8	44.4	15	65.1	22
1600	2/5	29.6	10	59.2	20	85.8	29
2000	1/2	35.5	12	71.0	24	106.6	36
2400	3/5	44.4	15	85.8	29	130.2	44
3000	3/4	53.3	18	106.6	36	159.8	54

(1) cl. = centiliters; used for emulsified rotenone. Ponds should be drained after 24 to 48 hours or more and refilled.

(Source: E.E. Brown and J.B. Gratzek, *Fish Farming Handbook;* 1980)

169

TABLE 7-24: NITROGEN FERTILIZERS FOR POND ENRICHMENT

SOURCE MATERIAL	CHEMICAL FORMULA	PERCENT NITROGEN	pH OF AQUEOUS SOLUTION
AMMONIUM METAPHOSPHATE	$(NH_4)_3PO_3$	17[1]	-
AMMONIUM NITRATE	NH_4NO_3	33.5	4.0
AMMONIUM PHOSPHATE	$(NH_4)_3PO_4$	11[2]	4.0
AMMONIUM SULFATE	$(NH_4)_2SO_4$	20	5.0
ANHYDROUS AMMONIA	$NH_3 \cdot H_2O$	82	-
AQUA-AMMONIA	$NH_3 \cdot H_2O$	40-50	-
CALCIUM CYANAMIDE	$CaCN_2$	22	
DIAMMONIUM PHOSPHATE	$(NH_4)_2HPO_3$	21[3]	8.0
UREA	H_2HCONH_2	46	7.2
SODIUM NITRATE	$NaNO_3$	16	7

[1] Also contains 73% P_2O_5

[2] Also contains 48% P_2O_5

[3] Also contains 48-52% P_2O_5

(Source: U.S. Fish and Wildlife Service, *Fish Hatchery Management*, 1982)

TABLE 7-25: FERTILIZATION REGIMES UTILIZED IN SHRIMP PONDS.

Key to abbreviations: (P) = pond preparation; (E) = exstensive ponds; N:P = nitrogen:phosphorus; U = urea; TSP = triple superphosphate; DAP = diammonium phosphate; AP = ammonium phosphate (16-20-0), FTC = "Ferticam" (37-6-3).

COUNTRY	FERTILIZERS (kg/ha)	N:P	SOURCE
COLOMBIA	U (15) + DAP (4)	9:1	Clifford (unpubl.)
ECUADOR	U (9-23) +TSP (0.9-2.3)	9:1	Villalon (1991)
ECUADOR (P)	U (16) + TSP (8)	4:1	Hirono (1989)
ECUADOR	U (8-12) + TSP (4-6)	4:1	Hirono (unpubl.)
ECUADOR (E)	U (20) + TSP (7)	6:1	Figueroa (1991)
ECUADOR (P)	U (22) + TSP (6)	8:1	Figueroa (1991)
ECUADOR	U (5.6) + TSP (2.2)	6:1	Figueroa (1991)
ECUADOR	U (15-30)+FTC (8.3-17)	45:1	Wigglesworth (1991)
INDONESIA(P)	U (IOO)+TSP(SO-100)	24:1	Chamberlain (1991)
INDONESIA (P)	U (150) + TSP (75)	3:1	Ahmad (1989)
INDONESIA	U (25)+TSP(10)	6:1	Ahmad(1989)
MEXICO(P)	U (15-35)+TSP(5-12)	7:1	Figueroa (1991)
PANAMA	U (20)+TSP(15)	3:1	DNA(1984)
PHILIPP. (E,P)	U (50)+ AP (50) [1]	7:1	Apud (1989)
PHILIPP. (E)	U (7) + DAP (65)	1:1	Subosa & Bautista
U.S.A. (P)	U (25) + TSP (15)	4:1	Jaenike (1989)
U.S.A.	U (45)		Figueroa (1991)
UNIVERSAL	-	20-30:1	ASEAN (1978)
UNIVERSAL (P)	U (7.9)+ TSP (2.6)	7:1	Cook (1991)
UNIVERSAL	U (2.5)+ TSP (0.5)	11:1	Cook (1991)
NOT SPECIFIED	U + TSP	15-30:1	Boyd (1989)

[1] also supplemented with 1-2 tons chicken manure/ha

(Source: H.C. Clifford, In: J. Wyban, *Proceedings of the World Aquaculture Society Special Session on Shrimp Farming*; 1992)

TABLE 7-26: AVERAGE ELEMENTAL COMPOSITION OF ORGANIC MANURES (VALUES ARE EXPRESSED AS % BY WEIGHT)[1]

MANURE	C:N RATIO	% MOISTURE-FREE BASIS		
		N	P	K
ANIMAL MANURES **FAECES/DUNG** [2]				
BUFFALO	19	1.23	0.55	0.69
CATTLE	19	1.91	0.56	1.40
SHEEP	29	1.87	0.79	0.92
GOAT & SHEEP (mixed)	-	1.50	0.72	1.38
HORSE	24	2.33	0.83	1.31
PIG	13	2.80	1.36	1.18
CAMEL	-	1.51	0.15	1.SO
ELEPHANT	43	1.29	0.33	0.14
TIGER	10	2.82	3.19	0.03
LION	9	3.60	3.21	0.04
HUMAN	8	7.24	1.72	2.41
POULTRY MANURE	9	3.77	1.89	1.76
DUCK MANURE	10	2.15	1.13	1.15
RABBIT MANURE	-	1.72	1.30	1.08
URINE				
BUFFALO	-	2.05	0.01	3.78
CATTLE	-	9.74	0.05	7.78
SHEEP	-	9.90	0.10	12.31
GOAT & SHEEP (mixed)	-	9.64	0.14	-
PIG	-	10.88	1.25	17.86
HORSE	-	13.20	0.02	10.90
HUMAN	0.8	17.14	1.57	4.86
MEALS				
BLOOD MEAL	3.5	11.12	0.66	
HORN AND HOOF MEAL	-	12.37	1.60	
BONE MEAL	8	3.36	10.81	-
FISH MANURE	4.5	7.5	2.82	0.8
PLANT MANURES **CROP RESIDUES**				
WHEAT STRAW	105	0.49	0.11	1.06
BARLEY STRAW	110	0.47	0.13	1.01
RICE STRAW	105	0.58	0.10	1.38
OATS STRAW	-	0.46	0.11	0.97
MAIZE STRAW	55	0.59	0.31	1.31
SOYBEAN STRAW	32	1.30	-	-
COTTON STALKS AND LEAVES	-	0.88	0.15	1.45
COTTONSEED MEAL	-	7.05	0.90	1.16
GROUNDNUT STRAW	19	0.59	-	-
GROUNDNUT HULLS	-	1.75	0.20	1.24
GROUNDNUT SHELLS	-	1.00	0.06	0.90
BEAN STRAW	-	1.57	0.32	1.34
COWPEA STEMS	-	1.07	1.14	2.54
COWPEA ROOTS	-	1.06	0.12	1.50
COFFEE PULP [3]	-	1.79	0.12	1.80
SUGARCANE TRASH	116	0.35	0.04	0.50

171

TABLE 7-26 (cont.):

MANURE	C:N RATIO	% MOISTURE-FREE BASIS		
		N	P	K
GRASS [4]	20	0.41	0.03	0.26
GREEN WEEDS	13	2.45	-	-
OIL PALM BUNCH ASH	-	-	1.71	32.50
OIL PALM PRESSED FIBER	-	1.24	0.10	0.36
OIL PALM SLUDGE CAKE	-	4.30	1.19	1.15
MOLASSES	-	2.09	5.30	1.99
COWPEA LEAVES	-	1.99	0.19	2.20
JUTE LEAVES	-	1.75	0.58	4.12
GROUNDNUT LEAVES	-	2.56	0.17	2.11
TREE LEAVES (general)	60	1.00	0.30	0.57
AQUATIC PLANTS AND ALGAE				
WATER HYACINTH	18	2.04	0.37	3.40
Azolla sp.	-	3.68	0.20	0.15
Lemna sp.	-	3.31	0.20	0.69
Chara vulgaris	-	1.27	0.19	0.84
Ceratophylum sp.	-	3.30	0.47	5.90
Elodia canadensis	-	3.29	0.51	3.26
Hydrilla sp.	-	2.70	0.28	2.90
Myriophyllum sp.	-	2.81	0.17	1.20
Pistia stratiotes	-	2.10	0.30	3.50
Potamogeton sp.	-	2.51	0.33	2.28
Typha sp	-	1.37	0.21	2.38
MARINE SEAWEEDS (air-dried) [5]	-	0.66	0.32	1.20
OILSEED CAKES				
CASTOR	-	4.89	0.80	1.04
COCONUT	-	3.07	1.23	1.57
COTTON-DECORTICATED	-	6.36	1.26	1.82
COTTON-UNDECORTICATED	-	3.95	0.81	1.35
LINSEED	-	5.48	0.60	0.99
NEEM	4.5	5.21	0.46	1.19
RAPE	-	5.08	0.8B	0.95
SAFFLOWER-DECORTICATED	-	7.88	0.97	1.59
SAFFLOWER-UNDECORTICATED	-	4.03	0.63	1.02
MUSTARD	-	4.93	0.53	0.65
SESAME	-	6.12	0.92	1.04
SOYBEAN	-	6.95	2.88	1.02
MISCELLANEOUS				
PEAT	80	1.08	0.02	0.08
ANIMAL/PLANT (MIXED) MANURES				
FARMYARD MANURE (general) [6]	-	0.80	0.21	0.68
RICE STRAW BEDDING	-	1.06	0.27	2.00
WHEAT STRAW BEDDING	-	1.09	0.17	1.40
LITTER BEDDING	-	1.13	0.20	2.03
STRAW	-	0.62	0.21	0.49
PEAT MOSS	-	0.88	0.16	0.85

172

TABLE 7-26 (cont.):

MANURE	C:N RATIO	% MOISTURE-FREE BASIS		
		N	P	K
EARTH BEDDING	-	0.48	0.14	0.40
RURAL COMPOSTS (general)	-	1.10	0.29	1.37
RAW MATERIAL				
STRAW	-	1.31	0.19	7.81
COW MANURE	-	0.37	0.10	0.08
BUFFALO MANURE	-	0.44	0.14	0.11
PIG MANURE	-	0.68	0.13	0.05
WATER HYACINTH	-	1.40	0.46	0.54
WATER HYACINTH [7]	13	2.05	0.48	2.10
COTTON STALKS	-	1.61	0.21	2.80
MIXED CROP RESIDUES	-	0.91	0.20	1.62
MULBERRY LEAVES	-	1.00	0.45	1.49
RICE STRAW	-	1.04	0.26	0.85
AZOLLA	-	1 00	1 43	3 53
PINE NEEDLES		o 99	0.63	2-93
URBAN REFUSE COMPOST	-	1.29	0.50	0.94
SEWAGE SLUDGE (general)	9	4.00	1 40	0 30
RAW SLUDGE	-	3.10	1.10	0.20
ANAEROBICALLY DIGESTED SLUDGE	10	3.30	1.60	0.67
AEROBIC ACTIVATED SLUDGE	-	6.00	1.40	0.80
RAW SAWDUST [8]	511	0.11	-	-
ROTTED SAWDUST [8]	208	0.25	-	-

[1] Adapted from Misra and Heese (1982)

[2] Mean moisture content of the fresh faeces of buffalo, cattle, sheep, horse, pig, poultry and humans is approximately 81%, 83%, 65%, 78%, 75%, 73% and 80% respectively.

[3] Data from Bressanik et al., (1975).

[4] Nitrogen figure reported is low:average nitrogen content of dried leafy grass is @ 4%.

[5] Gotaas (1956) reports C/N ratio and nitrogen content of dried seaweed as 19 and 1.9% respectively.

[6] Gotass (1956) reports C/N ratio and nitrogen content of dried farmyard manure as 14 and 2.15% respectively.

[7] Data from Little (1979).

[8] Data from Gotaas (1956)

(Source: FAO, *Nutrition and Feeding of Farmed Fish and Shrimp- Vol.ume 2: Nutrient Sources and Composition*; 1987)

TABLE 7-27: COMPOSITION OF FRESH MANURE FROM VARIOUS ANIMAL SPECIES.

COMPONENTS	MIXED DUNG	HORSE DUNG	CATTLE DUNG	SHEEP DUNG	PIG DUNG
WATER	75.0	71.3	77.3	64.6	72.4
ORGANIC MATTER	21.0	25.4	20.3	31.8	25.0
TOTAL NITROGEN (N)	0.50	0.58	0.45	0.83	0.45
PROTEINIC NITROGEN	0.31	0.35	0.28	-	-
AMMONIACAL NITROGEN	0.15	0.19	0.14	-	0.20
PHOSPHORUS (P_2O_5)	0.25	0.28	0.23	0.23	0.19
POTASSIUM (K_2O)	0.60	0.63	0.50.	0.67	0.60
CALCIUM (CAO)	0.35	0.21	0.49	0.33	0.18
MAGNESIUM (MGO)	0.15	0.14	0.11	0.18	0.09
SULPHURIC ACID (SO_3^{2-})	0.10	0.07	0.06	0.15	0.08
CHLORINE (CL)	-	0.04	0.10	0.17	0.17
SILICIC ACID	-	1.77	0.85	1.47	1.08
IRON AND ALUMINUM SESQUIOXIDES (R_2O_3)	-	0.11	0.05	0.24	0.07

(Source: T.V.R. Pillay, *Aquaculture Principles and Practices*; 1990)

TABLE 7-28: CHEMICAL OXYGEN DEMAND (COD)[1] AND BIOCHEMICAL OXYGEN DEMAND (BOD) OF AQUATIC MACROPHYTES COMPARED WITH ANIMAL WASTES.
(After Edwards, 1982; modified from Almazan and Boyd 1978; and Taiganides, 1977.)

AQUATIC MACROPHYTE	LIFE FORM	COD (mgO$_2$/mg dry weight)	$BOD_{0.5}$	BOD_1	BOD_2	BOD_3	BOD_4	BOD_5	RATIO OF COD BOD_5
Typha latifolia	Emergent	1.08	0.045	0.067	0.105	0.121	0.145	0.148	7.30
Eichhornia crassipes	Floating	0.88	0.044	0.079	0.109	0.128	0.145	0.157	5.61
Najas guadalupensis	Submersed	1.09	0.077	0.107	0.187	0.283	0.313	0.383	2.85
MEAN		1.02	0.055	0.084	0.134	0.177	0.201	0.229	5.25
PORK PIG WASTE									3.3
LAYING HENS WASTE									4.3

(columns across the top: BOD$_{(days)}$ (mg O$_2$/mg dry weight/day))

[1] All figures given for 30°C.

Chemical Oxygen Demand (COD) measures chemically the amount of oxygen required for the complete oxidation of a particular waste.

(Source: D. Little and J. Muir, *A Guide to Integrated Warm Water Aquaculture*; copyright © 1987 - with permission of University of Stirling Press)

TABLE 7-29: THE 24 HOUR BIOCHEMICAL OXYGEN DEMAND (BOD) FOR VARIOUS INPUTS INTO POND CULTURE OF FISH.
(After Almazan and Boyd, 1978; Edwards, 1980; and Schroeder, 1980.)

MATERIAL	% DRY MATTER	BOD g O_2/kg/24 hr at 30°C
GRAINS		
PELLETS (25% PROTEIN)	90.0	140
MILLED WHEATS/SORGHUM MIXTURE	90.0	96
WHEATGRAINS	91.0	40
SORGHUM GRAINS	88.0	18
MANURES		
CHICKEN MANURE	95.0	20-40
FIELD DRY MANURE	36.0	10
LIQUID COWSHED MANURE	12.5	7
LIQUID CALF MANURE	9.0	5
DRY HUMAN WASTES	26.5	35-50
HUMAN SEWAGE	2.0	2.5-3.0
FODDERS		
EMERGENT AQUATIC WEED	@8.0	5.4
FLOATING AQUATIC WEED	@8.0	6.3
SUBMERSED AQUATIC WEED	@8.0	8.6
TERRESTRIAL FODDER	@20.0	13.4

EXAMPLE:

A 1000 m^2 pond with average depth of 1 m holds 1000 m^3 of water.
If 100 kg of field dry manure is added to the pond, 24 hour BOD at 30°C is @ 10g/kg of manure/24 hours, for total oxygen requirement is 10 x 100 = 1000 g O_2/24hr.

Oxygen uptake is 1000g/1000 m^3 = 1 mg/l.

(Source: D. Little and J. Muir, *A Guide to Integrated Warm Water Aquaculture*; 1987)

TABLE 7-30: CRITICAL SECCHI DISC DEPTHS (in cm) FOR 1-METER DEEP FISH POND AS A FUNCTION OF WATER TEMPERATURE AND DISSOLVED OXYGEN CONCENTRATION (DO) AT DUSK [1] (from Boyd, 1979)

TEMP. °C	DO CONCENTRATION AT DUSK (ml/l)										
	2	3	4	5	6	7	8	9	10	11	12
20	37	S	S	S	S	S	S	S	S	S	S
21	58	26	S	S	S	S	S	S	S	S	S
22	79	42	21	S	S	S	S	S	S	S	
23	90	58	32	16	S	S	S	S	S	S	S
24	100	69	42	26	S	S	S	S	S	S	
25	100	79	53	37	21	S	S	S	S	S	S
26	100	85	63	48	32	16	S	S	S	S	S

175

TABLE 7-30 (cont.):

TEMP. °C	DO CONCENTRATION AT DUSK (mg/l)										
	2	3	4	5	6	7	8	9	10	11	12
27	100	90	69	53	37	26	S	S	S	S	S
28	100	95	74	58	45	32	21	S	S	S	S
29	100	95	79	63	53	40	29	18	S	S	S
30	100	100	85	69	58	45	34	26	16	S	S
31	100	100	87	74	63	50	40	32	21	S	S
32	100	100	90	79	66	55	45	37	29	18	S

[1] Data based on a pond pond containing 1,120 kg/ha (1,000 lb/acre) of channel catfish. Observed secchi disc values of less than the table values indicate that DO concentrations will drop below 2 mg/l by dawn. An (S) indicates that DO will not drop below 2 mg/l regardless of observed secchi disc values.

(Source: J.E. Lannan *et al., Principles and Practices of Pond Aquaculture;* 1986)

TABLE 7-31: PREDICTED GAINS (+) AND LOSSES (-) OF DISSOLVED OXYGEN BECAUSE OF DIFFUSION FROM A FISH POND DURING THE NIGHT BASED ON OBSERVED OXYGEN SATURATION AT DUSK
(Boyd, 1979; modified from Schroeder, 1975)

(These values are for a 1-m deep pond, for a 12-hour period)

DO CONCENTRATIONS AT DUSK (% of air saturation)	GAIN OR LOSS OF DO DURING THE NIGHT (mg/l)	DO CONCENTRATIONS AT DUSK (% of air saturation)	GAIN OR LOSS OF DO DURING THE NIGHT (mg/l)
50	+1.69	160	-1.64
60	+1.49	170	-1.82
70	+1.18	180	-1.98
80	+1.00	190	-2.11
90	+0.77	200	-2.37
100	0.44	210	-2.42
110	0.16	220	-2.54
120	-0.18	230	-2.67
130	-0.55	240	-2.76
140	-0.94	250	-2.91
150	-1.48		

(Source: J.E. Lannan *et al., Principles and Practices of Pond Aquaculture;* 1986)

VIII. Chemicals and Treatments

TABLE 8-1: SUMMARY OF CHEMICALS USED TO CONTROL DISEASES AND PARASITES OF WARMWATER FISHES[1]

Treatments specifically designed for channel catfish, but apply to other warmwater fishes

CHEMICAL	SUMMARY
OXYTETRACYCLINE	Commonly known as Terramycin[TM] - this antibiotic is used in controlling bacterial diseases. A premix for addition to feed is available when needed. Since oxytetracycline is usually expensive - it should be used only when needed. It can be used as a bath treatment at 15 ppm for 24 hours (or 0.57 gram per 10 gallons of water). (Some diseases are developing resistance to oxytetracycline. *Aeromonas* is one.)
COPPER SULFATE	Copper sulfate is often used to control external parasites and sometimes aquatic vegetation. The rate of usage ranges from 0.25-1.5 ppm - depending on water chemistry. Copper sulfate is more toxic in soft water than in hard water and a lower rate is used in soft water. Rates may be reduced in warm water. Plastic containers should be used with copper sulfate because of its corrosive effects. Galvanized containers should never be used. Copper sulfate will cause a die-off of plankton which may contribute to oxygen problems in ponds.
FORMALIN (formaldehyde)	Is used in treating for external parasites and fungal diseases. The rate of usage may range from 1-30 ppm in ponds to 150-250 ppm in tanks for one-two hours.
SALT	Common salt (NaCI) is used to cause external parasites to release from fish and fall to the bottom of the tank in which it is used. It is used as a 3% dip solution for up to 30 minutes. The fish should be removed when they begin to show signs of distress. It can be used at the rate 1,000-2,000 ppm in haul tanks (38-76 grams per 10 gallons).
POTASSIUM PERMANGANATE	Potassium permanganate is used for a number of purposes in fish culture. It is used to control external parasites at the rate of 2-8 ppm (this is 5.4-21.6 pounds per acre-foot)- depending an water chemistry. The effectiveness is related to water pH - hardness - and temperature.
TRICHLORFON (Dylox[TM] or Masoten[TM])	Monogenetic trematodes - copepods - anchor parasites have been controlled with trichlorfon on non-food fish. A rate of 0.25 ppm active ingredient is often used. Dylox breaks down rapidly in water with a high pH and at high temperatures.
CALCIUM HYDROXIDE OR CALCIUM OXIDE	Calcium hydroxide is commonly known as slaked or hydrated lime. It can be used as a disinfectant in drained ponds. It is used at the rate of 1,000 to 2,500 pounds per acre spread over the pond bottom.
FURANCE	Furance is used on a limited basis to treat bacterial infections. It is used at 0.25 ppm for one hour an three consecutive days in tank treatment. It can be fed at 100-200 milligrams active per 100 pounds of fish (catfish) for three to five days. (This is equivalent to 3.3-7.6 grams active per 100 pounds of feed.) Warning: Continued treatment may cause injury to skin.

(continued on page 178)

(continued from page 177)

[1] It is the responsibility of the person treating fish or water with chemicals to be sure that the chemical is legal for the purpose used. Always read the labels on containers.

From: *Principal Diseases of Farmed Catfish*, Southern Cooperative Series Bulletin 225. Alabama Agriculture Experiment Station, 1985.

TABLE 8-2: POSSIBLE TREATMENTS FOR COMMON PARASITES OF WARMWATER FISH[1]

Treatments described for channel catfish are applicable to many warmwater fish.

PARASITE	POSSIBLE TREATMENT
ICH DISEASE *ICHTHYOPHTHIRIUS MULTIFILLIS*	Repeated treatments with one or more of the following may be required: 2 ppm potassium permanganate copper sulfate (rate depends on water alkalinity) 15 ppm Formalin (One to three treatments at two-day intervals may be necessary.) *Ich* can be transferred on wet seines - boots - etc. - Disinfecting equipment will help prevent spread. The bottoms of dry ponds may be disked to destroy the encysted stage of *Ich* disease.
***TRICHODINA; CHILODONELLA COSTIA; TROCHOPHRYA; SCYPHIDIA;* PROLIFERATIVE GILL DISEASE**	Formalin - one application in ponds at 15 to 25 ppm - or 250 ppm in tanks for as long as fish will tolerate. Potassium permanganate in ponds at 2-3 ppm or as dictated by water chemistry. Broodstock should be treated before being transferred to spawning ponds to prevent spreading to fry. Copper sulfate - 0.25 - 0.5 ppm depending on water hardness. Copper sulfate is much more toxic in water with low alkalinity.
BLOOD SUCKERS (leeches)	Use 0.1% salt solution in holding tanks.
EPISTYLIS	Use 0.1-1.0 % salt solution in holding tanks.
FLUKES **MONOGENETIC TREMATODES** **DIGENETIC TREMATODES**	Formalin - 15-25 ppm in ponds or 250 ppm in tanks for one hour Potassium permanganate - 2.0 ppm in ponds. Control snails around ponds.
CESTODES (tapeworms and other intestinal worms)	Feed 125 milligrams of di-N-butyl tin oxide per pound of fish - or add 1% to the ration of feed for three days. The bottoms of empty ponds can be treated with calcium hydroxide (slaked lime) at the rate of 2 tons per acre.
COPEPODS	Repeated applications of Dylox at the rate of 0.25 ppm active ingredient.

[1] The possible treatments given here are not recommendations, but are those which have been used to aid in controlling fish diseases. Some of the chemicals listed have not been approved by the Food and Drug Administration of the United States for use with food fish.

From: *Principal Diseases of Farmed Catfish*, Southern Cooperative Series Bulletin 225. Alabama Agriculture Experiment Station, 1985.

TABLE 8-3: SOME CHEMICAL HERBICIDES WITH RECOMMENDED DOSAGES FOR AQUATIC USE.

COMPOUND	TARGET PLANTS	TREATMENT	COMMENTS
COPPER SULFATE	Algae (all types)	0.5-1.0 g/m³ (acid water); 1.5 g/m³ (alkalin water) as often as necessary	To avoid toxicity to fish treat only 1/3 to 1/2 of a pond at a time.
CUTRINE-PLUS (copper sulfate chelated with triethanolamine)	Algae (all types)	0.6 gallons per acre-foot (liquid form) or 60 lb/acre (66 kg/ha) (granular form) 2 or 3 times a year	Dilute liquid form at least 9 to 1.
COPPER OXYCHLORIDE	Filamentous algae (except *Chara*)	0.9 lb/acre (1 kg/ha)	-
MALACHITE GREEN	Filamentous algae (except *Chara*)	0.3 g/yd³	-
SUPERPHOSPHATE (powdered)	Filamentous algae (except *Chara*)	535 lb/acre (600 kg/ha)	-
QUICKLIME (powdered)	Floating algal masses	as necessary to temporarily raise the pH to 10.0 - 10.2.	Caution required; Danger of toxicity to fish.
SODIUM CHLORATE	Macrophytes	267 lb/acre (300 kg/ha)	Applied to dry pond bottom as a preventative.
CALCIUM CYANAMIDE	Macrophytes	668 lb/acre (750 kg/ha)	Applied to ponds filled with water but not stocked. Ponds so treated must be drained and refilled before stocking.
SODIUM ARSENITE	Macrophytes	-	To control weeds in stocked ponds.
AMMONIUM NITROGEN	*Elodea*	200 lb/acre (225 kg/ha)	Applied to dry pond bottom as a preventative.
AMMONIUM SULFATE	*Fontinalis*	200 lb/acre (225 kg/ha)	Applied to dry pond bottom as a preventative.
DIESEL FUEL [1]	Macrophytes	50 gal/acre	Sprayed on floating weeds
2,4-D [1]	Macrophytes	1-1.5 gal/50 gallons water per acre	Add 4 lb acid and 8 oz. detergent; mix and spray evenly
RODEO [1]	Floating or emersed macrophytes	1.5 gal/50 gallons water per acre	Mix and spray evenly
DIQUAT [1]	Floating or submersed macrophytes	1 gal/50 gallons water per acre	Do not use in muddy water
AQUATHOL [1]	Submersed macrophytes	2 gallons/acre	
FENAC (2,3,6 Trichlorophenyl acetic acid) [1]	Submersed macrophytes	20 lb/acre	
DALAPON [1]	Emersed macrophytes	10 lb/50 gallons of water per acre	Mix and spray

[1] Source: Mississippi Cooperative Extension Service, *Aquatic Weed Identification and Control*; 1983)

(Source: W. McLarney, *The Freshwater Aquaculture Book*, copyright © 1984 - with permission of Hartley & Marks, Publishers)

TABLE 8.-4: AGRICULTURE CHEMICAL TOXICITY TO SELECTED AQUATIC ANIMALS

Toxicity of various herbicides to species shown. The 96-hour LC_{50} is given in the ppm columns. The pounds of material needed per acre-foot of water (325,850 gallons) for the 96-hour LC_{50} is given in the lb columns [1].

HERBICIDE COMMON NAME	BLUEGILL		CHANNEL CATFISH		RAINBOW TROUT		CRAWFISH		FRESHWATER SHRIMP	
	ppm	lb	ppm	lb	ppm	lb	ppm	lb	ppm	lb
Endothall (AQUATHOL K)	343	932.9	150	408.0	230	625.0	-	-	-	-
Simazine (AQUAZINE)	16.0	43.5	-	-	2.8	7.6	-	-	-	-
Dicamba (BANVEL)	50.0	136.0	-	-	28	76.1			56	152.3
Copper sulfate [2]					+					
Dichlobenil (CASORON)	8.3	22.5	-	-	6.3	17.1	-	-	-	-
Copper ammnonium carbonate	3.28	8.9	-	-	0.02	0.05	-	-	-	-
Diquat (ORTHO)	245	666.4	-	-	10	27.2	-	-	-	-
Daiapon (DOWPON)	500	1,368.0	-	-	-	-	-	-	-	-
Endothall (HYDROTHOL 191)	0.94	2.5	0.5	1.3	0.5	1.5	-	-	0.05	0.1
Diuron (KARMEX)	8.2	22.3	-	-	4.9	13.3	-	-	-	-
Lime sulfur	49.0	133.3	-	-	8.0	21.7	-	-	-	-
Rotenone	0.02	0.06	0.002	0.007	0.03	0.08	-	-	-	-
Glyphosate (ROUNDUP)	5.6	15.2	13.0	35.3	8.3	22.5	-	-	-	-
Tebuthiuron (SPIKE)	112	304.6	-	-	144	391.6	-	-	-	-
Tribasic copper sulfate [2]										
Hexazinone (VELPAR)	370	1,006.4	-	-	320	870.4	-	-	56.0	152.3
2,4-D (WEEDAR 64)	0.6	1.60	0.3	0.8	0.2 ,0.6	1,389	3,778.1	0.1	0.4	
2,4-D (WEEDONE 170)	10.4	28.2	19.4	52.7	0.6	1.7	-	-	2.7	7.3

[1] Adapted from: Publication 1455. Extension Service of Mississippi State University, cooperating with U.S. Department of Agriculture. by Dr. Thomas L. Wellborn, Jr.. Leader, Extension Wildlife and Fisheries. Ruth Morgan. Associate Coordinator and Specialist-Pesticide Impact Assessment, and Geoffrey W. Guyton, Computer Programmer. Extension Entomology. All ppm values are given in active ingredient. When using a combination of chemicals, check each chemical in the combination to determine the toxicity of the combination. Always use pesticides according to the label directions which are on the product's purchased.

Tradenames are used only for the purpose of information, and the Mississippi Cooperative Extension Service does not guarantee or warrant the standard of the product, nor does it imply approval of the product to the exclusion of others that also may be suitable. The Federal Environmental Pesticide Control Act of 1972 stipulates: "The use of any registered pesticide in a manner inconsistent with labeling instructions is prohibited. Under current law, penalties may be levied against a purchaser who misuses pesticides." Pesticide salesmen, dealers, commercial applicators and university personnel also have a responsibility to use or recommend only EPA registered pesticides according to the label".

[2] Toxicity depends on total alkalinity of water. Can be very toxic in water with low alkalinity.

(Source: D.D. Thayer et al., Weed Control in Aquaculture and Farm Ponds, undated)

TABLE 8-5: SUGGESTED SAFE LEVELS FOR SOME PESTICIDES FOR AQUATIC LIFE [1]

CHEMICAL	SAFE LEVEL[2]	BIOACCUMULATION INDEX	PERSISTENCE INDEX
ALDRIN/DIELDRIN	0.003 mg/l	High	High
CHLORDANE	0.01 mg/l (F)	High	High
	0.004 mg/l (M)		
DDT	0.001 mg/l	High	High
DEMETON	0.01 mg/l	?	?
ENDOSULFAN	0.003 mg/l (F)	?	Low
	0.001 mg/l (M)		
ENDRIN	0.004 mg/l	Low	Low
GUTHION	0.01 mg/l	Low	Low
HEPTACHLOR	0.001 mg/l	High	High
LINDANE	0.01 mg/l(F)	Moderate	?
	0.004 mg/l(M)		
MALATHION	0.1 mg/l	Low	Low
METHOXYCHLOR	0.03 mg/l	Low	Low
MIREX	0.001 mg/l	High	High
PARATHION	0.04 mg/l	Low	Low
TOXAPHENE	0.005 mg/l	High	High
POLYCHLORINATED BIPHENYLS	0.001 mg/l	High	High

[1] Adapted from Environmental Protection Agency, 1976

[2] (F) = freshwater, (M) = marine

(Source: J.E. Lannan *et al.*, *Principles and Practices of Pond Aquaculture*; copyright © 1986 - with permission of Oregon State University Press)

TABLE 8-6: TOXICITY OF SELECTED CHLORINATED HYDROCARBON INSECTICIDES TO AQUATIC LIFE

PESTICIDE	96-hr LC50[1] (µg/liter)	SAFE LEVEL (µg/liter)
ALDRIN/DIELDRIN	0.20 - 16	0.003
BHC	0.17 - 240	4000
CHLORDANE	5 - 3,000	0.010
DDT	0.24 - 2	0.001
ENDRIN	0.13 - 12	0.004
HEPTACHLOR	0.10 - 230	0.001
TOXAPHENE	1 - 6	0.005

[1] LC$_{50}$ = Lethal Concentration for 50% of population

(Source: C. E. Boyd, *Water Quality in Ponds for Aquaculture*, 1990)

TABLE 8-7: MAXIMUM PERMISSIBLE PESTICIDE CONCENTRATIONS WHICH MAY BE TOLERATED BY FISH.

PESTICIDE	CONCENTRATION (mg/l)
ORGANOCHLORINE PESTICIDES	
ALDRIN	0.01
DDT	0.003
DIELDRIN	0.005
CHLORDANE	0.004
ENDRIN	0.003
LINDANE	0.02
TOXAPHENE	0.01
ORGANOPHOSPHATE INSECTICIDES	
DIAZINON	0.002
DURSBAN	0.001
MALATHION	0.008
PARATHION	0.001
TEPP	0.3
CARBAMATE INSECTICIDES	
CAEBARYL	0.02
ZECTRAN	0.1
HERBICIDES; FUNGICIDES; DEFOLIANTS	
AMINOTRIAZOLE	300.0
DIQUAT	0.5
DIURON	1.5
2,4"D	4.0
SILVEX	2.0
SIMAZINE	10.0
BOTANICALS	
PYRETHRUM	0.01
ROTENONE	10.0

(Source: T.V.R. Pillay, *Aquaculture Principles and Practices*, 1990)

TABLE 8-8: TOXICITY OF SELECTED HEAVY METALS TO AQUATIC LIFE

METAL	96-hr LC50[1] (µg/liter)	SAFE LEVEL (µg/liter)
CADMIUM	80-420	10
CHROMIUM	2,000 - 20,000	100
COPPER	300-1,000	25
LEAD	1,000 - 40,000	100
MERCURY	10 - 40	0.10
ZINC	1,000 - 10,000	100

[1] LC_{50} = Lethal Concentration for 50% of the population.

(Source: C. E. Boyd., *Water Quality in Ponds for Aquaculture*, 1990)

REFERENCES

ADCP. 1983. *Fish Feeds and Feeding in Developing Countries*. United Nations Development Programme and Food and Agriculture Organization of the United Nations. ADCP/REP/83/18. FAO, Rome.

Ahmad, T. and C.E. Boyd. 1988. Design and performance of paddle wheel aerators. Aquaculture Eng., 7:39-62.

Ahmad, T. 1989. Shrimp aquaculture in Indonesia. In: D.M. Akiyama (Ed.) *Proceedings of the S.E. Asia Shrimp Farm Management Workshop*. American Soybean Assocation, Singapore.

Akiyama, D.M. and N.L. M. Chwang. 1989. Shrimp feed requirements and feed management. In: D.M. Akiyama, (Ed.) *Proceedings of the S.E. Asia Shrimp Farm Management Workshop*. American Soybean Assocation, Singapore.

Aleem, M.I.H. 1959. The physiology and chemo-autrophic metabolism of *Nitrobacter agilis*. Thesis. Cornell University, New York, NY.

Alimov, A.F. and N.V. Shadrin. 1977. The calorific value of some representatives of freshwater benthos. *Hydrobiol. J.*, 13:68-73.

Allen, M. M. 1968. Simple conditions for growth of unicellular blue-green algae on plates. *J. Phycol.* 4: 1-4.

Allen, R.D., 1984. *Ingredient analysis table: 1984 Edition. Feedstuffs* 56(30):22-30.

Almazan, G. and C.E. Boyd. 1978. Effects of nitrogen levels on rates of oxygen consumption during decay of aquatic plants. *Aquat. Bot.*, 5:119-126.

American Association for Vocational Instructional Materials (AAVIM). 1973. Planning for an Individual Water Sytem. AAVIM Engineering Center, Athens, GA.

Anderson, R.J., E.W. Kienholz, and S.A. Flickinger. 1981. Protein requirements of smallmouth bass and largemouth bass. *J. Nutr.*, 111:1085-1097.

Andrews, J.Y. and Y. Matsuda. 1975. Influence of various culture conditions on the oxygen consumption of channel catfish. *Trans. Amer. Fish. Soc.*, 104:322-327.

Andrews, J.W., L.V. Sick, and G.J. Baptist. 1972. The influence of dietary protein and energy levels on growth and survival of penaeid shrimp. *Aquaculture*, 1:341-347.

Appler, H.N. and K. Jauncey. 1983. The utilization of a filamentous green alga (*Cladophora glomerata* (L. Kutzin) as a protein source in pelleted feeds for *Sarotherodon* (Tilapia) *niloticus* fingerlings. *Aquaculture*, 30:21-30.

Apud, F.D. 1989. *Recent Developments in Prawn Pond Culture*. Aquaculture Extension Pamphlet No. 1,

Aquaculture Dept. SEAFDEC, Iloilo, Philippines.

AQUACOP. 1977. Reproduction in captivity and growth of *Penaeus monodon* Fabricus in Polynesia. In: *Proc. of the 8th Ann. Workshop of the World Mariculture Society.* J.W. Avault (Ed.) Louisiana State University Press, Baton Rouge, LA., USA, pp. 927-945.

ASEAN. 1978. *Manual on Pond Culture of Penaeid Shrimp.* ASEAN 77/SHR/CUL3. ASEAN National Coordinating Agency of the Phillipines. Manila, Philippines.

ASHRAE. 1985. *ASHRAE Handbook - 1985 Fundamentals: Inch-Pound Edition.* American Association of Heating, Refrigerating, and Air-Conditioning Engineers, Inc. Atlanta, GA.

Austreng, E. and T. Refstie. 1979. Effect of varying dietary protein level in different families on rainbow trout. *Aquaculture,* 18:145-156.

Backmann, R.W. 1962. Evaluation of a modified C-14 technique for shipboard estimation of photosynthesis in large lakes. Great Lakes Research Publ. No. 8.

Balazs, G.H. and E. Ross. 1976. Effect of protein source and level on growth and performance of the captive freshwater prawn *Macrobrachium rosenbergii. Aquaculture,* 7:299-313.

Balazs, G.H., E. Ross, and C.C. Brooks. 1973. Preliminary studies on the preparation and feeding of crustacean diets. *Aquaculture,* 2:369-377.

Ball, R.O. and H.J. Cambell, Jr. 1974. Static aeration systems problems and performance. Proc. of the 29th Purdue Industrial Waste Conference, Purdue University, 328-337.

Barlow, S.M. and M.L. Windsor. 1984. Fishery by-products. International Association of Fish Meal Manufacturers, Technical Bulletin No. 19, September 1984.

Bates, R.G., 1975. pH Scales for Seawater. In: E.D. Goldber (Ed.), *The Nature of Seawater,* Dahlem Konferenzen, Berlin. pp. 313-338.

Bath, D. *et. al.* 1984. Composition of by-products and unusual feedstuffs. *Feedstuffs,* 56(30):32- 36.

Beamish, F.W.H. 1964. Respiration of fishes with special emphasis on standard oxygen consumption II. Influence of weight and temperatue on respiration of several species. *Can. J. Zool.,* 42:177-194.

Benijts, F., E. Van Voorden, and P. Sorgeloos. 1976. Changes in the biochemical composition of the early larval stages of the brine shrimp, *Artemia salina* L., In Proc. 10th Eur. Symp. on Mar. Biol., Vol. I., Persoone, G. and E. Jaspers (Eds.), Universa Press, Wetteren, Belgium, I.

Benson, B.B. and D. Krause. 1984. The concentration and isotopic fractionation of oxygen dissolved in freshwater and seawater in equilibrium with the atmosphere. *Limn. Oceanogr.* 29:630-632.

Bernhard, M., A. Zattera and P. Filesi, 1966. Suitability of various substances for use in the culture of marine organisms. *Publ. Sta. Zool. Napoli,* 35: 89-104.

Bernier, C.J. 1962. Measurement technique for the radiant energy requirements of growing plants. Illumination Engineering Society Conference paper (unpublished).

184

Beveridge, M.C.M. 1987. *Cage Aquaculture*, Fishing News Books, Surrey, England.

Bhaskar, T.I. and S. Ahamed Ali. 1984. Studies on the protein requirements of post larvae of the prawn *Penaeus indicus* H. Milne Edwards using purified diets. *Ind. J. Fish.*, 31:74-81.

Bickford, E.D., and S. Dunn. 1972. *Lighting for Plant Growth*. Kent State University Press. 221 pp.

Bischoff, H. W. and H. C. Bold. 1963. *Phycological Studies. IV. Some algae from Enchanted Rock and related algal species*. The Univ. of Texas Pub. No. 6318.

Blankley, W.F. 1979. Toxic and inhibitory materials associated with culturing. In: J.R. Stein (Ed.) *Handbook of Phycological Methods: Culture Methods and Growth Measurements*. Cambridge University Press. pp. 207-233.

H.C. Bold and M.J. Wynne, 1978. *Introduction to the Algae*. Prentice-Hall, Inc. Englewood Cliffs, NJ.

Bolton, W. and R. Blair. 1977. *Poultry nutrition*. Bull. Ministr. Agric. Fish. Food G.B. (174):134.

Boon, B. and H. Laudelout. 1962. Kinetics of nitrite oxidation by *Nitrobacter winogradskyi*. *Biochem. J.* 85:440-447.

Boonyaratpalin, M., and M.B. New. 1982. Evaluation of diets for *Macrobrachium rosenbergii* reared in concrete ponds. In: *Giant Prawn Farming*, M.B. New (Ed.), Elsevier Science Publ. Corp., Amsterdam-Oxford-New York, pp. 249-256.

Bose, A.N., S. N. Ghosh, C.T. Yang and A. Mitra. 1991. *Coastal Aquaculture Engineering*, Hodder & Stoughton Publishing Co., New York, NY.

Bowen, H.J.M. 1966. *Trace elements in Biochemistry*. Academic Press, New York, 241 pp.

Boyd, C.E. 1968. Fresh-water plants: a potential source of protein. *Econ. Bot.*, 22:359-368.

_____. 1969. The nutritive value of three species of water weeds. *Econ. Bot.*, 23:123-127.

_____. Chemical analysis of some vascular aquatic plants. *Arch. Hydrobiol.*, 67:78-85.

_____. 1976. Lime requirement and application in fish ponds. *FAO Tech. Conf. on Aquaculture*, Fishing News Books, Ltd., Farnham, England: 120-122.

_____. 1979. Aluminum sulfate (Alum) for precipitating clay turbidity from fish ponds. *Trans. Amer. Fish. Soc.*, 108:307-313.

_____. 1990. *Water Quality in Ponds for Aquaculture*. Alabama Agricultural Experiment Station, Auburn University. Birmingham Publ. Co., Birmingham, AL.

Boyd, C.E., and C.P. Goodyear. 1971. Nutritive quality of food in ecological systems. *Arch. Hydrobiol.*, 69:256-270.

Boyd, E.E. and D.J. Martinson. 1984. Evaluation of propeller-aspirator-pump aerators. *Aquaculture*, 36:283-292.

185

Boyd, C.E., T. Ahmad, and Z. La-fa. 1988. Evaluation of plastic pipe, paddle wheel aerators. Aquacult. Eng., 7:63-72.

Bozniak, E. 1969. Laboratory and field studies of phytoplankton communities. Ph.D. Dissertation. Washington University, St. Louis, Missouri.

Branckaert, R., S. Tessema and R.S. Temple. 1976. The use of local by-products for formulating diets in tropical African countries. In: *Proceedings of the Symposium on Water Quality and Management through Biological Control*. Gainesville, U. of Florida.

Bressanik R., *et al.* 1975. The use of coffee processing waste as animal feed. In: *Proceedings of Conference on Animal Feeds of Tropical and Subtropical Origin*. London School of Pharmacy, London, 1-5 April 1974, Tropical Products Institute.

Brett, J.R. and C.A. Zala. 1975. Daily pattern of nitrogen excretion and oxygen consumption of sockeye salmon (*Oncorhynchus nerka*) under controlled conditions. *J. Fis. Res. Board Can.*, 32:2479-2486.

Brown, E.E. and J. B. Gratzek. 1980. *Fish Farming Handbook: Food, Bait, Tropicals and Goldfish*. Van Nostrand Reinhold, New York, NY.

Bruggeman, E. P. Sorgeloos, and P. Vanhaecke. 1980. Improvements in the decapsulation technique of *Artemia* cysts, In: The Brine Shrimp *Artemia*. Vol. 3. G. Persoone, P. Sorgeloos, O. Roels, and E. Jaspers (Eds.), Universa Press, Wetteren, Belgium.

Buckman, H.O. and N.C. Brady. 1960. *The Nature and Property of Soils*. 6th ed., Macmillan Co., New York, NY.

Burns, R.L. and A.C. Mathieson. 1972. Ecological studies of economic red algae. II. Culture studies of *Chondrus crispus* Stackhouse and *Gigartina stellata* (Stackhouse) Batters. J. Exp. *Mar. Biol. Ecol.* 8:1-6.

Busch, R.L., C.S. Tucker, J.A. Steeby, and J.E. Reames. 1984. An evaluation of three paddlewheel aerators used for emergency aeration of channel catfish ponds. *Aquacult. Eng.*, 3:59-69.

Buswell, A.M., T. Shiota, N. Lawrence, and I. Van Meter. 1953. Laboratory studies on the kinetics of the growth of *Nitrosomonas* with the relation to the nitrification phase of the BOD test. *Applied Microbiology* 2:21-25.

Carraro, S. 1983. *Etude de l'influence de defferents regimes alimentaires a base d'azolla sur la croissance   de Sarotherodon niloticus. Memoir presente par Simonetta Carraro en view de l'obtention du grade de Licencia en Sciences*, Universite Catholique de Louvain, Faculte des Sciences, Louvain-la-Neuve, pp. 81-83.

Chamberlain, G.W. 1991. Shrimp farming in Indonesia: I - Growout Techniques. *World Aquaculture* 22(2):12-27.

Chemtrol. 1979. *Plastic Piping Handbook*. Celanese Piping Systems, P.O. Box 1032, Louisville, Kentucky.

Chen, L.C.M., T. Edelstein, and J. McLachlan. 1969. *Bonnemaisonia hamifera* Hariot in nature and in culture. *J. Phycol.* 5: 211-220.

Chesness, J.L. and J.L. Stephens. 1971. A model study of gravity flow aerators for catfish raceway systems.

186

Trans. Amer. Soc. Agricult. Eng., 14:11678-1169; 1174.

Chesness, J.L., J.L. Stephens, and T.K. Hill. 1972. Gravity flow aerators for raceway culture systems. Research Report 137, Agricultural Experiment Station, University of Georgia.

Chesness, J.L., L.J. Fussell, and T.K. Hill. 1973. Mechanical efficiency of a nozzle aerator. *Trans. Amer. Soc. Agricult. Eng.*, 16:68,71.

Choubert, G., and P. Luquet. 1983. Utilization of shrimp meal for rainbow trout (*Salmo gairdneri* Rich.) pigmentation. Influence of fat content of the diet. *Aquaculture*, 32:19-26.

Chong, C.V.Y. 1977. *Properties of Materials.* McDonald and Evans, Ltd., Plymouth, UK.

Christiansen, J.E. 1935. *Measuring Water for Irrigation.* California Agricultural Experiment Station Bulletin 588.

Chu, S.P. 1942. The influence of the mineral composition of the medium on the growth of planktonic algae. I. Methods and culture media. *J. Ecol.* 30:284-325.

Chuapoehuk, W. and T. Pothisoong. 1985. Protein requirements of catfish fry, *Pangasius sutchi*, Fowler. In: C. Y. Cho, C.G. Cowey and T. Watanabe (Eds.) *Finfish Nutrition in Asia: Methodologial Approaches to Research and Development.* Ottawa, Canada, IDRC-233e.

Clifford, H.C., III. 1992. Marine Shrimp Pond Management: A Review. In: J. Wyban (Ed.) *Proc. of the Special Session on Shrimp Farming.* World Aquaculture Society, Baton Rouge, LA.

Colinvaux, L.H., K.M. Wilbur, and N. Watabe. 1965. Tropical marine algae: growth in laboratory culture. *J. Phycol.* 1:69-78.

Collins, M.T., J.B. Gratzek, D.L. Dawe, and T.G. Nemetz. 1975. Effects of parasiticides on nitrification. J. *Fish. Res. Board Can.* 32:2033-2037.

Collins, et al. 1976. Effects of antibacterial agents on nitrification in an aquatic recirculating system. *J. Fish. Res. Board Can.* 33:215-218.

Colt, J. and H. Westers. 1982. Production of gas by aeration. *Trans. Amer. Fish. Soc.*, 111:342-360.

Colt, J. and C. Orwicz. 1991. Aeration in intensive culture. In: D.E. Brune and J. R. Tomasso (Eds.), *Aquaculture and Water Quality.* World Aquaculture Society, Baton Rouge, LA.

Colt, J.E. and G. Tchobanoglous. 1981. Design of aeration systems for aquaculture. In: L.J. Allen and E.C. Kinney (Eds.)*Proc. Bioeng. Symp. Fish Cult.*, Traverse City, MI, 16 -18, October, 1979. *Fish Culture Section of the American Fisheries Society*, Bethesda, MD.

Colt, J. 1984. Computation of dissolved gas concentration in water as functions of temperature, salinity and pressure. Amer. Fish Soc. Spec. Publ. No. 14.

Colvin, L.B. and C.W. Brand. 1977. The protein requirement of penaeid shrimp at various life-cycle stages in controlled environmental systems. Proc. 8th Ann. Meet. World Maric. Soc.

Colvin, P.M. 1976. Nutritional studies on penaeid prawns: Protein requirements in compounded diets for juvenile *Penaeus indicus*. *Aquaculture*, 7:315-326.

Conover, R.J. 1978. Transformation of organic matter. In: O. Kinne (Ed.,) *Marine Ecology IV: Dynamics*. John Wiley & Sons, New York, NY.

Cook, H.L. 1991. Improving low shrimp production due to acid or potentially acid soils. In: M. C. Zendejas and G.W. Chamberlain (Eds.) *Taller Sobre Cultivo de Camaron, Proceedings of a Shrimp Farming Workshop in Mazatlan, Mexico on 17-19 July 1991*. Industrias Purina S.A. de C.V., Mexico City, Mexico.

Cooley, M.L. 1976. Feed ingredients guide. In: H.B. Pfost and D. Pickering (Ed.) *Feed Manufacturing Technology*. Feed Production Council, American Feed Manufacturers Assoc., Inc. Arlington, Virginia.

Connell, J.J. and P.F. Howgate. 1959. The amino acid composition of some British food fishes. *J. Sci. Food Agric.* 10:241-244.

Corbin, J.S., M. M. Fujimoto, and T. Y. Iwai, Jr. 1983. Feeding practices and nutritional considerations for Macrobrachium rosenbergii culture in Hawaii. In: J.P. McVey (Ed.), *Handbook of Mariculture: Vol 1. Crustacean Aquaculture*. CRC Press, Boca Raton, FL.

Cornacchia, J.W. and J.E. Colt. 1984. The effects of dissolved gas supersaturation on larval striped bass *Morone saxatilis* (Walbaum). *J. Fish Dis.*, 7:15-27.

Cowey, C.B. and J.R. Sargent. 1972. Fish nutrition. *Adv. Mar. Biol.* 10:383-492.

Cowey, C.B. et. al. 1971. Studies on the nutrition of marine flatfish. Growth of the plaice *Pleuronectes platessa* on diets containing proteins derived from plants and other sources. *Mar. Biol.* 105:145-153.

Creswell, D.C. and I.P. Kompaing. 1981. Studies on snail meal as a protein source for chickens. 1. Chemical composition, metabolizable energy, and feeding value for broilers. *Poultry Science*, 60(8):1854-1860.

Cummins, K.W. and J.C. Wuycheck. 1971. Caloric equivalents for investigations in ecological energetics. *Mitt. Int. Ver. Limnol.*, 18:1-158.

Dabrowski, K. 1977. Protein requirements of grass carp fry (*Ctenopharyngodon idella* Val.). Aquaculture, 12:63-73.

Dabrowski, K. and M. Rusiecki. 1981. Content of total and free amino acids in zooplanktonic food of fish larvae. *Aquaculture*, 41:333-344.

D'Agostina, A. 1975. Antibotics in culture of invertebrates. In: Smith, W.L. and M. H. Chanley (Eds.). *Culture of Marine Invertebrate Animals*. Plenum Press, New York, NY.

Darden, W.H. 1966. Sexual differentiation in *Volvox aureus*. *J. Protozool.* 13:239-255.

Davis, A.T. and R.R. Stickney. 1978. Growth responses of *Tilapia aurea* to dietary protein quality and quantity. *Trans. Amer. Fish. Soc.*, 107:479-483.

Davis, E.A., J. Dedrick, C.S. French, H.W. Milner, J. Myers, J.H.C. Smith and H.A. Spoehr. 1953.

Laboratory experiments on *Chlorella* culture at the Carnegie Institution of Washington. Department of Plant Biology. In: Burlew, J.S. (Ed.), *Algal Culture from Laboratory to Pilot Plant*, Carnegie Inst. Washington Publ. 600.

Davies, E.M., G.L. Rumsey and J.G. Nickum, Egg-processing wastes as a replacement protein source in salmonid diets. *Prog. Fish Cult.*, 38:20-22.

Dawes, C.J., J.W. LaClaire, and R.E. Moon. 1976. Culture studies on *Eucheuma nudum* J. Agardh, a carrageen producing red alga in Florida. *Aquaculture* 7:1-9.

Deppe, K.J. and M.S. Engel. 1960. *Untersuchunger liberidie temperaturatuvabhang del nitratbildung durch Nitrobacter wimogradski buch Bei undehem und gehemmtem washtum. Zentlol. Bakt. Parasitkde II:P* 113:561-568.

Deshimaru, O. and K. Shigeno. 1972. Introduction to the artificial diet for prawn *Penaeus japonicus. Aquaculture.* 1:115-133.

Deshimaru, O. and K. Kuroki. 1974. Studies on a purified diet for prawn -1. Basal composition of diet. *Bull. Jap. Soc. Fish.*, 40:413-419.

Deshimaru, O. and Y. Yone. 1978. Optimum level of dietary protein for prawn. *Bull. Jap. Soc. Sci. Fish.*, 44:1395-1397.

Deshimaru, O., *et. al.* 1985. Nutritional quality of compounded diets for prawn *Penaeus monodon. Bull. Jap. Soc. Sci. Fish.*, 51:1037-1044.

De Silva, S.S. and M.K. Perera. 1985. Effects of dietary protein level on growth, food conversion, and protein use in young *Tilapia nilotica* at four salinities. *Trans. Amer. Fish. Soc.*, 114:584-589.

Devendra, C. 1979. *Malaysian feedstuffs.* Serdang, Malaysian Agricultural Research and Development Institute (MARDI).

Deyoe, C.W., O.W. Tiemeier, and C. Supper. 1968. Effects of protein, amino acid levels and feeding methods on growth of fingerling channel catfish. *Prog. Fish. Cult.*, 30:187-195.

Dijkstra, F., H. Jennekins, and P. Nooren. 1979. The development and application of water-jet aeration for waste water treatment. *Prog. Water. Tech.*, 11:181-191.

Direccion Nacional de Acuicultura (DNA). 1984. *Manual de cria de camarones penaeidos.* Ministerio de Desarollo Agropecuario, Santiago de Verguas, Panama.

Doty, M.S. and M. Oguri. 1959. The carbon-fourteen technique for determining primary plankton productivity. *Publ. Sta. Zool Napoli*, 31 (suppl) 70-94.

Driver, E.A., L.G. Sugdew, and R.J. Kovach. 1974. Calorific, chemical and physical values of potential duck foods. *Freshwater Biol.*, 4:281-292.

Dyer, D.L. and D.E. Richardson. 1962. Materials of construction in algal culture. *Appl. Microbiol.* 10:129-132.

Edwards, P. 1980. The production of microalgae on human wastes and their harvest by herbivorous fish. In: G. Shelef and G. J. Soeder (Eds.) *Algae Biomass*. Elsevier/North Holland Biomedical Press, Netherlands.

_____. 1982. Integrated fish farming in Thailand. ICLARM Newsletter., 5, 3.

Edwards, P., M. Kamai and K.L. Wee. 1985. Incorporation of composted and dried water hyacinth in pelleted feed for the tilapia *Oreochromis niloticus* (Peters). *Aquacult. Fish. Management*, 16(3)233-248.

Elmslie, L.J. 1982. Snails and snail farming. *World Animal Review*, 41:20-26.

Emerson, K., R.C. Russo, R.E. Lund and R.V. Thurston., 1975. Aqueous ammonia equilibrium calculations: Effects of pH and temperature. *J. Fish. Res. Board Can.*, 32:2379-2383.

Engle, M.S. and M. Alexander. 1958. Growth and autotrophic metabolism of *Nitrosomonas europea*. *J. Bact.* 76:217-222.

EPA (Environmental Protection Agency) 1968. Report of the Committee on Water Quality Criteria. Federal Water Pollution Control Administration. US Department of the Interior.

FAO. 1970. Amino acid content of foods and biological data on proteins. Food Policy and Food Science Service, Nutrition Division. FAO Food and Nutrition Serices - Collection FAO No. 21.

FAO. 1987. *Feeding and Nutrition of Farmed Fish and Shrimp - A Training Manual. Volume 1: Essential Nutrients*. FAO Field Document No. 2. GCP/RLA/075/ITA.

_____. 1987. *Feeding and Nutrition of Farmed Fish and Shrimp - A Training Manual. Volume 2: Nutrient Sources and Composition*. FAO Field Document No. 5. GCP/RLA/075/ITA.

Figueroa, J.L.A. 1991. *Tecnicas de fertilizacion en el cultivo de camaron*. In: M.C. Zendejas and G.W. Chamberlain (Eds.) *Taller Sobre Cultivo de Camaron, Proceedings of a Shrimp Farming Workshop in Mazatlan, Mexico on 17-19 July 1991*. Industrias Purina S.A. de C.V., Mexico City, Mexico.

Finstein, M.S. and C.C. Delwiche. 1965. Molybdenum as a micro nutrient for *Nitrobacter*. *J. Bact.*, 89:123-128.

Foltz, J.W. 1982. A feeding guide for single cropped channel catfish (*Ictalurus punctatus*). *J. World Maricult. Soc.*, 13:274-281.

Forster, J.R. M. and T.W. Beard. 1973. Growth experiments with the prawn *Palaemon serratus* Pennant fed with fresh and compound foods. M.A.F.F. Fisheries Investigations Series 11 27(7): 16.

Fortier, S. and F.C. Scobey. 1926. Permissible canal velocities, *Transactions American Society of Civil Engineers*.

Gaastra, P. 1959. Photosynthesis of crop plants as influenced by light, carbon dioxide, temperature and stomatal diffusion resistance. Mededel, Landbouwhogeschool Wageningen 59:1-68.

Gallagher, M. and W. D. Brown. 1975. Composition of San Francisco Bay brine shrimp (*Artemia salina*). *Agric. Food. Chem.*, 23:1631-1635.

Garling, D.L. and R.P. Wilson. 1976. Optimum dietary protein to energy ratio for channel catfish fingerlings (*Ictalurus punctatus*). *J. Nutr.*, 106:1368-1375.

Godin, V.J. and P.C. Spensley. 1971. *Crop Product Digests No. 1 Oils and Oilseeds*. Tropical Products Institute, London.

Gohl, B. 1981. Tropical feeds. *FAO Animal. Prod. Health Ser.*, 12:529.

Gotaas, H.B. 1956. *Composting: Sanitary Disposal and Reclamation of Organic Wastes*. World Health Organization Monograph Series, No. 31., WHO, Geneva.

Guillard, R.R.L. and J.H. Ryther. 1962. Studies of marine planktonic diatoms. I. *Cyclotella nana* (Hussedt) and *Detonula conferoaces* (Cleve) Gran. *Can. J. Microbiol.*, 8:229-239.

Guillard, R.R.L. 1973. Division rates. In: J.R. Stein (Ed.) *Handbook of Phycological Methods*. Cambridge University Press. Cambridge.

Guillard, R.R.L. 1975. Culture of phytoplankton for feeding marine invertebrates. In: Smith, W.L. and M. H. Chanley (Eds.), *Culture of Marine Invertebrate Animals*. Plenum Press, New York, NY.

Hackney, G.E. and J.E. Colt. 1982. The performance and design of packed column aeration systems for aquaculture. *Aquacult. Eng.*, 1:275-295.

M.J. Hammer. 1975. *Water and Waste-Water Technology*. John Wiley & Sons., New York, NY.

Hardison, T. B. 1977. *Fluid Mechanics for Technicians*. Reston Publishing Co., Inc. Reston, VA.

Haskell, D.C., R.O. Davies, and J. Reckahn. 1960. Factors in hatchery pond design. *New York Fish and Game Journal.*, 7:112-129.

Hastings, W.H. 1974. Study of pelleted fish foods stability in water. In: J.L. Gaudet (Ed.), *Report of the Ⓐ Workshop on Fish Feed Technology and Nutrition*, Warmwater Fish Culture Laboratories, Stuttgart, Arkansas, USA, 7-19 September, 1970. Washington, D.C., US Government Printing Office, EIFAC/Bureau of Sport Fisheries and Wildlife Resource Publication 102.

Hastings, W.H. 1973. *Experience relative a la preparation d'aliments des poissons et a leur alimentation. Repport prepare pou le Projet Regional de Recherche et de Formation Piscicoles*, FAO FI:DP/RAF/66/-054/1.

Hepher, B. 1988. *Nutrition of Pond Fishes*. Cambridge University Press, New York, NY.

Herwig, N. 1979. *Handbook of Drugs and Chemicals Used in the Treatment of Fish Diseases*. Charles C. Thomas, Publisher, Springfield, IL.

Hickling, C.F. 1971. *Fish Culture*. Faber and Faber, London.

Hilton, J.W. 1983. Potential of freeze-dried worm meal as a replacement for fish meal in trout diet formulations. *Aquaculture*. 32:277-283.

Hirono, Y. 1989. Shrimp farm management in Ecuador. In: D.M. Akiyama (Ed.) *Proceedings of the S.E. Asia*

Shrimp Farm Management Workshop. American Soybean Association, Singapore.

Hnath, J.G. 1983. Hatchery disinfection and disposal of infected stocks. In: F.P. Meyer, J.W. Warren and T.G. Carey (Eds.). *A Guide to Integrated Fish Health Management in the Great Lakes Basin.* Special Publication, 83(2)121-33. Great Lakes Fishery Commission, Ann Arbor, Michigan.

Hoffman, T. and H. Lees. 1952. The biochemistry of the nitrifying organisms. *Biochem.* 52:140-142.

Hollaender, A. 1956. Editor. *Radiation Biology, Vol. III, Visible and Near Visible Light.* McGraw-Hill Book Company, Inc., New York, NY.

Horvath, L., G. Tamas, and I. Tolg. 1984. *Special Methods in Pond Fish Husbandry.* Halver Corporation, Seattle, USA.

Hughes, E.O., P.R. Gorham, and A. Zehnder. 1958. Toxicity of a unialgal culture of *Microcystis aeruginosa*. *Can. J. Microbiol.* 4:346-348.

Huguenin, H.E. and J. Colt. 1989. *Design and Operating Guide for Aquaculture Seawater Systems.* Developments in Aquaculture and Fisheries Sciences, 20. Elsevier, New York, NY.

Hulbert, R. and D. Feben. 1933. Hydraulics of rapid filter sand. *Journal of the American Water Works Assoc.* 25(1)20-65.

Hur, S.B. 1991. The selection of optimum phytoplankton species for rotifer culture during cold and warm seasons and their nutritional value for marine finfish larvae. In: W. Fulks and K. L. Main (Ed.) *Rotifer and Microalgae Culture Systems.* Oceanic Institute, HI.

Imada, O., *et. al.* 1979. Development of a new yeast as a culture medium for living feeds used in the production of fish seed. Bull. Jap. Soc. Sci. Fish., 45:955-959.

Jackson, A.J., A.K. Kerr and C.B. Cowey. 1984. Fish silage as a dietary ingredient for salmon. 1. Nutritional and storage characteristics. *Aquaculture* 38:211-220.

Jansen, J.J.A. 1981. *Bamboo in building structures.* Unpublished Ph.D. thesis, University of Eindhoven.

Jaenike, F. 1989. Management of a shrimp farm in Texas. In: D.M. Akiyama (Ed.) *Proc. of the S.E. Asia Shrimp Farm Management Workshop.* American Soybean Association, Singapore.

Janssen, J. 1985. *Elevage du poisson - chat africain Clarias lazera (Cuv & Va., 19840) En Republique Centrafricaine.* IV - Alimentation haut Commissariat charge du Tourisme, des Eauz, Forets, Chasses et Peches, Project FAO GCP/CAF/007/NET, Bangui, Document Technique No. 23.

Jauncey, K. 1981. The effects of varying dietary composition on mirror carp (*Cyprinus carpio*) maintained in thermal effluents and laboratory recycling systems. In: *Proc. of World Symposium on Aqaculture in Heated Effluent and Recirculation Systems: Vol. II.* I.H. Heenemann Gmby and Co., Berlin.

Jauncey, K. 1982. The effects of varying dietary composition on mirror carp (*Cyprinus carpio*) maintained in thermal effluents and laboratory recycling systems. In: *Proceedings of World Symposium on Aquaculture in Heated Effluent and Recirculation Systems, Vol. II,* I.H. Heenemann Gmby and Co., Berlin.

Jobling, M. and A. Wandsvik. 1983. Quantitative protein requirements of Artic charr *Salvelinus alpinus* (L.) *J. Fish Biol.*, 22:705-712.

Kanazawa, A. *et. al.* 1980. Nutritional requirements of the puffer fish: purified test diet and optimum protein level. *Bull. Jap. Soc. Sci. Fish.*, 46:1357-1361.

Kaushik, S.J. and P. Luquet. 1980. Influence of bacterial protein incorporation and of sulphur amino acid supplementation to such diets on growth of rainbow trout, *Salmo gairdneri* Richardson. *Aquaculture*, 19:163-175.

Kawai, A. Y. Yoshida and M. Kimata. 1965. Biochemical studies on the bacteria in aquarium with circulating system. I. Changes of the qualities of the breeding water and bacterial population of the aquarium during fish cultivation. *Bull. Jap. Soc. Sci. Fish.* 31(1):65-71.

Kawai, A. Y. Yoshida and M. Kimata. 1965. Biochemical studies on the bacteria in the aquarium with circulating system. II: Nitrifying activity of the filter sand. *Bull. Jap. Soc. of Sci. Fish.* 31:65.

Kay, D.E. 1973. *Root crops. Tropical Products Institute Crop and Product Digest No. 2.* Tropical Products Institute, London.

Kay, D.E., 1979. *Crop and product digest, No. 3, Food legumes.* Tropical Products Institute, London.

Kelly, C.B. 1974. *The Toxicity of Chlorinated Waste Effluents to Fish and Considerations of Alternative Process for the Disinfection of Waste Effluents.* Virginia State Water Control Board.

Kevgor Aquasystems. 1990. *Aquaculture Facilities.* Vancouver, BC, Canada.

Kholdebarin, B. and J.J. Oertli. 1977. Effect of pH and ammonia on the rate of nitrification of surface water. *Jour. of the Water Poll. Control. Fed.*, 49:1688-1692.

Khoo, K.H., C.H. Culberson and R.C. Bates, 1977. Thermodynamics of the dissociation of ammonium ion in seawater from 5 to 40° C. *J. Sol. Chem.*, 6:281-290.

Kleschin, A.F. 1960. *Die Pflanze und das Licht.* Akademie-Verlag. Berlin.

Klust, G. 1982. *Netting Materials for Fishing Gear.* Fishing News Books, Ltd., Farnham, Surrey, UK.

Klust, G. 1983. *Fibre Ropes for Fishing Gear.* Fishing News Books, Ltd., Farnham, Surrey, UK.

Kongkeo, H. 1991. An overview of live feeds in Thailand. In: W. Fulks and K. L. Main (Eds.) *Rotifer and Microalgae Culture Systems.* Oceanic Institution, Inc., Honolulu, HI.

Kuwa. M. 1983. Corrosion and protection of fish culturing floating cage made of wire netting. *Bull. Jap. Soc. Sci. Fish.*, 49, 165-175.

Lannan, J.E. , R. O. Smitherman, and G. Tchobanoglous. *Principles and Practices of Pond Aquaculture.* Oregon State University Press, Corvallis, OR.

Laudelout, H. and L. Van Tichelen. 1960. Kinetics of the nitrite oxidation by *Nitrobacter winogradsky. J. Bacteriology* 79:39-42.

Lee, D.L. 1971. Studies on the protein utilization related to growth in *Penaeus monodon* Fabricus. *Aquaculture*, 1:1-3.

Lee, J. S. 1991. *Commercial Catfish Farming*. Interstate Publishers, Inc., Danville, IL.

Lees, H. 1952. The biochemistry of the nitrifying organism. *Biochem. J.*, 52:134-139.

Leger, Ph., D.A. Bengston, K. L. Simpson and P. Sorgeloos. 1986. The use and nutritional value of *Artemia* as a food source. *Oceanogr. Mar. Biol. Ann. Rev.*, 24: 521-623.

Leitritz, E. and R.C. Lewis. 1976. Trout and salmon culture (hatchery methods). California Department of Fish and Game, Fish. Bulletin 164.

Levine, G. and T.L. Meade. 1976. The effects of disease treatment on nitrification in closed system aquaculture. In: J.W. Avault, Jr. (Ed.). *Proc. 7th Ann. Meet. World Mariculture Soc.*

Lewin, J., 1966. Physiological studies of the boron requirment of the diatom, *Cylindrotheca fusiformis*. Reimann and Lewis. *J. Exp. Bot.*, 17:473-479.

Liao, I.C., H.M. Su and J.H. Lin. 1983. Larval foods for penaied prawns. In: McVey, J.P. (Ed.) *Handbook of Mariculture: Vol 1. Crustacean Aquaculture*. CRC Press, Boca Raton, FL.

Liao, P.B. 1971. Water requirements for salmonids. *Prog. Fish. Cult.*, 33,210-215.

Lieder, U. 1965a. *Das Eiweiss der Nahrung der Karpfen. Dtsch. Fisch. Ztg.*, 12:16-26.

D. V. Lightner. 1983. Diseases of cultured penaeid shrimp. In: McVey, J.P. (Ed.) *Handbook of Mariculture: Vol 1. Crustacean Aquaculture*. CRC Press, Boca Raton, FL, pp. 289-43-71.

Lim, E., S. Sukhawongs, and F.P. Pascual. 1979. A preliminary study on the protein requirement of *Chanos chanos* (Forskal) fry in a controlled environment. *Aquaculture*, 17:195-201.

Linn, J.G. *et al.* 1975. Nutritive value of dried or ensiled aquatic plants. 1. Chemical composition. *J. Anim. Sci.*, 41:601-609.

Ling, S.W., 1967. Feeds and feeding of warm-water fishes in ponds in Asia and the Far East. *FAO Fish. Rep.*, 44(3):291-309.

Little, E.C.S. 1979. *Handbook of Utilization of Aquatic Plants*. FAO Fish. Tech. Pap. 187.

Little, D. and J. Muir. 1987. *A Guide to Integrated Warm Water Aquaculture*. Institute of Aquaculture Publications, University of Stirling, Scotland.

Little, E.C.S. and I.E. Hensen. 1967. The water content of some important tropical water weeds. *PANS* (c), 13(3):223-227.

Loveless, J.E. and H.A. Painter. 1969. The influence of metal ion concentration and pH value on the growth of a nitrosomonas strain isolated from activated sludge. *J. Gen. Microbiol.* 52:1-14.

Lovell, R.T. 1972. Protein requirement of cage-cultured channel catfish. *Proc. 26th. Ann. Conf. of S.*

Eastern Assoc. of Game and Fish Comm. 1972.

_____. 1975. Fish feeds and nutrition: How much protein in feeds for channel catfish. *The Commercial Fish Farmer*, March-April: 40-41.

_____. 1979, Nutritional value of catfish processing waste. *Commercial Fish Farmer and Aquaculture News*, May/June, 5(4):2.

_____. 1989. *Nutrition and Feeding of Fish.* Van Nostrand Reinhold, New York, NY.

Lundegaard, G. 1985. *Keeping Marine Fish.* Blandford Press, Dorset, England.

Luning, K. 1971. Seasonal growth of *Laminaria hyperborea* under recorded underwatr light conditions near Helgoland. In: D.J. Crips (Ed.), *Proc. 4th European Mar. Biol. Symp.*, Cambridge Univ. Press, Cambridge, England.

MacDonald, P., R.A. Edwards and J.F.D., Greenhalgh. 1977. *Animal Nutrition.* Longman, London and New York.

Malone, R. 1991. *Design of Recirculating Blue Crab Shedding System.* Louisiana Sea Grant.

Marek, M. 1975. Revision of supplementary feeding tables for pondfish. *Bamidgeh*, 27(3):57-64.

Mathias, J.A., *et. al.* 1982. Harvest and nutritional quality of *Gammarus lactustris* for trout culture. *Trans. Am. Fish. Soc.*, 111:83-89.

Matty, A.J., and K.P. Lone. 1985. Hormonal control of protein deposition. In: B. Cowey, A.M. Mackie and J.G. Bell (Eds.), *Nutrition and Feeding in Fish.* Academic Press, London.

Matty, A.J. and P. Smith. 1978. Evaluation of a ycast, a bacterium and an alga as a protein source for rainbow trout. 1. Effect of protein level on growth, food conversion efficiency and protein conversion efficiency. *Aquaculture*, 14:235-246.

Mazid, M.A., *et. al.* 1979. Growth response of *Tilapia zilli* fingerlings fed iso-calorific diets with variable protein levels. *Aquaculture*, 18:115-122.

McLarney, W. 1984. *The Freshwater Aquaculture Book: A Handbook for Small Scale Fish Culture in North America.* Hartley & Marks, Inc. Point Roberts, WA.

McLaughlin, J. 1964. Some considerations of the growth of marine algae in artificial media. *Can. J. Microbiol.*, 10:769-782.

McLachlan, J. 1973. Growth media - marine. In: J.R. Stein (Ed.). *Handbook of Phycological Methods: Culture Methods and Growth Measurements.* Cambridge University Press.

McVey, J.P. 1983. *Handbook of Mariculture: Volume 1. Crustacean Aquaculture.* (Ed.), CRC Press, Boca Raton, FL.

_____. 1983. *Handbook of Mariculture: Volume 2. Finfish Aquaculture.* (Ed.) CRC Press, Boca Raton, FL.

McLusky, D.S. 1981. *The Estuarine Ecosystem*. Blackie.

Metcalf and Eddy, Inc. 1979. *Wastewater Engineering: Treatment, Disposal, Reuse*. McGraw-Hill, New York, NY.

Meyerhof, O. 1917. *Untersuchunger uber den atmungsuourgong nitrifizierenden bakterien. Pflugers Archges Physiol.* 166:240-280.

Meyers, S.P. 1986. Utilization of shrimp processing wastes. *INFOFISH - Marketing Digest*, No. 4/86. pp. 18-19.

Meyers, S.P. 1987. Crawfish - total product utilization. *INFOFISH-Marketing Digest*, No. 3/87. pp. 31-32.

Miller, J.W. 1976. Fertilization and feeding practices in warm-water pond fish culture in Africa. *CIFA Tech. Pap.* (84) Suppl. 1:512-541.

Millikin, M.R. 1982. Qualitative and quantitative nutrient requirements of fishes: A review. *Fish. Bull.* 80:655-686.

_____. 1983. Interactive effects of dietary protein and lipid on growth and protein utilization of Age-0 striped bass. *Trans. Amer. Fish. Soc.*, 112:185-193.

Millikin, M.R., A.R. Fortner, and L.V. Sick. 1980. Influence of dietary protein concentration on growth, food conversion ratio and general metabolism of juvenile prawn (Macrobrachium rosenbergii). *Proc. World Maric. Soc.*, 11:382-391.

Milne, P.H. 1970. Fish farming: a guide to the design and construction of net enclosures. *Mar. Res.*, 1:31.

Milne, P.H. 1972. *Fish and Shellfish Farming in Coastal Waters*. Fishing News Books, Ltd., Farnham, Surrey, UK.

Misra, R.V. and P.R. Hesse. 1982. Comparative analyses of organic manures. FAO/UNDP Regional Project RAS/75/004 (Improving soil fertility through organic recycling). Project Field Document No. 24.

Moe, M.A., Jr. 1989. *The Marine Aquarium Reference: Systems and Invertebrates*. Green Turtle Publications, Plantation, FL.

Mohsen, A.F., A.H. Nasr, and A.M. Metwalli. 1973. Effect of different light intensities on growth, reproduction, amino acid synthesis, fat and sugar contents in *Ulva fasciata* Delile. *Hydrobiologia* 43:125-235.

Moore, J.M. and C.E. Boyd. 1984. Comparison of devices for aerating inflow of pipes. *Aquaculture*, 38:89-96.

Muller-Fuega, A., J. Petit, and T.J. Sabaut. 1978. The influence of temperature and wet weight on the oxygen demand of rainbow trout (*Salmo gairdneri* R.) in freshwater. *Aquaculture* 14:355-363.

Murai, T., M.A. Fleetwood, and J.W. Andrews. 1979. Optimum levels of dietary crude protein for fingerling American Shad. *Prog. Fish. Cult.*, 41:5-6.

Murai, T., *et. al.* 1985. Effects of dietary protein and lipid levels on performance and carcass composition of fingerling carp. *Bull. Jap. Soc. Sci. Fish.*, 51:605-608.

Murray, A.P. and R. Marchant. Nitrogen utilization in rainbow trout (*Salmo gairdneri* Richardson) fed mixed microbial biomass. *Aquaculture*, 54:263-275.

Murray, S.N. and P.S. Dixon. 1973. The effect of light intensity and light period on the development of thallus form in the marine red alga *Pleononsporium squarrulosum* (Harvey) Abbott (Rhodophyta: Ceramiales). I. Apical cell division - main axes. *J. Exp. Mar. Biol. Ecol.* 13:15-27.

_____. 1975. The effect of light intensity and light period on the development of thallus form in the marine red alga *Pleononsporium squarrulosum* (Harvey) Abbott (Rhodophyta: Ceramiales). II. Cell enlargment. *J. Exp. Mar. Biol. Ecol.* 19:165-176.

Nagy, Z. 1979. The air-lift aerator and its application in sewage treatment. *Prog. Water. Tech.*, 11:101-109.

National Research Council (NRC). 1981. *Nutrient Requirements of Coldwater Fishes.* National Academy Press, Washington, D.C.

_____ 1982. *United States - Canadian Tables of Feed Composition.* Committee on Animal Nutrition, Washington, D.C., National Academy Press.

_____. 1983. *Nutrient Requirements of Warmwater Fishes and Shellfishes.* National Academy Press, Washington, D.C.

Nelson, D.H. 1931. Isolation and characteriation of *Nitrosomonas* and Nitrobacter. *Zenthol Bakt Parasit. Abt. II* 83:280-311.

Newton, G.L., *et. al.* 1977. Dried *Mermetia illucens* larvae meal as a supplement for swine. *J. Anim. Sci.*, 44(3):395-400.

Nichols, H.W. 1973. Growth media-freshwater. In: Stein, J. R. (Ed.) *Handbook of Phycological Methods.* Cambridge University Press. Cambridge.

Nichols, H.W., and H.C. Bold. 1965. *Thrichosarcina polymorpha gen. et sp. nov. J. Phycol.* 1:34-38.

Nigrelli, R.F. 1936. Life history of *Oodinium ocellatum. Zoologica* (NY) 21: 129-164, 9 pl.

Nigrelli, R.F. and G.D. Ruggieri. 1966. Enzootics in the New York Aquarium caused by *Cryptoracaryon irritans* Brown, 1951 (= *Ichthyophthirius marinus* Sikama, 1961), a histophagous ciliate in the skin, eyes and gills of marine fishes. *Zoologica* (NY) 51:97-102, 7 pl.

Nomura, M. and T. Yamazaki. 1977. *Fishing Techniques.* Japanese International Cooperation Agency, Tokyo.

Nose, T. and S. Arai. 1973. Optimum level of protein in purified diet for eel, *Anguilla japonica. Bull. Freshwater Fish. Res. Lab.* (Tokyo), 22: 145-155.

Ogino, C. and K. Saito. 1970. Protein nutrition in fish. 1. The Utilization of dietary protein by young carp. *Bull. Jap. Soc. Sci. Fish.*, 36:250-254.

Ogino, C., C.B. Cowey, and J.Y. Chiou. 1978. Leaf protein concentrate as a protein source in diets for carp and rainbow trout. *Bull. Jap. Soc. Sci. Fish.*, 44:49-52.

Ogino, C. 1963. Studies on the chemical composition of some natural foods of aquatic animals. *Bull. Jap. Soc. Sci. Fish.*, 29:459-462.

_____. 1980. Protein requirements of carp and rainbow trout. *Bull. Jap. Soc. Sci. Fish.*, 46:385-388.

Oppenheimer, C.H. and G.S. Moreira. 1980. Carbon, nitrogen and phosphorous content in the developmental stages of the brine shrimp Artemia. In: Persoone, G. P. Sorgeloos, O. Roels, and E. jaspers (Eds.), *The Brine Shrimp Artemia, Vol.2.* Universa Press, Wetteren.

Orme, L.E. and C.A. Lemm. 1973. Use of dried sludge from paper processing wastes in trout diets. *Feedstuffs*, Dec. 10:28-29.

Osborne, K. 1983. Full spectrum lighting in the planted aquarium. Part 1. *FAMA (Freshwater and Marine Aquarium Magazine)*, 5(12):14.

Ott, F. D. 1965. Synthetic media and techniques for the xenic culture of marine algae and flagellates. *Virginia J. Sci.* (N.S.) 16:205-218.

Oyenuga, V. A., 1975. The nutritional value of conophor seed (*Tetracarpidium conophorum* Welw.). *Proceedings of the Conference on Animal Feeds of Tropical and Subtropical Origin.* London School of Pharmacy, 1-5 April 1974.

Page, J.W. and J. W. Andrews. 1973. Interactions of dietary levels of protein and energy on channel catfish (*Ictalurus punctatus*). *J. Nutr.*, 103:1339-1346.

Parsons, T.R. , K. Stephans, J.D.H. Strickland. 1961. On the chemical composition of eleven species of marine phytoplankters. *J. Fish. Res. Board Can.*, 18(6):1001-1016.

Papaperaskeva-Papoutsoglou, E. and M.N. Alexis. 1986. Protein requirements of young grey mullet, *Mugil capito*. *Aquaculture*, 52:105-115.

Phillips, G.B. and E. Hanel, Jr. 1960. *Use of Ultraviolet Radiation in Microbiological Laboratories*. Technical Report BL 28, Revision of Special Report 211. U.S. Army Chemical Corps, Biological Laboratories, Fort Detrick, Md.

Pillay, T.V.R. 1990. *Aquaculture Principles and Practices.* Fishing News Books,Ltd., Oxford, U.K.

Piper, R.G., I.B. McElwain, L.E. Orme, J.P. McCracken, L.G. Fowler, and J.R. Leonard. 1982. *Fish hatchery management*. US Dept. of the Interior, Fish and Wildlife Service, Washington, D.C.

Platt, B.S. 1962. *Tables of representative values of foods commonly used in tropical countries*. Privy council, Medical Research Council Special Report Series No. 302 (Revised edition of Special Report No. 253), London, HMSO.

Prather, E.E. and R.T. Lovell. 1973. Response of intensively fed channel catfish to diets containing various protein-energy ratios. *Proc. of 27th. Ann. Conf. of South Eastern Assoc. of Game and Fish Comm.*

Pringsheim, E. G. 1964. *Pure Cultures of Algae, Their Preparation and Maintenance.* University Press, Cambridge.

Provasoli, L. 1963. Growing marine seaweeds. In: Devirville, D. and J. Feldmann (Eds.), *Proc. Int. Seaweed Symp.*, 4:9-17. Pergamon Press, Oxford.

Provasoli, L. 1968. Media and prospects for the cultivation of marine algae. In: Watanabe, A. and A. Hattori (Eds.) *Cultures and collection of Algae.* Proc. US-Japan Conf. Hakone, Sept. 1966. Jap. Soc. Plant Physiol.

Pullin, R.S.V. and G. Almazan. 1983. *Azolla* as a fish food. *ICLARM Newsletter*, January 1983.

Ree, W.O. 1949. Hydraulic characteristics of vegetation for vegetated waterways. *Agricultural Engineering* 30(4):184-187,189.

Reece, D.L., *et. al.* 1975. A blood meal-rumen contents blend as a partial or complete substitute for fish meal in channel catfish diets. *Prog. Fish Cult.*, 37:15-19.

Reinemann, D.J. and M.B. Timmons. 1989. Prediction of oxygen transfer and total dissolved gas pressure in airlift pumping. *Aquaculture Eng.*, 8:29-46.

Richman, S. 1958. The transformation of energy by *Daphnia pulex. Ecol. Monogr.*, 28:273-291.

Rodhe, W. 1948. Environmental requirements of freshwater plankton algae. Symbol. Bot. Upsaliensis 10(1):5-145.

Ross, B. and L.G. Ross. 1984. The oxygen requirements of *Oreochromis niloticus* under adverse conditions. In: *Proc. Int. Symp. on Tilapia in Aquaculture*, Nazareth, Israel. May 8-13, 1983. Tel Aviv University, Israel.

Ross, L.G. and B. Ross. 1984. Anaesthetics and sedative techniques for fish. Institute of Aquaculture, University of Stirling, Scotland.

Rumsey, G.L., *et. al.* 1981. Dairy-processing wastes as a replacement protein source in diets of rainbow trout. *Prog. Fish Cult.*, 43:86-88.

Ryther, J.H. and R. R. Guillard. 1962. Studies of marine planktonic diatoms. II. Use of *Cyclotella nana* Hustedt for assays of vitamin B-12 in seawater. *Can. J. Microbiol.*, 8:437-445.

Sabaut, J.J. and P. Luquet. 1973. Nutritional requirements of the gilthead bream *Chrysophrys aurata.* Quantitative protein requirements. *Mar. Biol.*, 18:50-54.

Salonen, K., J. Sarvala, I. Hakala, and M.L. Viljanen. 1976. The relation of energy and organic carbon in aquatic invertebrates. *Limnol. Oceanogr.*, 21:724-730.

Santiago, C.B., M. Banes-Aldaba, and M.A. Laron. 1982. Dietary crude protein requirement of *Tilapia nilotica* fry. *Phillipp. J. Biol.*, 11:255-265.

Satia, B.P. 1974. Quantitative protein requirements of rainbow trout. *Prog. Fish. Cult.*, 36:80-85.

Schaeperclaus, W. 1933. *Textbook of Pond Culture*. Trans. W.P.A., Project 50-11861, Stanford University.

Schauer, P.S., D.M. Johns, C.E. Olney and K.L. Simpson. 1980. International study on Artemia. IX. Lipid level, energy content and fatty acid composition of the cysts and newly hatched nauplii from five geographical strains of *Artemia*, in *The Brine Shrimp Artemia, Vol. 3*, G. Persoone, G. P. Sorgeloos, O. Roels and E. Jaspers, (eds.) Universa Press, Wetteren, Belgium.

Schindler, D.W., A.S. Clark, and J.R. Gray. 1971. Seasonal calorific values of freshwater zooplankton, as determined with Phillipson bomb calorimeter modified for small samples. *J. Fish. Res. Bd. Can.*, 28:559-564.

Schroeder, G.L. 1975. Nighttime material balance for oxygen in fish ponds receiving organic wastes. *Bamidgeh*, 27:65-74.

_____. 1980. Fish farming in manure-loaded ponds. In: R.S.V. Pullin and Z.H. Shehadeh (Eds.) *Integrated Agriculture-Aquaculture Farming Systems*. ICLARM Conf. Proc. 4,73-86.

Schultz, E. and H.J. Oslage. 1976. Composition and nutritive value of single-cell protein (SCP). *Anim. Feed Sci. Tech.*, 1:9-24.

Segedi, R. and W.E. Kelley. 1964. A new formula for artificial sea water. In: J.R. Clark and R.L. Clark (Eds.) *Sea-water Systems for Experimental Aquariums: A Collection of Papers*. Res. Rept. 63, Bur. Sport Fish. Wildl., Washington, DC.

Seidel, C.R., *et. al.* 1980. Culture of Atlantic silversides fed on artificial diets and brine shrimp nauplii. *Bull. Jap. Soc. Sci. Fish.*, 46:237-245.

Seidel, C.R., J. Kryznowek, and K.L. Simpoon. 1980. International study on Artemia. XI. Amino acid composition and electrophoretic protein patters of *Artemia* from five geographical locations. In: *The Brine Shrimp Artemia, Vol. 3*. Persoone, G., P. Sorgeloos, O. Roels, and E. jaspers, (Eds.) Universa Press, Wettern, Belgium.

Simpson, Klein-MacPhee and A.D. Beck. 1982. Zooplankton as a food source. In: G.D. Pruder, C.J.Langdon and D.E. Conklin (Eds.) *Second International Conference on Aquaculture Nutrition*, Rehobeth Beach, Delaware, Special Publication No. 2, World Mariculture Society, Louisiana State University, Baton Rouge, LA.

Siriwardene, J.A., S.S.E. Ranawana and G. A. Piyasena. 1970. Study of the feeding value of *Salvinia auriculata* for growing pigs. *Trop. Agric., Colombo*, 126(1):31-34.

Skinner, F.A. and N. Walker. 1961. Growth of *Nitrosomonas europea* in batch and continuous culture. *Archiv. for Microbiologie* 38:339-349.

Smith, L.L. *et. al.* 1985. Growth and digestion by three sizes of *Penaeus vannamei* Boone: effects of dietary protein level and protein source. *Aquaculture*, 46:85-96.

Smith, R.H. and R. Palmer. 1976. A chemical and nutritional evaluation of yeast and bacteria as dietary protein sources for rats and pigs. *J. Sci. Food Agric.*, 27:763-770.

fungi as protein sources for simple stomached animals. *J. Sci. Food Agric.*, 26:785-795.

Smith, W.L. and M. H. Chanley. 1975. *Culture of Marine Invertebrate Animals*. Plenum Press, New York, NY.

Soeder, C.J. 1981. Chemical composition of microbial biomass as compared to some other types of single cell protein (SCP). In: *Wastewater for Aquaculture, Proceedings of a Workshop on Biological Production Systems and Waste Treatment*. J.V. Grobbelda, C.J. Soeder and D.F. Toerien (Eds.). Bloemfontein, South Africa, 12-13 March 1980, U.O.F.S. Publ., Series C.

Sorgeloos, P. P. Lavens, P. Leger, W. Tackaert and D. Versichele. 1986. *Manual for the Culture and Use of Brine Shrimp Artemia in Aquaculture*.

Speece, R.E. 1970. Design of U-tube aeration systems. *J. Sanit. Eng. Div. ASCE*, 96:715-725.

Speece, R.E., M. Madrid, and K. Needham. 1971. Downflow bubble contact aeration. *J. Sanit. Eng. Div.*, ASCE, 97:433-441.

Speece, R.E., J.L. Adams, and C.B. Wooldridge. 1969. U-tube aeration operating characteristics. *J. Sanit. Eng. Div.*, ASCE, 95:563-574.

Spinelli, J. 1980. Unconventional feed ingredients for fish feed. In: *Fish Feed Technology*, Rome, UNDP/FAO, ADCP/REP/80/11:187-214.

Spotte, S. 1979. *Fish and Invertebrate Culture: Water Management in Closed Systems*. John Wiley & Sons, New York, NY.

_____. 1979. *Seawater Aquariums: The Captive Environment*. John Wiley & Sons, New York, NY.

Springhall, J.A. 1969. Composition of a number of tropical and sub-tropical feedstuffs. *PNG Agric. J.*, 20 (3-4):85-88.

Stafford, E.A. and A.G.J. Tacon. 1984. Nutritive value of the earthworm *Dendrodrilus subrubicundus*, grown in domestic sewage, in trout diets. *Agric. Wastes*, 9:249-266.

Stafford, E.A. 1985. The nutritional evaluation of dried earthworm meal (*Eisenia foetida*, Savigny, 1826) included at low levels in production diets for rainbow trout, *Salmo gairdneri* Richardson. *Aquaculture Fish. Management*, 16:213-222.

Stanley, R.W. and L. B. Moore. 1983. The growth of *Macrobrachium rosenbergii* fed commercial feeds in pond cages. *J. World. Maric. Soc.*, 14:174-184.

Starr, R.C., 1964. The culture collection of algae at Indiana University. *Am. J. Bot.* 51: 1013-1044.

_____. 1969. Structure, reproduction and differentiation in *Volvox carteri f. nagariensis* Iyengar, strains HK9 and 10. *Arch. Protistenk*. 111:204-222.

Stein, J.R. 1966. Growth and mating of *Gonium percorale* (Volvocales) in defined media. J. Phycol. 2:23-28.

_____. 1973. Handbook of Phycological Methods: Culture Methods & Growth Measurements. Cambridge University Press, New York, NY.

Stosch, H. A. von, 1964. Wirkungen von Jod und Arsenit auf Meeresalgen in Kultur. *Proc. Intern. Seaweed Symp.* 4: 142-150.

Strasburg, D.W. 1964. An aerating device for salt well water. In: J.R. Clark and R.L. Clark (Eds.), *Seawater Systems for Experimental Aquariums.* T.F.H. Publications, Jersey City, NJ.

Subosa, P.F. and M.N. Bautista. 1991. Influence of stocking density and fertilization regime on growth, survival and gross production of *Penaeus monodon* fabricus in brackishwater ponds. *Israeli Journal of Aquaculture- Bamidgeh*, 43(2):69-76.

Tacon, A.G.J. 1982. Utilization of chick hatchery waste: The nutritional characteristics of day-old chicks and egg shells. *Agric. Wastes*, 4:335-343.

_____. 1983. Replacement of marine fish protein in salmonid diets. Report to the Chief Scientist Group, Ministry of Agriculture, Fisheries and Food, June 1983, University of Stirling.

_____. 1986. Papua New Guinea; Development of carp feeds. FI:TCP/PNG/4503 Field Document 3 July, 1986.

_____. 1986a. Larval shrimp feeding - Crustacean tissue suspension: Pracitical alternative for shrimp culture. Rome, UNDP/FAO, ADCP Report No. ADCP/MR/86/23.

_____. 1987. *The Nutrition and Feeding of Farmed Fish and Shrimp - A Training Manual: 1. The Essential Nutrients.* FAO Field Document 2, Brasilia, Brazil.

_____. 1987. *The Nutrition and Feeding of Farmed Fish and Shrimp - A Training Manual: 2. Nutrient Sources and Composition.* FAO Field Document 5, Brasilia, Brazil.

Tacon, A.G.J. and P.N. Ferns. 1979. Activated sewage sludge, a potential animal foodstuff 1. Proximate and mineral content: seasonal variation. *Agric. Environm.*, 4:257-269.

Tacon, A.G.J., E.A. Stafford and C.A. Edwards. 1983. A preliminary investigation of the nutritive value of three terrestrial lumbricid worms for rainbow trout. *Aquaculture* 35: 187-199.

Tacon, A.G.J., J.L. Webster, and C.A. Martinez. 1984. Use of solvent extracted sunflower seed meal in complete diets for fingerling rainbow trout (*Salmo gairdneri* Richardson). *Aquaculture*, 43:381-389.

Taiganides, E.P. 1977. Bioengineering properties of feedlot wastes. In: E.P. Taiganides (Ed.) *Animal Wastes.* Applied Science Publ. Ltd., Essex, UK.

Tang, Y. A. 1982. *Planning, Design and Construction of a Coastal Fish Farm.* Advances in Aquaculture. FAO, by Fishing News Book Ltd.

Tatterson, I.N. and M.L. Windsor. 1974. Fish silage. *J. Sci. Food Agric.*, 25:369-379.

Tebbutt, T.H.Y. 1972. Some studies on reaeration in cascades. *Wat. Res.*, 6:297-304.

Teshima, S., G. Gonzalez, and A. Kanazawa. 1978. Nutritional requirements of Tilapia: Utilization of dietary protein by *Tilapia zilli*. *Mem. Fac. Fish.*, Kagoshima Univ., 27:49-57.

Teshima, S., A. Kanazawa, and Y. Uchiyama. 1985. Optimum protein levels in casein-gelatin diets for *Tilapia nilotica* fingerlings. *Mem. Fac. Fish.*, Kagoshima Univ., 34:45-52.

Thayer, D.D. (undated) Weed Control in Aquaculture and Farm Ponds.

Thomas, W.A., H.A. Spalding and J. Pavlovich. 1967. *The Engineers Vest Pocket Book*. National Book Store, Philippines.

Toerber, E.D. and M.G. Mandt. 1979. Greater oxygen transfer with jet aeration system. *Water and Sewage Works*, 126:71-75.

Toledo, J., J.A. Cisneros, and O. Ortiz. 1983. Requirementos nutricionales en alevines de *Oreochromis aureus* (*T. nilotica*). 1. Nivel optimo de proteina con dietas purifacadas. *Rev. Lat. Acui*. Lima, Peru, No. 18-1-48, Dec. 1983, pp. 8-12.

Trama, F.B. 1957. The tranformation of energy by an aquatic herbivore, *Stenonema pulchellum*. Ph.D. Thesis. Univ. Michigan. Ann Arbor. Mich., USA.

U.S. Department of Agriculture (USDA). 1982. *Ponds - Planning, Design Construction*.

U.S. Department of the Interior. 1981. Water Measurement Manual. Washington, D.C.

U.S. Fish and Wildlife Service. 1982. *Fish Hatchery Management*. U.S. Department of the Interior, Washington, D.C.

Van Baalen, C. 1967. Further observations on growth of single cells of coccoid blue-green algae. J. Phycol. 3:154-157.

Vanhaecke, P. and P. Sorgeloos. 1983. International Study on *Artemia*. XIX. Hatching data for 10 commercial sources of brine shrimp cysts and re-evaluation of the "hatching efficiency" concept. *Aquaculture*, 30:43.

Vanhaecke, P. 1980. International Study on Artemia. IV. The biometrics of *Artemia* strains from different geographical origin. In: *The Brine Shrimp Artemia, Vol. 3*, G. Persoone, P. Sorgeloos, O. Roels, and E. Jaspers (Eds.,) Universa Press, Wetteren, Belgium.

Venkataraman, L.V., W.E. Becker, and T.R. Shamala. 1977. Studies on the cultivation and utilization of the alga *Scenedesmus acutus* as a single cell protein. *Life Science*, 20:223-234.

Venkataramia, T., G.J. Lakshmi, and G. Gunter. 1975. Effect of protein level and vegetable matter on growth and food conversion efficiency of brown shrimp. *Aquaculture*, 6:115-125.

Villalon, J.R. 1991. *Practical Manual for Semi-intensive Commercial Production of Marine Shrimp*. TAMU-SG-91-501, TAMU Sea Grant College Program, College Station, Texas.

Viola, S. and G. Zohar. 1984. Nutrition studies with market size hybrids of *Tilapia* (*Oreochromis*) in intensive culture: 3. Protein levels and sources. *Bamidgeh*, 36:3-15.

Vijverberg, J. and Th.H. Frank. 1965. The chemical composition and energy content of copepods and cladocerans in relation to size. *Freshwater Biol.*, 6:333-345.

Waaland, J.R. 1973. Experimental studies on the marine algae *Iridaea* and *Gigartina*. *J. Exp. Mar. Biol. Ecol.* 11:71-80.

Walne, P.R. 1974. *Culture of Bivalve Molluscs*. The Whitefriars Press, Ltd. London and Tondridge.

Wang, K.W., T. Takeuchi, and T. Watanabe. 1985. Optimum protein and digestible energy levels in diets for *Tilapia nilotica. Bull. Jap. Soc. Sci. Fish.*, 51:141-146.

Wannigama, N.D., D.E.M. Weerakoon, and G. Muthukumarana. 1985. Cage culture of *Sarotherodon niloticus* in Sri Lanka: Effect of stocking density and dietary crude protein levels on growth. In: C.Y. Cho, C.B. Cowey and T. Watanabe (Eds.), *Finfish nutrition in Asia: Methodological approaches to research and development*. IDRC-233e, IDRC, Ottawa, Canada.

Waris, H. 1953. The significance for algae of chelating substances in the nutrient solutions. *Physiol. Plant.*, 6:538-543.

Watanabe, T., T. Arakawa, C. Kitajima, K. Fukusho and S. Fujita. 1978a. Proximate and mineral composition of living feeds used in seed production of fish. *Bull. Jap. Soc. Sci. Fish.*, 44:979-984.

Watanabe, T., T. Arakawa, C. Kitajima, and S. Fujita. 1978b. Nutritional evaluation of proteins of living feeds used in seed production of fish. *Bull. Jap. Soc. Sci. Fish.*, 44:985-988.

Watanabe, T., C. Kitajima, and S. Fujita. 1983. Nutritional value of live organisms used in Japan for mass propagation of fish: A review. *Aquaculture*, 34:115-143.

Weast, R.C., 1987. *Handbook of Chemistry and Physics*. CRC Press, Inc. Boca Raton, FL.

Wee, K.L., N. Kerdchuen and P. Edwards. 1987. Use of waste grown tilapia silage as feed for *Clarias batrachus. J. Aqua. Trop.* 1987.

Wee, K.L. and A.G.J. Tacon. 1982. A preliminary study on the dietary protein requirement of juvenile snakehead. *Bull. Jap. Soc. Sci. Fish.*, 48:1463-1468.

West, J.A. 1967. *Pilayella littoralis F. rupincola* from Washington: the life history in culture. J. Phycol. 3:150-153.

_____. 1974. Controlling *Rhodochorton* production. *Carolina Tips* 37(1):1-2.

Wheaton, F.W. 1985. *Aquacultural Engineering*. John Wiley & Sons, New York, NY.

Wheaton, F., J. Hochheimer, and G. Kaiser. 1991. Fixed film nitrification filters for aquaculture. In: D.E. Brune and J. R. Tomasso (Eds.) *Aquaculture and Water Quality*. World Aquaculture Society, Baton Rouge, LA.

Whitfield, M. 1974. The hydrolysis of ammonium ions in seawater - a theoretical study. *J. Mar. Biol. Assoc. U.K.*, 54:565-580.

Wigglesworth, J.M. 1991. The use of chemical fetilizers to manipulate algal based food webs in commercial shrimp production ponds. *Abstract, 22nd World Aquaculture Society*. San Juan, Puerto Rico.

Wikfors, G.H. 1986. Altering growth and gross chemical composition of two microalgal molluscan food species by varying nitrate and phosphate. *Aquaculture*, 59:1-14.

Wilson, R.P. and C.B. Cowey. 1985. Amino acid composition of whole body tissue of rainbow trout and Atlantic salmon. *Aquaculture* 48:373-376.

Wilson, R.P., D.W. Freeman and W.E. Poe. 1984. Three types of catfish offal meals for channel catfish fingerlings. *Prog. Fish Cult.*, 46:126-132.

Windell, J.T., R. Armstrong and J.R. Clineball. 1974. Substitution of brewers single-cell protein into pelleted fish feed. *Feedstuffs*, 46:22-23.

Winfree, R.A. and R.R. Stickney. 1981. Effects of dietary protein and energy on growth, feed conversion efficiency and body composition of *Tilapia aurea*. *J. Nutr.*, 111:1001-1012.

Winograsdsky, S. and H. Winogradsky. 1933. Etudes sur la microbiologie clu sol. *Annls. Inst. Pasteur, Paris* 50:350-432.

Wissing, T.E. and A.D. Hasler. 1968. Calorific values of some invertebrates in Lake Mendota, Wisconsin. *J. Fish. Res. Board Can.*, 25:2515-2518.

Wissing, T.E. and A.D. Hasler. 1971. Intraseasonal changes in caloric content of some freshwater invertebrates. *Ecology*, 52:371-371.

Wood, J.F., B.S. Capper and L. Nicolaides. 1985. Preparation and evaluation of diets containing fish silage, cooked fish preserved with formic acid and low-temperature-dried fish meal as protein sources for mirror carp (*Cyprinus carpio*). *Aquaculture*, 44:27-40.

Woods Hole Engineering Associates, Inc. 1984. *Design Guide for Use of Copper Alloy Expanded Metal Mesh in Marine Aquaculture*. Contract Report Technology for the Copper Industry. INCRA Proj. 268B. Woods Hole Engineering Associates, Inc., Woods Hole, MA.

J. Wyban and J. Sweeney. 1991. *Intensive Shrimp Production Technology*. Oceanic Institute, Honolulu, Hawaii.

Yoshida, M. and H. Hoshii. 1978. Nutritional value of earthworms for poultry feed. *Jap. J. Poult. Sci.*, 15:308-310.

_____. 1980. Nutritive evaluation of 16 samples of single cell protein grown on agricultural waste materials by growing chicks. *Agric. Biol. Chem.* 44:2671-2676.

Yurkowski, M. and J.L. Tabachek. 1978. Proximate and amino acid composition of some natural fish foods. Paper presented at the First International Symposium on Finfish Nutrition and Feed Technology, Hamburg, 20-23 June, 1978, Paper E/33.

Zendejas, M.C. 1991. Alimentos para camaron y sistemas de alimentaction. In: M.C. Zendejas and G.W. Chamberlain (eds.) *Taller Sobre Cultivo de Camaron, Proceedings of a Shrimp Farming Workshop in Mazatlan, Mexico on 17-19 July, 1991*. Industrias Purina S.A. de C.V., Mexico City, Mexico.

Zein-Eldin, Z.P. and J. Corliss. 1976. *The Effect of Protein Levels on Growth of Penaeus aztecus*. FAO Technical Conference on Aquaculture, Kyoto, Japan, 26 May - 27 June, 1976. Fishing News Books Ltd., London.

Zeitoun, I.H., *et al.* 1973. Influence of salinity on protein requirements of rainbow trout (*Salmo gairdneri*) fingerlings. *J. Fish. Res. Board Can.*, 30:1867-1873.

Zerbe, W.B. and C. B. Taylor. 1953. *Sea Water Temperature and Density Reduction Tables*. Spec. Publ. 198. U. S. Coast Geodetic Survey.

Zimmerman, D.R. and S.B. Tegbe. 1977. Evaluation of a bacterial SCP for young pigs and rats. *J. Anim. Sci.*, 46:469-475.

Zohar, G. 1986. Improved feeding chart for tilapias in high stocking rates. (In Hebrew with English summary). *Fisheries and Fishbreeding in Israel*, 19:28-33.

Printed in the United States
By Bookmasters